# Microscopic and Spectroscopic Imaging of the Chemical State

# PRACTICAL SPECTROSCOPY
## A SERIES

### Edited by Edward G. Brame, Jr.

*The CECON Group*
*Wilmington, Delaware*

# Microscopic and Spectroscopic Imaging of the Chemical State

edited by
Michael D. Morris
*University of Michigan*
*Ann Arbor, Michigan*

CRC Press
Taylor & Francis Group
Boca Raton London New York

CRC Press is an imprint of the
Taylor & Francis Group, an **informa** business

CRC Press
Taylor & Francis Group
6000 Broken Sound Parkway NW, Suite 300
Boca Raton, FL 33487-2742

First issued in paperback 2019

© 1993 by Taylor & Francis Group, LLC
CRC Press is an imprint of Taylor & Francis Group, an Informa business

No claim to original U.S. Government works

ISBN-13: 978-0-8247-9104-9 (hbk)
ISBN-13: 978-0-367-40233-4 (pbk)

**Library of Congress Cataloging-in-Publication Data**

Microscopic and spectroscopic imaging of the chemical state / edited by Michael
    D. Morris.
        p.   cm. -- (practical spectroscopy; 16)
    Includes bibliographical references and index.
    ISBN 0-8247-9104-5 (acid-free)
    1. Spectrum analysis.   2. Microscope and microscopy.   I. Morris,
Michael D.   II. Series: Practical spectroscopy; v. 16.
    QD95.M47   1993
    539'.6'0287--dc20                                                    93-13593
                                                                           CIP

Visit the Taylor & Francis Web site at
http://www.taylorandfrancis.com

and the CRC Press Web site at
http://www.crcpress.com

# Preface

As chemists whose research is centered on individual molecules, we often treat the larger world as essentially homogeneous. The structure of a molecule is important to us. The dynamic behavior of ensembles of molecules matters, of course, but we often ignore the distribution of molecules across a surface or through a solid. Yet spatial distributions are often the keys to understanding functions of complex systems, including living organisms. Mechanisms of formation, transformation, and damage are also addressed by observing distributions of key components.

Mapping or imaging approaches to problems are familiar to our colleagues in the life sciences, earth sciences, and materials science. They have long understood the power of light microscopy and electron microscopy to relate structure to function. Few chemists take full advantage of either family of techniques. Fewer still use imaging based on their own well-established and powerful spectroscopies.

It is time, then, to provide the chemist with a general introduction to techniques that image the *chemical state* of a sample. We define these as techniques for mapping distributions of molecules or their structural units. The images are simply plots of the intensity of the molecule-specific signal as a function of two or three coordinates. Generally, but not invariably, the plots are spatial distributions. Most, but not quite all, of the imaging methods described in this volume are based on spectroscopies that are familiar to the modern chemist. Most are microscopies and are useful on distance scales ranging from individual atoms to about 1 mm.

The range of available imaging techniques is broader than any individual can handle competently. This volume draws on the skills of chemists who are leaders in major areas of spectroscopic imaging. Their chapters summarize general principles and review important applications and chosen techniques: light microscopies based on electronic and vibrational transitions, x-ray microscopies, magnetic resonance microscopies, scanned probe microscopies, and mass spectrometric microscopies. In addition, tutorial chapters introduce the chemist to the principles of light microscopy and scanned probe microscopy. Another tutorial chapter outlines image processing and image enhancement for the chemist. Because the volume has been written for practicing chemists who are familiar with modern spectroscopies, the underlying spectroscopies are discussed only to the extent necessary to understand their uses in imaging.

The field of chemical state imaging, as much as any area of experimental chemistry, is almost completely dependent on the revolution in computer technology of the last 15–20 years. Almost all the microscopies we discuss employ small computers for data acquisition, data storage, and data display. Most of the illustrations in the book were generated by computers. Yet computers are infrequently mentioned in the volume, except, of course, in the discussion of image processing. Perhaps that is the strongest evidence for their ubiquity. Like telephones, computers are simply there.

In any volume the editor must choose which topics to include and exclude. With one exception, we have tried to include the most important contrast generation principles in current use, as well as some emerging techniques. With apologies to our electron microscopy colleagues, we have deliberately excluded electron microscopy from our coverage. The electron microscopy literature is already large. Some treatises have a pronounced chemical flavor, which is lacking in the review literature of most of the topics of this book. The omission is an acknowledgment of the widely understood importance of electron microscopy to the chemical sciences.

Without the expertise, enthusiasm, and literary skills of the chapter authors, this book would not exist. Other people have made invaluable contributions without the hope of the modest fame of authorship. Graduate students and postdoctoral fellows in the editor's research group served as semivoluntary reviewers and technical consultants. Ms. Lynn Phaneuf applied her formidable word processing skills to much of the manuscript, greatly speeding the editor's work. Leslie Morris, whose expertise is in the much maligned fields of political organization and government, provided invaluable advice on the organization of an enterprise involving almost two dozen busy professionals scattered across North America. The editor hopes that the result of these combined efforts is a volume that will prepare readers to use chemical state imaging in their own work.

*Michael D. Morris*

# Contents

# Contributors

**Thomas P. Beebe, Jr.** Department of Chemistry, University of Utah, Salt Lake City, Utah

**Gareth R. Eaton** Department of Chemistry, University of Denver, Denver, Colorado

**Sandra S. Eaton** Department of Chemistry, University of Denver, Denver, Colorado

**Barry M. Gordon** Department of Applied Science, Brookhaven National Laboratory, Upton, New York

**Lynn W. Jelinski** College of Engineering, Cornell University, Ithaca, New York

**Keith W. Jones** Department of Applied Science, Brookhaven National Laboratory, Upton, New York

**Raoul Kopelman** Departments of Chemistry and Physics, University of Michigan, Ann Arbor, Michigan

**Duane A. Krueger** Analytical Sciences Laboratory, The Dow Chemical Company, Midland, Michigan

**Michael D. Morris**   Department of Chemistry, University of Michigan, Ann Arbor, Michigan

**Robert A. Morris**   Department of Mathematics and Computer Science, University of Massachusetts at Boston, Boston, Massachusetts

**Robert W. Odom**   Charles Evans & Associates, Redwood City, California

**Bradford G. Orr**   Department of Physics, University of Michigan, Ann Arbor, Michigan

**David M. Pallister**   Department of Chemistry, University of Michigan, Ann Arbor, Michigan

**David L. Patrick**   Department of Chemistry, University of Utah, Salt Lake City, Utah

**Joan F. Power**   Department of Chemistry, McGill University, Montreal, Quebec, Canada

**Weihong Tan**   Department of Chemistry, University of Michigan, Ann Arbor, Michigan

**Patrick J. Treado**   Department of Chemistry, University of Pittsburgh, Pittsburgh, Pennsylvania

**Marjorie S. Went**   College of Engineering, Cornell University, Ithaca, New York

# 1
# Fundamentals of Light Microscopy

**David M. Pallister and Michael D. Morris**  *University of Michigan, Ann Arbor, Michigan*

The light microscope is perhaps the one instrument that links the chapters in this book. Visual observation can provide important structural and morphological information, of course. Yet, even where other microscopies are used to examine the sample, the light microscope is almost always used for preliminary examination and for positioning the sample in the instrument. Consequently, an understanding of the basic principles and operation of the light microscope are important to researchers contemplating almost any form of modern chemical state imaging.

## I. HISTORICAL DEVELOPMENT OF THE MICROSCOPE

The earliest microscopes date back to the middle of the fifteenth century. These simple devices were one-lens loupes used to examine small specimens at low magnification. The compound microscope using two or more lenses in combination came soon after. Early lenses were not achromatic, and the images produced by these instruments were surrounded by fringes of color. During the early nineteenth century true achromatic lenses were being produced. The work of the English microscopist Lister in 1830 put achromatic design on a sound theoretical basis and led to the development of high power microscope objectives [1,2].

The towering figure of modern microscope design is Ernst Abbe (1840–1905), a German physicist [3,4]. Working at the firm of Carl Zeiss, he standardized the production of the microscope. His research there, particularly on the design of objectives, underlies all modern microscope design. Abbe

**1**

introduced the concept of numerical aperture to define the resolution of an objective. He established the diffraction theory of image formation and then began the development of modern Fourier optics.

His technological contributions include the oil immersion lens and the Abbe condenser. Collaboration with glass maker Otto Schott enabled Abbe to develop high quality oil immersion objectives using fluorite and low dispersion glasses developed by Schott Glass. By the turn of the century, Abbe had developed a whole series of excellent, highly corrected objectives and eyepieces. They fully realized the theoretical resolving power possible with the wide field microscope.

Microscope design has evolved since this era, and many new contrast principles have been introduced. The phase contrast microscope and the scanning confocal microscope, two major twentieth-century developments, owe much to the pioneering efforts of Ernst Abbe.

## II. GENERAL FEATURES OF THE LABORATORY MICROSCOPE

Formation of an image from the light microscope is caused by diffraction of light by a specimen. The diffraction pattern produced by an object is gathered by a lens system and focused to produce an magnified interference pattern at the image plane. Image formation is, therefore, a process of double diffraction, first by the object and then by the lens, to produce the observed image. Absorption of light may also occur, and will provide color.

Laboratory compound microscopes utilize the same basic optical components: a condenser to illuminate the object, an objective to gather light from the object, and an eyepiece to form the final image at the viewer's eye or camera. The mechanical appearance can vary, but the microscope will usually incorporate all the features shown in Fig. 1 [5].

The body tube positions the objective and the eyepiece at the proper distance. Usually the body tube incorporates a rotating nosepiece to mount several objectives. To focus the microscope, either the stage or the objective is moved. The fine adjustment of the focusing mechanism must be able to control motion with a resolution comparable to the depth of field of the highest power objectives, typically 0.25–0.5 micrometers ($\mu$m).

The substage carries the illumination optics (condenser). The light source provides illumination for the condenser system. Usually, 20–50 W quartz–halogen lamps are used.

The condenser provides bright, even illumination. It includes a focusing mechanism, numerical aperture adjustment, and a diaphragm to match the circle of illumination to the size of the viewing field to reduce stray light.

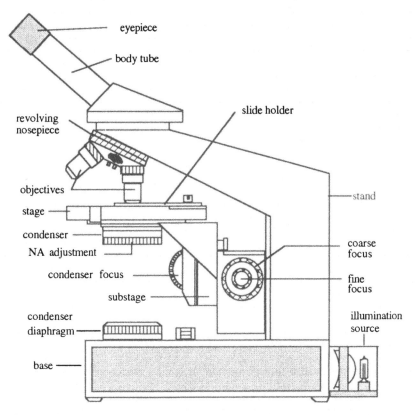

**Figure 1** The research microscope. This upright microscope, equipped for visual observation, is a composite of features from several manufacturers.

## III. MAGNIFICATION AND IMAGE FORMATION IN THE FIXED TUBE LENGTH MICROSCOPE

Most of the specimen magnification is performed by the objective, which is a carefully designed combination of lenses intended to provide achromatic, diffraction-limited imaging. For most purposes, the magnification properties of the microscope objective can be modeled as a single thin lens. In the thin lens approximation, magnification of an objective can be approximated by Eq. (1).

$$M = \frac{-i}{o} \tag{1}$$

where magnification $M$ is given by the ratio of the image distance $i$ to the object distance $o$. These distances are defined in Fig. 2, which also shows a worked example.

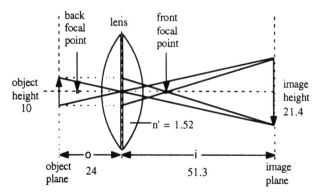

**Figure 2** Magnification and the conjugate points of a lens: $n'$ is the refractive index of the lens glass; $o$ is the object distance, and $i$ is the image distance.

The conjugate points of the object and image planes are determined by the focal length of the lens through the Gaussian thin lens formula, given in Eq. (2).

$$\frac{1}{f} = \frac{1}{i} + \frac{1}{o} \tag{2}$$

In Eq. (2), $f$ is the focal length of the lens. It is the distance at which an image of an object located infinitely far from the lens is brought to focus.

As shown in Fig. 3, the standard microscope has fixed distances between its components; the standard values are given in the figure. This design is often called the Abbe or fixed tube length design. It is also called the 160 mm tube length design, because the standard tube length is 160 mm. There is, however, another design, called the infinity-corrected design, in which the tube length is not fixed. This microscope is discussed in Section V.

In the fixed tube length design, the image focal plane is at a fixed distance from the objective seating plane, 150 mm in the 160 mm tube length design. The magnification of the objective is determined by its focal length and the tube length of the microscope. Objectives are designed to be parfocal. That is, they focus at the same distance, usually 45 mm, from the seating plane (7).

A real image (Fig. 3) is formed at the focal plane 10 mm from the end of the body tube. The eyepiece, in combination with the lens of the eye, forms a real image on the retina. This produces the effect of viewing an upright virtual image of the object. Normal viewing of an object through the microscope will produce a virtual image that is approximately 250 mm from the eyepiece.

The overall magnification of the fixed tube length microscope is given by Eq. (3).

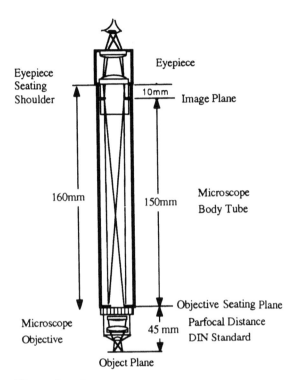

**Figure 3** Microscope body tube components and dimensions, and location of object and image plane.

$$\text{magnification} = \frac{\text{optical tube length}}{\text{focal length of objective}} \times \frac{250 \text{ mm}}{\text{focal length of eyepiece}} \quad (3)$$

Equation (3) is only approximate, because the optical system also includes the lenses of the user's eyes [5].

## IV. THE MICROSCOPE OBJECTIVE

### A. Properties of Objectives

Microscope objectives are classified by the degree of correction of both monochromatic (Seidel) and chromatic optical aberrations, especially chromatic aberrations. Monochromatic aberrations are image distortions caused by the use of spherical surfaces for the lenses of the objective. Chromatic aberration is caused by refractive index dispersion of glass. The focal length of a lens depends on the refractive index of the glasses of which it is fabricated, and it changes with wavelength.

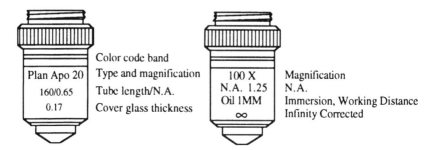

| Plan Apo 20 | Color code band | 100 X | Magnification |
| 160/0.65 | Type and magnification | N.A. 1.25 | N.A. |
| 0.17 | Tube length/N.A. | Oil 1MM | Immersion, Working Distance |
| | Cover glass thickness | ∞ | Infinity Corrected |

**Figure 4**   Microscope objectives, with identifying information.

Generally an objective is labeled with information about its intended use. This information usually includes the magnification, the numerical aperture, the body tube length, and the thickness of the coverslip. For an immersion objective, the type of the immersion liquid will be given. Infinity-corrected objectives will have the infinity symbol and the tube lens focal length, which gives the objective the specified magnification [7]. Metallographic objectives will have the inscription "no cover" or will lack a coverslip thickness inscription, indicating that the objective is designed to be used without a coverslip [5]. Examples are shown in Fig. 4.

The working distance is the distance from the end of the objective to the object plane. Low numerical objectives can have working distances of 2–15 mm. High magnification, high numerical aperture objectives will have working distances less than 0.2 mm [8].

The numerical aperture NA is the product of the refractive index of the immersion media and the half-angle of the cone of light accepted by the objective $\theta$, given in Eq. (4).

$$NA = n \sin (\theta) \tag{4}$$

Numerical aperture is a measure of the resolution and light-gathering power of an objective. The maximum angle of light acceptance is 90°, so the maximum NA depends on the refractive index of the immersion medium used with the objective.

## B.  Types of Objectives

Standard objectives are corrected for spherical aberration, coma, and astigmatism at one wavelength. The degree of correction of longitudinal chromatic aberration varies with the type of objective [7,8]. Table 1 lists some standard objectives and their properties.

Achromats are corrected for chromatic aberration at a red wavelength (644 or 657 nm), and a blue wavelength (480 nm). At all other wavelengths, the

**Table 1**  German Standard (DIN) Microscope Objectives: 45 mm Parfocal Distance

| Magnification | Focal length (mm) | Field of view (mm) | Working distance (mm) | Numerical aperture | Color code |
|---|---|---|---|---|---|
| 4× | 30.60 | 4.50 | 15.80 | 0.10 | Red |
| 10× | 16.60 | 1.80 | 6.30 | 0.25 | Yellow |
| 20× | 8.78 | 0.90 | 1.50 | 0.40 | Light blue |
| 40× | 4.50 | 0.45 | 0.45 | 0.65 | Green |
| 63× | 3.05 | 0.30 | 0.28 | 0.85 | Dark blue |
| 100× | 1.86 | 0.18 | 0.13 | 1.25 | |

focal points will be different, resulting in a secondary fringes around the focused image.

In apochromats longitudinal chromatic aberration is removed for three wavelengths, one each in the red, blue, and green. As a consequence, there is only weak chromatic aberration at all other wavelengths. It is perceptible only in the off-axis region of view. Apochromats are also corrected for spherical aberration at two wavelengths. In general, they form better images than achromats.

The amount of correction in the semiapochromat is between that of achromats and apoachromats. Longitudinal chromatic aberration is reduced for the green line, but not minimized. This type of objective is sometimes called a fluorite objective, if fluorite is used in its construction.

Plan apochromats or achromats are used for videomicroscopy, photomicroscopy, or whenever a flat field of view is necessary. A field-flattening element is incorporated into the objective to minimize the Petzval field curvature of the image front.

In neoachromats correction for field flatness is between that of achromats and plan achromats. Neochromatic designs commonly incorporate glasses with high refractive index and low dispersion.

Immersion objectives offer the highest magnification and highest numerical apertures available for the light microscope. Although the maximum light-gathering angle remains 90°, the use of a fluid of refractive index of 1.3–1.5 increases the amount of light brought into the objective, resulting in numerical apertures above 1.0. The large values of numerical aperture, in turn, provide the highest resolutions and useful magnification available from the light microscope. Most immersion objectives are designed for one fluid. Usually this is an oil of $n = 1.52$. Objectives designed for water immersion are available, but are less common. The Zeiss Plan-Neofluars are designed to work with different immersion fluids, such as oil, water, or glycerin. A correction collar adjusts the focus for immersion fluids of different refractive index [9].

Many manufacturers of microscope objectives compensate for aberrations in an objective series by overcorrecting the eyepiece aberrations. In combination the optics work well together, but mismatching optics can degrade the performance of the microscopic.

For example, chromatic aberration free (CF) optics are available from both Nikon and Zeiss. Longitudinal chromatic aberration and chromatic difference of magnification are corrected in both the objectives and eyepieces. The correction is made independently for each optical component. These designs completely eliminate chromatic aberration across the entire field of view.

## C.  Reflecting Objectives

Reflecting objectives employ spherical mirrors instead of lenses. They are used for infrared or ultraviolet microscopy or wherever the choice of transmissive materials is limited. The most common design is the Schwarzchild objective, which uses a large concave gathering mirror and a small convex focusing mirror.

Mirrors are inherently free of chromatic aberrations, but they do exhibit spherical aberration, coma and astigmatism. The Seidel aberrations can be minimized only by increasing the size of the secondary mirror, thus reducing the effective aperture [10,13]. Consequently high numerical aperture Schwarzchild objectives are not available.

The working distance of a reflecting objective is much greater than that of a similar numerical aperture glass objective. The working distance of a 0.4 NA Schwarzchild objective is 14.5 mm. The equivalent glass objective will have a working distance of about 1.5 mm [10].

## V.  THE INFINITY-CORRECTED MICROSCOPE

In the infinity-corrected microscope, the objective is a lens system operating at an infinite conjugate ratio. The object is positioned at the focal plane of the objective, and the image is focused at infinity. A tube lens, also optimized to work at an infinite conjugate ratio, is used to focus the image at a finite distance. The optical train is shown in Fig. 5. The magnification of this microscope is the ratio of the objective focal length to the tube lens focal length, as given in Eq. (5).

$$M = \frac{f_{\text{tube lens}}}{f_{\text{objective}}} \tag{5}$$

The same kinds of objective are available for infinity-corrected and finite tube length microscopes. However, finite tube length objectives are not designed to work with a tube lens and are not interchangeable with infinity-corrected objectives.

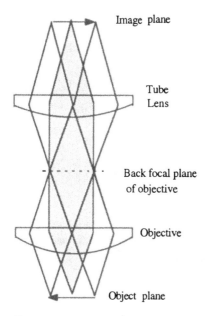

Image plane

Tube
Lens

Back focal plane
of objective

Objective

Object plane

**Figure 5** The optical schematic of the infinity-corrected microscope. The image from the object is focused at infinity. The tube lens is used to form a real image at the image plane.

In an infinity-corrected system the tube length is arbitrary. Several optical devices such as beam splitters, polarizers, or filters can be placed between the objective and the tube lens without affecting the position of the image plane. This configuration is especially helpful in reflection, fluorescence, and laser microscopies, where beam splitters and other devices must be present in the optical path of the microscope.

## VI. THE COVERSLIP

Many biological specimens require the use of a coverslip, and many microscope objectives are designed for use with a coverslip. Typically the thickness is 0.17 mm. Coverslip objectives with numerical aperture greater than 0.4 should be equipped with a correction collar to adjust for differences in coverslip thickness or refractive index to avoid spherical aberration and degradation of contrast. If a coverslip is not used with an objective designed for one, the aberrations eliminated by the optical design will be reintroduced.

Figure 6 illustrates the polarization effects of the coverslip. Polarized light measurements of samples under a coverslip will include systematic errors from Fresnel reflection losses. The polarization problem of light at the air–coverslip

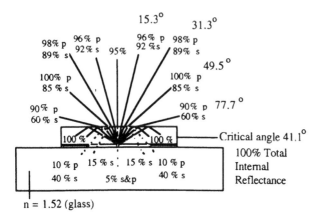

**Figure 6**  Polarization effects at a coverglass–specimen interface. The percentages indicate the reflection back into the specimen of light polarized parallel (p) and perpendicular (s) to the plane of incidence, for various angles.

interface can be eliminated with a homogeneous immersion system. If the immersion liquid refractive index matches the refractive index of glass, the interface is transparent to image formation [8].

Coverslips are rarely used in reflected light microscopy, because specimen reflectance is often about the same as coverslip reflectance, and contrast would be poor [5].

## VII. EYEPIECES

The widely used Huygens eyepiece consists of two plano–convex lenses, both of the same glass. This eyepiece, which is made achromatic by choosing the spacing and the focal lengths of the lenses properly, produces a negative magnification of the image and is essentially free from lateral chromatic aberration.

The Ramsden eyepiece consists of two plano–convex lenses. This positive magnification, achromatic eyepiece has the advantage that the real image is not between the two lenses. A cross hair or reticule in the system will have the same magnification ratio as the real image from the objective.

The Kellner eyepiece is a modified Ramsden eyepiece in which an achromat replaces the second plano–convex lens. The achromat is chosen to provide a better field of view and eye relief for the user. The achromat also improves color correction and reduces spherical aberration and field curvature.

Variations of the basic eyepiece designs provide better field or view, highly corrected field flatness, or variable magnification. Special eyepieces are available for the user with eyeglasses. Because video cameras and photographic film require a real image focused at the film or sensor plane, special amplifying or projection eyepieces are available which provide a flat image field outside the body tube.

The objective is the critical component in the microscope. It provides magnification at high numerical aperture, maximizing the resolution of the intermediate real image in the body tube. High magnification of this intermediate image decreases performance because residual aberrations of the objective are also magnified. Consequently eyepiece magnification should be moderate and total magnification never should exceed 1000 NA [9]. This topic is discussed in Section XI.

## VIII. VIDEO MICROSCOPY

Since the early 1980s video cameras have become increasingly common for viewing microscope images. Both color and monochrome cameras are used. Although the typical video camera cannot provide the image definition of photographic film, let alone the human eye, video observation is convenient and has the advantage of allowing images to be stored accessibly in analog (tape) or digital form. Where laser illumination is used, video observation protects the user against accidental exposure to intense laser light.

Cameras employing interline transfer charge-coupled devices (CCDs) as image sensors have largely replaced the earlier tube-based cameras. CCD cameras, compact and relatively inexpensive, provide adequate image definition in both monochrome and color versions. Although broadcast video cameras often use three sensors for color separation, laboratory cameras use a single sensor and an integrated filter system. Consequently color images have lower resolution (but not 1/3!) than monochrome images.

Images can be viewed on free-standing monitors. Computer boards are available that convert the camera signals to formats (such as VGA) used by computer monitors, allowing observation of the image directly on the computer screen. This approach is often called live video.

With a specialized system called a frame grabber, video images can be digitized for storage or processing. A frame grabber synchronizes analog–digital (A/D) conversion with the raster scan of the camera. The digitized image is usually has 8 bits (256 levels) of color intensity in monochrome, or 8 bits of each of three colors for color. There are several competing standards for file formats. Monochrome and color frame grabbers are available as inexpensive boards for standard laboratory computer systems. A frame grabber may include, or be designed to work in conjunction with, live video conversion hardware.

Slow scan, or frame transfer, CCD cameras are used for low light level microscopy, such as fluorescence or Raman microscopy. They provide still, monochrome images, but with 12–16 bits of intensity levels. The low noise amplifiers and high resolution A/D converters used with these devices operate at low frequencies, and image digitization may take several seconds. The cameras must be operated at subambient temperatures, usually between $-40$ and $-140°C$. At these temperatures dark current is negligible and integration times of several seconds to several hours are possible.

## IX. IMAGE BRIGHTNESS

Image brightness is proportional to the square of the numerical aperture and inversely proportional to the square of the lateral magnification, as shown in Eq. (6) [8].

$$\text{image brightness} = \left( \frac{NA_{obj}}{\text{magnification}} \right)^2 \tag{6}$$

For low light level microscopy, immersion objections are advantageous because they maximize numerical aperture at any magnification. Excessive magnification simply increases optical aberrations and degrades image contrast and resolution, as well as reducing brightness. As discussed below, video cameras do not require the high magnification that human vision needs. They can be especially useful at low light levels.

## X. MICROSCOPE ILLUMINATION

### A. Koehler Illumination

It is important to use an even source of illumination for examination of specimens. In the laboratory microscope, this is usually provided by the Koehler illumination system. The image of the light source is positioned at the back focal point of the objective. This configuration provides incoherent illumination. Figure 7 shows the component arrangement: in this epi-illumination version, used in fluorescence, reflectance, and Raman microscopy, the objective functions as a condenser. In transmission, a separate condenser is used, as illustrated in Fig. 1.

The Koehler illumination system provides adjustment for numerical aperture and an aperture diaphragm. Microscope resolution depends on both the NA of the condenser or illumination source and the NA of the objective, as in Eq. (7), where $\lambda$ is the wavelength of light.

$$\text{resolution} = \frac{1.22\lambda}{NA_{condenser} + NA_{objective}} \tag{7}$$

To realize the full resolution of the microscope, the condenser NA must be approximately equal to the objective NA. A condenser numerical aperture slightly

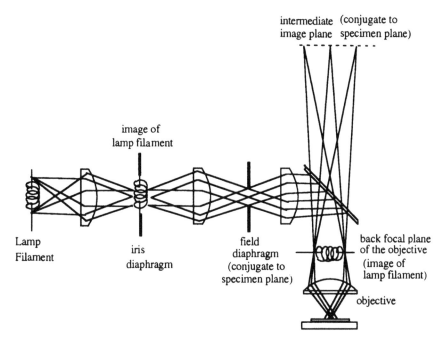

**Figure 7** Koehler illumination: the epi-illumination version. In transmission or epi-illumination the lamp filament is imaged at the back focal plane of the objective. Light at the specimen is completely out of focus and is uniform across the field.

less than that of the objective works best [6,12]. The lower condenser NA reduces stray light entering the microscope.

A condenser diaphragm is used to adjust the field of illumination to the field of view of the imaging optics of the microscope. When the magnification is changed, the condenser diaphragm should be adjusted to match the field of illumination with the field of view to reduce stray light.

## B. Koehler Epi-Illumination

In the epi-illumination, or vertical illumination, form of Koehler illumination, the objective also acts as the condenser. A beam splitter is used to reflect the illumination light from the source to the object. In fluorescence microscopy, a dichroic mirror is usually used as the beam splitter, to provide some spectral filtering [17].

Epi-illumination systems for finite tube length microscopes require a Galilean telescope at the entrance of the epi-illumination attachment to move the intermediate image plane a distance equivalent to the height of the attachment. This

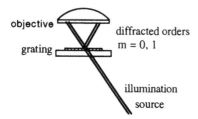

objective

grating

diffracted orders
m = 0, 1

illumination
source

**Figure 8**   Oblique illumination.

telescope adds aberrations and reduces transmission. Addition of a beam splitter
to the finite tube length system results in distortion of the image plane. Infinite
tube length systems do not require use of Galilean magnifiers to add components
to the microscope, so added image distortion is negligible.

## C.   Oblique Illumination

Oblique illumination (Fig. 8) can provide a twofold increase in resolution over
axial illumination. This technique is simple to implement with the laboratory
microscope. Addition of a cardboard annulus on top of the condenser diaphragm
aperture provides this type of illumination. Oblique illumination is required for
phase contrast and dark field microscopies [7,9,16].

## D.   Dark Field Illumination

Special dark field condensers provide oblique illumination and ensure that no
illumination light enters the objective. A field can be observed only when a sample diffracts the illumination source. Dark field methods are particularly well
adapted for microscopic viewing of cracks in structures and other details not
observed under normal bright field illumination [9]. Reflected light dark field
illumination is an important way to improve contrast in reflected light microscopy. It is especially valuable for specimens that exhibit very high specular reflection [12].

## XI.   IMAGE RESOLUTION

Increasing the magnification of an objective in itself does not directly improve
the resolution of the microscope. Rather, resolution is determined by the numerical aperture of the objective and the condenser. However, numerical aperture
and magnification are not independent, and high numerical aperture is found in
high magnification objectives.

## A. Lateral Resolution

The theoretical limit of lateral resolution is set by diffraction. A point source is imaged by a lens of circular cross section as a central disk and concentric bright rings, the Airy pattern. The irradiance distribution of the Airy disk therefore determines the theoretical limit of resolution for two point sources close to one another. Equation (8), the Rayleigh criterion, is the conventional definition of resolution when objective and condenser NA are equal. The resolution is the separation at which the maximum of one Airy disk function just touches the first minimum of an adjacent Airy disk. The criterion is same as the spectroscopist's Rayleigh criterion for resolution of two adjacent lines.

$$\text{resolution} = \frac{1.22\lambda}{2\text{NA}} \tag{8}$$

Equation (8) strictly applies to objectives completely free from monochromatic and chromatic aberration. Modern objectives can come quite close to diffraction-limited performance.

As the spectroscopist will realize, the Rayleigh criterion is an operational definition of resolution, not a fundamental one. There are several ways to obtain higher resolution with a conventional light microscope. The most common one is confocal scanning microscopy (Section XII). Computer image processing also can be used to increase the resolution of digitized images by deconvolution of the objective's point spread function, which describes the diffraction of a point source.

## B. Depth of Field

An observed image actually consists of a convolution of contributions from those objects at the object plane and the objects within a small axial displacement, or depth of field. The axial Airy disk function determines the resel or resolution limit in the axial direction. An approximate value for the longitudinal resolution limit is given by Eq. (9) [11,15]. This equation is a good estimate for low numerical aperture.

$$\text{axial resolution} = \frac{2\lambda}{\text{NA}^2} \tag{9}$$

Magnification also decreases the depth of field. To minimize the depth of field, the user should select the highest practical numerical aperture and magnification [8].

Depth of field decreases with increasing spatial frequency. The depth of field is minimized when an object is focused, therefore. Defocusing decreases the spatial frequency content of an image and increases the depth of field.

A narrow depth of field is not always a benefit. In some applications it may be necessary to compare features that vary in depth in an object. Low magnification, low NA objectives, such as long working distance objectives, improve focus and resolution of axially displaced objects [5].

## C. Magnification for Visual Observation and Videomicroscopy

A minimum magnification is necessary for human vision to discern objects at the resolution limit of a microscope [9]. At the visual reference for the microscope, 250 mm, human vision can distinguish features 0.15 mm apart. This corresponds to 2 minutes of arc, or 0.034°. For light in the middle of the visible spectrum, 589 nm, the minimum magnification necessary for human vision to discern two points that are just resolved by the objective is given by Eq. (10).

$$509 \times NA = \frac{2 \times NA \times 1.5 \times 10^{-4} \text{ meter}}{5.89 \times 10^{-7} \text{ meter}} \tag{10}$$

A minimum magnification of about 500 NA is necessary for the human eye to perceive objects at the resolution limit of an objective.

Video cameras can be analyzed by a similar argument. For a typical slow scan CCD camera, the pixel width is 20 $\mu$m. At least 3 pixels are necessary to discern a minimum between two points. Accordingly, a magnification of 200 NA or greater is necessary for such a camera to respond to the resolution limit of the objective. Video rate interline CCD cameras with 12 mm wide sensors have pixel widths of about 10 $\mu$m. A magnification of 100 NA or more is necessary.

For any viewing device, little is gained by increasing the magnification above the critical value. A magnification perhaps twice the value of Eq. (10), or its analogs for video or photographic cameras, is useful for direct observation. Further magnification adds no detail and is usually called "empty." For digital deconvolution, further magnification may be needed to define small differences of intensity, however.

## XII. CONFOCAL MICROSCOPY

Confocal microscopy uses a form of critical coherent illumination to provide superresolution from the optical microscope. Not only is the axial resolution improved, but the spatial filtering inherent in the technique also removes axial out-of-focus light from the image. Because they lack flare from out-of-focus light, confocal images are sharper than conventional microscope images.

The objective and condensers are focused on or convolve the same point on the object. The focused image of the point always appears at the same point in the image plane of the objective. A pinhole or slit at this point is used to aperture

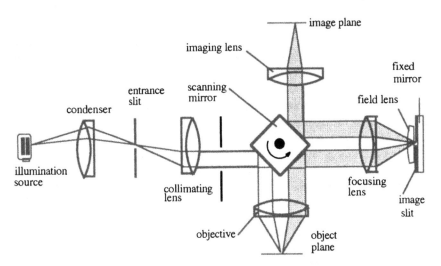

**Figure 9**  The scanning confocal microscope. Here a slit is used as a spatial filter, and a line is scanned rather than a point. The line of source light is brought to focus on the specimen, and this line is imaged on the slit by the field lens. The image is built up by rotating the scanning mirror system, so that the slit is scanned across the specimen.

or remove extraneous portions of the lateral *and* axial Airy disk diffraction pattern at the image plane [18]. The confocal geometry really images just one point on the object plane. Some means must be included to systematically scan this point around the specimen, or to scan the specimen under the fixed point in the object plane. Because only one point is measured at a time, a computer or video system is usually used to build up and display the image. Figure 9 shows a typical confocal system, in which a moving mirror system is used to scan the specimen and bring the light back to a fixed-slit spatial filter.

Resolution of the confocal microscope depends on the size of the aperture or slit. Surprisingly, a slit performs almost as well as a pinhole. Convolution of the objective and condenser point source response results in a spread function that is essentially the square of the usual Airy disk response. This convolution improves both lateral resolution and depth discrimination [18].

Equation (11) gives the confocal lateral resolution of a confocal microscope. Equation (12) describes the axial resolution.

$$\text{confocal lateral resolution} = 0.36\,\frac{\lambda^2}{\text{NA}^2} \tag{11}$$

$$\text{confocal axial resolution} = 4\,\frac{\lambda^2}{\text{NA}^4} \tag{12}$$

Depth discrimination is the property for which confocal microscopy is usually used. The fourth-power dependence on numerical aperture means that with high NA ($>$ 0.9) objectives, the confocal microscope provides image "slices" much thinner than 1 µm.

Many scanning mechanisms are used to produce confocal images. Various mechanical, acousto-optic and electro-optic devices are used to scan the beam across a specimen. The microscopes are usually used in fluorescence or reflectance, with epi-illumination. Generally, a low power argon ion or helium–neon laser is used as the light source, and the systems are generally called laser scanning confocal microscopes.

Aside from its greater complexity, the confocal microscope has another disadvantage. The images it forms are much dimmer than those from conventional (wide field) microscopes. This is because most of the light passed by a wide field microscope originates from above or below the region of best focus. This light, which can be 80–95% of the total, is rejected by the spatial filter. There is an inevitable tradeoff between image sharpness and image brightness.

It is also possible to generate multiple confocal conjugate points using a rotating disk, called a Nipkow disk, containing an array of apertures that define multiple illumination points and spatial filters at the conjugate point. The Nipkow disk works well with incoherent illumination. However, the holes cover only 1–5% of the disk area, and the images can be quite dim. Other moving devices containing arrays of slits or circular apertures are also used. Laser scanning methods are too slow to produce real-time confocal images [19]. The Nipkow disk, or tandem scanning, confocal microscope can produce real-time confocal images directly to the viewer eyepiece.

The scanning slit microscope, illustrated in Fig. 9, has higher throughput than aperture-based devices and is simpler to design and operate. Although it is confocal in only one lateral direction, its images are sufficiently sharp for most purposes. It can employ higher power lasers than point-scanned systems, because the laser power can be focused to a slit rather than a point.

Although it is still uncommon in chemistry laboratories, the confocal microscope has found wide application in biomedical and materials sciences laboratories. The uniquely sharp images can be as valuable to chemists as to other scientists.

## REFERENCES

1. Ford, B. J. (1985). *The Story of the Simple Microscope, Single Lens*, Harper & Row, New York.
2. Jacker, C. (1966). *Window on the Unknown, A History of the Microscope*. Scribner's, New York.
3. Volkman, H. (1966). Ernst Abbe and his work, *Appl. Opt. 5*: 1720.
4. Lipson, S. G., and Lipson, H. (1969). *Optical Physics*, Cambridge University Press, New York.

5. Mason, C. W. (1983). *Handbook of Chemical Microscopy*, Vol. 1, 4th ed., Wiley, New York.
6. Melles Griot Corp. (1988). *Optics Guide 4*, Irvine, CA.
7. Abramowitz, M. (1985). *Microscope, Basics and Beyond*, Vol. 1. Olympus Corp., Precision Instruments Division, Lake Success, NY.
8. Pluta, M. (1988). *Advanced Light Microscopy*, Vol. 1, *Principles and Basic Properties*. Elsevier, New York.
9. Determann, H., and Lepusch, F. (no date). *The Microscope and Its Applications*. Ernst Leitz Wetzlar GmbH, Wetzlar, Germany.
10. Fracon, M. (1961). *Progress in Microscopy*, Vol. 9, Row, Peterson, New York.
11. Born, M., and Wolf, E. (1980). *Principles of Optics, Electromagnetic Theory of Propagation, Interference and Diffraction of Light*, 6th ed., Pergamon Press, Elmsford, NY.
12. Abramowitz, M. (1990). *Reflected Light Microscopy, An Overview*. Vol. 3, Olympus Corp., Precision Instruments Division. Lake Success, NY.
13. Beck, J. L. (1969). A new reflecting microscope objective with two concentric spherical mirrors, *Appl. Opt. 8*: 1503.
14. Hopkins, H. (1955). The frequency response of a defocused optical system, *Proc. R. Soc. London, A231*:91.
15. Richardson, M. (1990). Confocal microscopy and three-dimensional visualization. *Am. Lab. 22* (17): 73.
16. Abramowitz, M. (1990). *Contrast Methods in Microscopy, Transmitted Light*, Vol. 2, Olympus Corp., Precision Instruments Division, Lake Success, NY.
17. Pluta, M. (1989). *Advanced Light Microscopy*, Vol. 2, *Specialized Methods*. Elsevier, New York.
18. Wilson, T., and Sheppard, C. (1984). *Theory and Practice of Scanning Optical Microscopy*, Academic Press, New York.
19. Pawley, J. B. (1990). *Handbook of Biological Confocal Microscopy*, Plenum Press, New York.

# 2
# Applications of Light Microscopy

**Duane A. Krueger**   *The Dow Chemical Company, Midland, Michigan*

## I. INTRODUCTION

This chapter presents examples of and references for the industrial, nonbiological applications of various light microscopical techniques used to delineate chemical information. Mason and Chamot [1,2] published a classical, definitive treatise on chemical microscopy. Other reference texts [3–13], as well as biennially [14–20] and periodically published journals [21–23, 50] contain valuable resources for applications not discussed in this review. Since numerous light microscopical techniques (including Nomarski or differential interference contrast [DIC], Hoffman modulation contrast, and normal and oblique reflected light) distinguish primarily material morphologies, they are omitted from the following discussion. The transmitted or reflected light microscopical techniques treated, therefore, include fluorescence, polarized light, phase contrast, and coaxial illumination. Ancillary equipment associated with the use of these techniques includes confocal laser scanning microscopy (CLSM), hot stage, light filtration, and infrared microscopy, and documentation using videomicrography and photomicrography.

## II. FLUORESCENCE MICROSCOPY

Most microscopical identifications of materials require the observation of differences in refractive indices of components to be observed. Materials that exhibit fluorescence reveal chemical differences which refractive index alone cannot. In fact, many seemingly homogeneous materials reveal strikingly clear information that was not visible under normal illumination. The technique is used extensively by biologists (usually with artificially stained specimens) and in forensic science, but it is becoming more widely used in the industrial laboratory. A variety of polymeric materials, inorganic materials, solvents, and solvent residues that are used in layered composites, multicomponent mixtures, and coatings fluoresce under appropriate excitation wavelengths [24–31]. In addition, contaminants and/or contaminant residues generated on processing may exhibit characteristic fluorescence [30]. In many cases, little or no sample preparation time is required if the materials are fluorescent or exhibit different fluorescence colors.

Excitation of a sample generates fluorescent light of lower energy and longer wavelength than the impinging radiation. Hence, unless the chemical structure generates a more complex fluorescence, ultraviolet excitation will normally generate a bluish image, whereas blue excitation usually generates a yellowish image. The human eye responds well to bright objects on a black field, so this condition is sought for greatest sensitivity. For fluorescence microscopy, excitation wavelengths are filtered out by a barrier filter for sensitivity and safety reasons. Fluorescence microscope manufacturers usually employ a mercury arc light source that has a wide range of wavelengths for excitation and is equipped with a convenient filter system. The Nikon cube and filter combinations summarized in Table 1 are typical.

The filter system provides a black field if no excitation of the specimen occurs and differing bright colors depending on the material excited and the filter chosen. For optimal work, especially in the ultraviolet (UV), dry and immersion UV optics, fluorites, or planapochromats should be used. For work in the low UV (wavelengths < 350 nm), special quartz UV microscope optics, quartz slides, and quartz coverslips are required. When components in the material do not fluoresce, selective staining with appropriate dyes may be required to enhance component feature(s) of interest. The sensitivity and advantage of fluorescence microscopy, then, depends on the nature of the materials studied as they are observed in the microscope. Quantification of fluorescent components must be cautiously undertaken, since the coefficients of extinction of materials vary widely, and flare from bright components may obscure less intensely emit-

**Table 1** Nikon Fluorescence Filter Cube Combinations

| Excitation mode | Wavelength (nm) | Excitation filter | Dichroic mirror | Eyepiece filter |
|---|---|---|---|---|
| UV | 365 | UV330–380 | DM400 | 420K |
| Blue | 495 | IF420–485 | DM510 | 520–560 |
| Violet | 405 | IF395–425 | DM455 | 470K |
| Green | 546 | IF335–550 | DM580 | 580W |

**Table 2** Selected List of Fluorescing Materials

| Material | Observed color | |
| --- | --- | --- |
| | UV (365 nm) excitation | Blue (495 nm) excitation |
| Diamond | Various | Various |
| Zinc oxide | Blue-white | Yellow |
| Starch | Blue-white | Yellow |
| Degraded PVC | Blue | None |
| Ion-exchange beads | Blue | Yellow |

ting species. In addition, for transparent specimens, the use of epifluorescence (reflected light excitation) allows the option of projecting fluorescence over the transmitted light image in order to map components in the specimen. For materials that are mounted in epoxies for polishing or microtoming prior to analysis, care must be taken to use nonfluorescing or contrasting fluorescent media. The best sensitivity occurs with absolutely black background fields.

Identification of materials and contaminants may be determined in bulk samples by positive or negative fluorescence, as compared to the surrounding matrix. Most neat polymers do not fluoresce with excitation below 400 nm. However, additives or degradation products of the polymer system may exhibit characteristic fluorescence. In other materials, some natural fibers, dyed synthetic fibers, and corrosion or add-mixtures of components are observable. Table 2 lists some common materials that fluoresce and gives the colors observed under UV and blue excitation.

Fluorescence data can be, therefore, extremely useful for the study of materials in their natural state or under thermal stress.

## A. Fluorescence Applications

Figures 1 and 2 illustrate the application of the fluorescence technique to layer thickness of an expensive, weatherable acrylonitrile–ethylpropyldiene–styrene (AES) coated acrylonitrile–butadiene–styrene (ABS) base polymer.

Since the refractive indices of the two pigmented polymers are nearly identical, the layers are indistinguishable by direct examination by either normal transmitted or reflected bright field illumination. In these figures only microtome artifacts, but not layer delineation, were observed with either bright field, polarized light or phase contrast techniques or reflected bright field illumination for rough-cut and microtome-polished specimens. Infrared (IR) analysis is also ineffective due the invisibility of the layers. Since both polymers stain very poorly, transmission electron microscopy (TEM) may provide weak contrast differences to distinguish the polymers and then only after time-consuming and

(a)

**Figure 1** Use of fluorescence to observe the coating on a microtomed tan-pigmented acrylonitrile–butadiene–styrene (ABS) polymer capped with acrylonitrile–ethylpropyl-enediene–styrene (AES). The AES coating is invisible under (a) bright field illumination (250×), (b) phase contrast microscopy (250×), (c) polarized light microscopy (250×), or (d) reflected bright field illumination (25×). The weaker coating fluorescence (e) is visible under blue epifluorescence (250×).

complex staining procedures with ruthenium tetraoxide or osmium tetraoxide vapors, followed by microtoming. By applying fluorescence microscopy, the less fluorescent AES layer is quite measurable in either polished or rough-cut sample preparations of the material.

Ceramic materials fluoresce to advantage [32,33]. Figure 3 illustrates the rapid location of contamination on a ceramic electronic component, using blue excitation. The ceramic itself, which is white under normal reflected bright field illumination, fluoresces red under UV excitation.

Figure 4 shows striking color differences in ceramic wafers using only normal long wavelength UV lamp excitation at one-to-one magnification. Since these materials have nearly identical elemental composition, but different

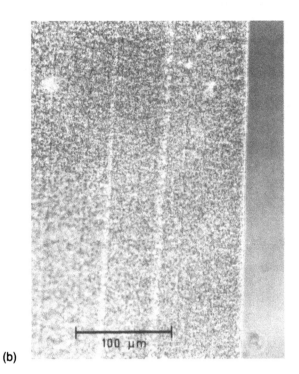

(b)

100 µm

100 µm

(c)

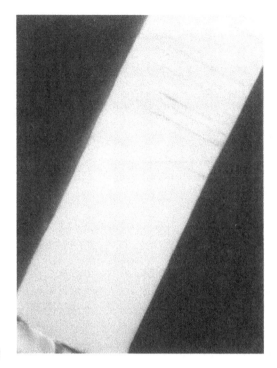

(d)

(e)

**Figure 1** *continued*

(a)

**Figure 2** Use of fluorescence to observe the coating on a rough-cut, black-pigmented ABS polymer capped with AES. (a) The coating is invisible under reflected bright field illumination (70×). (b) The nonfluorescent (dark) coating is visible under blue epifluorescence (130×).

(b)

**Figure 2** *continued*

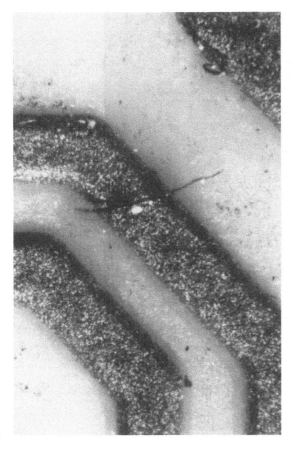

(a)

**Figure 3** Detection of contamination on a ceramic computer part. (a) Reflected bright field illumination (80×). (b) Blue, 495 nm epifluorescence (50×). The contaminant is visible as bright regions on a dark background.

fluorescence properties, the technique allows the identification of unique ceramic structures. As a quality control technique, the method may be used to aid in the determination of consistency in composition of these materials.

Asphalt is a difficult material to analyze because of its variable texture, blackness, and opacity. It usually exhibits little or no fluorescence. To make asphalt a more durable product, compatible polymer components may be added to it to achieve desired properties. Compositional analysis may be accomplished by tedious thin-sectioning or solvent extraction of the asphaltic phase followed by TEM examination, but the information is more readily obtained using fluorescence microscopy.

(b)

**Figure 3** *continued*

Figure 5 illustrates the distribution of polymer in asphalt at various mixing stages using UV excitation. Not only is the distribution observable, but with application of confocal scanning microscopy, the sample can be optically sectioned to reveal the three-dimensional dispersion structure.

Polymer contamination by black specks of material is ubiquitous. Metal or fiber contamination is relatively simple to determine. A speck of a degraded polymer goes through visible color change stages of yellow, amber, red-brown, and black as it proceeds ultimately to carbon under anaerobic conditions. This contamination is difficult to identify by infrared spectroscopy, because IR is insensitive to carbon and nonoxidative degradation products. Many polymers may be studied using the hot stage to follow the degradation progress over time. Some polymers and their degradation products do not fluoresce. In other polymer systems, degradation products fluoresce under both UV and blue excitation.

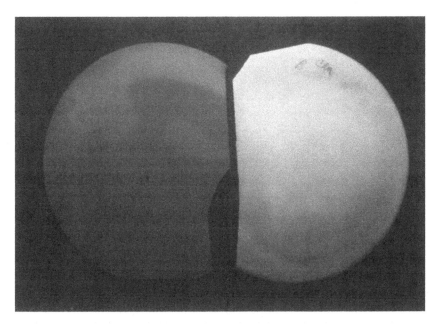

**Figure 4**   Ceramic wafer fluorescence using long wavelength UV (365 nm) excitation. The two materials have nominally the same composition, but the wafer on the left fluoresces blue, while the one on the right fluoresces magenta.

In yet other polymers, some degradation products fluoresce differently under UV and blue excitation. Figures 6 and 7 illustrate fluorescence changes as two polymeric materials, polyvinyl chloride and ion-exchange beads, respectively, are altered by heating.

Figure 6 shows beads of neat, unheated polyvinyl chloride and the polymer as it appears after heating. Little information is provided by transmitted light or polarized light microscopy. Obvious differences are detected under reflected bright field, but more characteristically, there is fluorescence generated by UV excitation but not by blue excitation. Examination of samples under different wavelengths of excitation can provide detection schemes for individual polymer systems.

The ion-exchange beads of Fig. 7 show few discernible differences by normal transmitted or reflected illumination, but under either blue or UV excitation, the dark, nonfluorescing heated beads are quickly distinguished from the fluorescent unheated, neat sample. By using fluorescence, degradation of certain polymers may be rapidly detected and, when this technique is combined with spectrophotometric analysis, the composition of the polymer's degradation products may be elucidated.

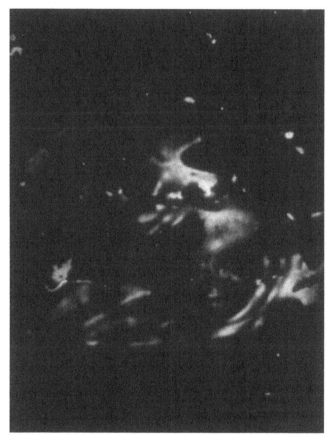

(a)

**Figure 5** Polymer distribution in asphalt during mixing, as seen by UV (365 nm) excited fluorescence: (a) start of mixing (150×) and (b) after mixing (260×).

Fluorescence also can be used to aid structural analysis via micro-IR. For contamination in which the defect is indistinguishable from the bulk matrix under normal illumination, the analyst can mount a specimen on pinhole apertures. For such materials, concentration centers of high fluorescence, rather than diffuse regions, may be accurately placed over the pinhole masks to allow cleaner spectra of the material of interest to be identified with reduced matrix interference.

## B.  Spectrophotometry

The progress of thermal degradation processes and the characterization of degradation products may be followed spectrophotometrically using a system such as the Zeiss UMSP, or by attaching photomultiplier tubes (PMTs) or photo-

(b)

diodes to the microscope via a port (usually at the photographic trinocular of the microscope). The success of this analysis depends on the optical construction of the microscope, the existence of aperturing ability to eliminate interfering components, the dynamic range of the detection system, and the spectral interpretation software. Fully integrated systems, such as the UMSP, may cost $225,000, a price that includes microscope with quartz optics, computer, software, scanning stage with 0.25 µm resolution at 2500 µm/s maximum speed, aperture diaphragms of $\geq 0.5$ µm, grating monochromators, and UV–visible–near-infrared detectors. The UMSP 80 has a spectral range of 240–2100 nm with quartz optics. Less costly systems are usually limited by the UV transmission of microscope optics and the scanning abilities of the excitation and detector components. Simple, less costly photomonitors with digital readout may be added to a fluorescence microscope. These accessories are inexpensive but are limited to simple intensity measurements.

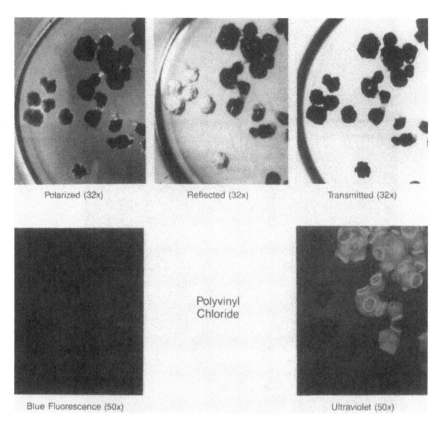

**Figure 6** Identification of neat and thermally degraded polyvinyl chloride (PVC). Top row: only the reflected light image shows differences between neat PVC (left) and partially decomposed material. Bottom row: the characteristic fluorescence of PVC degradation products is observed under UV excitation but not under blue excitation.

## C.  Cautions for Interpretation of Fluorescence Images

Illumination effects must be evaluated. A sample is subjected to high intensity photon irradiation during observation. The relatively low fluorescent yield of many materials requires extended irradiation during photographic documentation. In addition, higher magnification brings the objective closer to the specimen surface, increasing the exposure of the material, according to the law of inverse squares. As a result, the thermal and bleach stability of the sample must be established.

An important effect in some materials, such as polymers, especially darkly pigmented formulations, and asphalt, can be alteration by melting, flow, or dis-

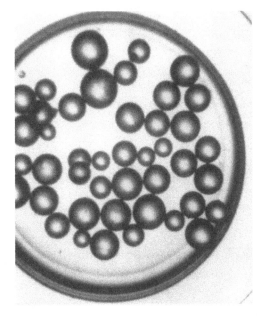

(a)

**Figure 7**  Fluorescence identification of partially decomposed (heated) ion-exchange beads. (a) Transmitted bright field illumination (18×) does not indicate which beads have been heated. (b) UV (365 nm) excitation shows fluorescence (50×) of unheated beads; decomposed beads remain dark.

tortion of the sample during extended examination. Not only is the sample perturbed, but there is the potential of contamination of the objective. A second effect can be time-dependent bleaching of the fluorophore or dye. Inorganics and some pigment additives or polymers exhibit good resistance to bleaching, but biological materials and dyes are more susceptible. In this case, one should compose the field of interest using longer working distance, lower magnification objectives before choosing a higher magnification objective for observation. Bleaching is an even more serious impediment to photomicrographic documentation of these samples.

## D.  Documentation of Fluorescence Data

Caution is necessary when photomicrographic documentation of the data is required, as it is with dark field and polarized light microscopy. Automatic exposure meters underintegrate fields when a few bright spots are generated on a predominantly dark field. The result is an overexposed image with poor detail. In addition, the typical low intensity of fluorescence requires fast film and extremely stable microscope conditions to reduce exposure times and prevent

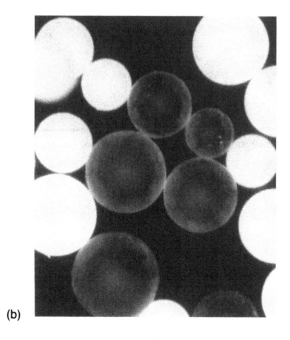

(b)

**Figure 7** *continued*

blurring. With color photomicrography, reciprocity failure with long exposures and low emulsion sensitivity cause problems with obtaining "true" observed hues and intensity. Test exposures and standard conditions for the films used by the laboratory will help establish consistent documentation. For additional hints on photography, see Delly [34].

For videomicroscopical documentation, one must consider the additional factors of camera sensitivity and stability. Initial observation of eye of materials under common "black light" or UV lamps (long and short wavelength) obtained from scientific supply companies may reveal fluorescence differences among materials. The differences may suggest potential image analysis or "on-line" analysis schemes. For extremely low magnification work, a macro lens attached to a video camera may give disappointing video results by comparison to the initial visual examination under the same conditions. The eye is insensitive to the UV, while the video camera detects the excitation wavelength along with fluorescence, obscuring the information. The insertion of barrier filters in the light path between the object and the camera lens in this situation, therefore, is critical. Barrier filtration is taken for granted in the use of normal fluorescence microscope optics, of course. In addition, the illumination detection capabilities and image enhancement features of the camera will determine the success of video documentation [35].

## III. POLARIZED LIGHT MICROSCOPY

## A. Basics of Polarized Light Microscopy

Polarized light microscopy (PLM) is a powerful tool for the rapid characterization and identification of many materials: minerals, fibers, inorganics, organics, polymers, ceramics, polymorphs, and so on. Strain-free optical components for the microscope are essential for this method. Subset techniques of PLM include dispersion staining, reflection polarization, absorption polarization, crossed polarizers, partially crossed polarizers, first-order red plate, and quartz wedge. Structural information concerning materials may be derived from orthoscopic (normal observation) and conoscopic (using Bertrand lens or observation after removal of occular, magnified with a focusing telescope) techniques. Refractive indices, crystallographic axial orientation, crystal sign, crystal family (isotropic, uniaxial, or biaxial), birefringence, optical angle ($2E$, $2V$, or $2H$) are among the quantifiable data used for identification by PLM.

Bulk and microelemental analyses of samples and composites alone often are ineffective in distinguishing among glass, asbestos, fiber glass, quartz, sand, and diatoms. Figure 8 illustrates the dramatic difference in the form taken by some of the silica-based compounds that chemical composition alone cannot detect.

All these materials (and about 30 others) are predominantly $SiO_2$. In the art world, pigments from different forms of $TiO_2$ may indicate the era in which a painting originated. Scanning electron microscopy (SEM) combined with energy-dispersive spectroscopy (EDS) can characterize materials elementally. The elemental detection range is dependent on the sensitivity of the detector and its window thickness (if present). SEM also can characterize some of these materials by size, shape, and surface characteristics, but it cannot determine internal structure or optical properties. X-ray diffraction is often useful but cannot identify noncrystalline particles, and it cannot differentiate particles differing only by shape. SEM cannot identify many organic compounds, unless they are of biological origin or exhibit characteristic external forms (e.g., pollen grains). TEM and electron diffraction require extensive sample preparation and suffer from many of the limitations mentioned earlier. Qualitative and quantitative morphological information obtained with PLM can easily distinguish a wide variety of organic and inorganic materials. By application of dispersion staining techniques, compounds with the same chemical composition, such as silicate glasses, are rapidly differentiated.

Minerals and polymers exist in many different forms, including fibers. The term ''fiber'' can be applied to any small elongated piece of matter, regardless of the source or composition of the material. Fibers are generated from biological and nonbiological, natural and man-made sources. They are essential to the

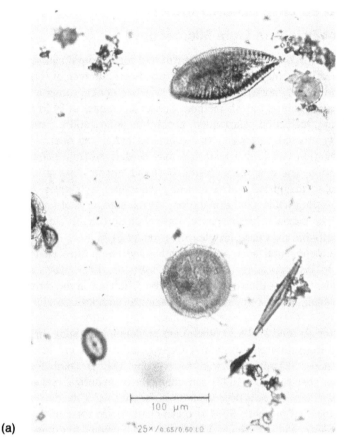

(a)

**Figure 8** Morphology of silicate materials in bright field and polarized light microscopy: (a) diatoms, bright field, (250×); (b) quartz, PLM plus first-order red plate (50×);
(c) fiberglass, PLM plus first-order red plate (100×); and (d) chrysotile asbestos, PLM
plus first-order red plate (250×).

human body, clothing, and structural components and are ubiquitous in our environment, especially as common dust. Some fibrous forms are strictly regulated in industry and/or the environment for health reasons. Examples include
asbestos because of asbestosis and textile fibers because of white lung. In examination of an area of asbestos using the phase contrast method, for example,
any fiber with an aspect ratio (length-to-width ratio) greater than 3:1, regardless
of the actual composition, is considered to be asbestos. In most forensic cases,
product contamination cases, and occupational health exposure questions, the
exact material composition is critical.

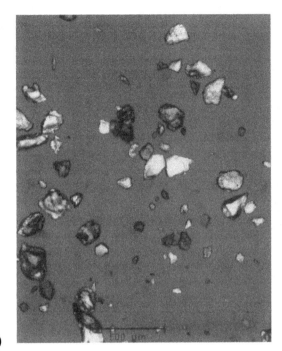

(b)

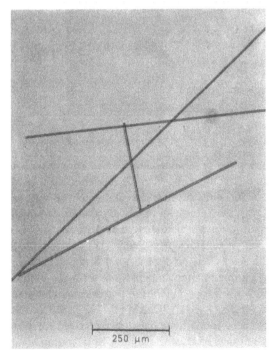

(c)

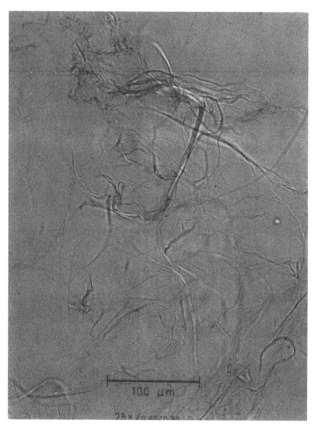

(d)

**Figure 8** *continued*

Many minerals have the same or similar chemical building blocks in their crystalline or noncrystalline structure. For example, common silicate minerals are characterized by tetrahedral units of a central silicon ion ($Si^{4+}$) associated with four oxygen ions ($O^{2-}$). As silicate minerals crystallize, tetrahedra may join by sharing oxygen atoms to form structures with widely varying Si:O ratios (e.g., single-chain 1:3 ratio pyroxenes, double-chain 4:11 ratio amphiboles, tetrahedral sheet 2:5 ratio micas or clays, and tetrahedral network 1:2 quartz). In addition, other elements (aluminum, iron, magnesium, etc.) are exchanged in the crystal lattice to form other series of minerals (e.g., the tremolite–actinolite–ferroactinolite series: $Ca_2Mg_5Si_8O_{22}(OH)_2$-$Ca_2(Mg_{0.2} \ Fe_{0.8})_5Si_8O_{22}(OH)_2Ca_2$-$Fe_5Si_8O_{22}(OH)_2$. Many of these forms have been characterized for systematic identification, as in *Dana's System of Mineralogy* [36], by quantitative measurements of optical properties (both transmitted and reflected light, depending on the transparency of the material) correlated with x-ray diffraction. Some of the compounds, such as the mineral species chrysotile and lizardite, both $Mg_3Si_2$-$O_5(OH)_4$, are polymorphs and have a similar elemental composition during polymerization; their three-dimensional crystal structures, however, are different. The fibrous polymorph chrysotile is the very common form of asbestos that is regulated by statute.

Natural and synthetic nonsilicate minerals also form fibers [37]. Sulfur and metallic elements (e.g., tellurium, arsenic, gold) form fibers near volcanic vents, as a result of their sublimation as volcanic gases escape the earth. Sulfur-containing minerals (sphalerite and wurtzite, both polymorphs of ZnS; gypsum, $CaSO_4 \cdot 3H_2O$; livingstonite, $HgSb_4S_8$; uranopilite, $(UO_2)_6(SO_4)(OH)_{10} \cdot 12H_2O$ and sulfates), vanadite, $(Pb_5(VO_4)_3Cl)$, phosphates, and arsenates are naturally occurring and have fibrous forms. Synthetic fibers are classified into man-made mineral fibers (MMMF) and man-made vitreous fibers (MMVF). MMVFs have glasslike structures, while MMMFs crystalline rather than amorphous structures.

Further distinctions are made with carbon-based compounds, such as nylons, and celulosics. When these fibers are defined less than 25 μm in diameter, they are usually called whiskers. Ceramic fibers may be defined as silicon-based fibers with higher concentrations of other elements (e.g., boron, calcium, magnesium) or these elements in combination with light and/or transition atoms (boron carbides/nitrides, aluminum nitrides, graphite alone or doped with traces of La, Y, Ce, etc.). Each of these fibers has its own distinguishing characteristics that allow its identification with the light microscope.

The light microscope is essential in the compositional analysis of mortar and cement. Method C457 (point count analysis) of the American Society for Testing and Materials (ASTM) uses direct low magnification, stereo microscopic examination, under reflected bright field illumination, to generate percentage concentrations of cement paste and aggregate and the size distribution of voids. Figure 9 depicts the hexagonal, platy calcium hydroxide deposits at the edges of the voids exhumed from a freshly cleaved sample of cement. This view provides information about the relative amounts and distribution of calcium hydroxide present in the cement material. Cleavage study also reveals the strength of the mortar, since higher strength material fractures through quartz grains, while weaker mortars fracture around these grains. Application of calculations from such data reveals the specific surface and spacing factor for the sample, using the formula

$$\bar{L} = \frac{3}{\alpha} (1.4P/A + 1)^{1/3} - 1$$

where $\bar{L}$ = spacing factor, $P$ = % paste, $A$ = % total air, $\alpha$ = specific surface (4/b), and $b$ = average width of air void.

ASTM Method C856 (petrographic examination) employs both low and high magnification with transmitted illumination through thin, polished sections

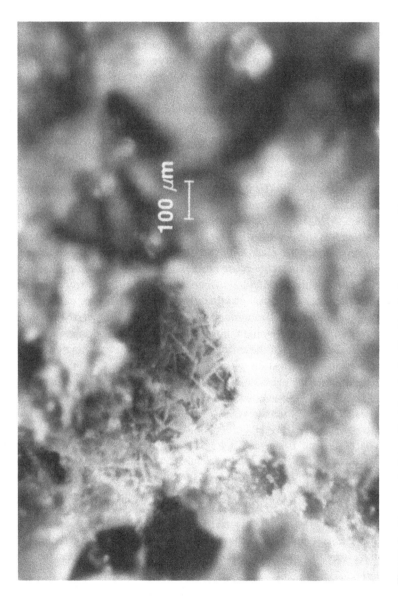

**Figure 9**  Hexagonal plates of calcium hydroxide deposits in a cement void.

(<30 μm thick). Bright field, dark field, and polarized light (crossed, slightly crossed, with/without first-order red plate) techniques provide chemical information. At low magnification (about 4×), dark field reveals corrosion products, some additives with conjugated double bonds, and homogeneity of the cement mixture. Using bright field and polarized light at magnifications of 4–64×, the state of hydration, extent of carbonation, and aggregate composition can be estimated. Clinker structures, illustrated in Fig. 10, suggest the extent of hydration in cement. In curing, normal calcium silicate reacts with water to form a donut ring boundary in the particle.

The further the ring progresses toward the center of the particle, the more hydrated the structure. In addition, hydrated particles are isotropic or, at most, weakly birefrigent. Carbonation of calcium hydroxide from exposure to air is revealed by polarized light, as shown in Fig. 11. Because calcium carbonate is more birefringent under polarized light, it appears as a lighter region in the specimen. The effect is nearly invisible using normal lighting conditions.

Individual minerals in the sample are readily detected by PLM. Examples are quartz (clear birefringent crystals with definite extinction), limestone (fine-grained aggregate that appears to be milky white without extinction), quartzite (variety of fine, randomly oriented quartz grains within a particle), feldspar (twinning lines), and marble dust (fine, birefringent shards). Using magnifications from 100× to 1000×, finer detail of the cement paste is determined. Higher magnification clarifies the busy, yellowish mottle detail in the paste to suggest substantial conversion of calcium hydroxide to calcium carbonate, while quantification of hydration in clinkers and determination of marble dust distribution is possible. At areas of crack formation or at potential fracture sites, such as that viewed in Fig. 12, the fine needle structure characteristic of ettringite $\{Ca_6[Al(OH)_6]_2 \cdot 24H_2O\}(SO_4)_3 \cdot 2H_2O$, may be observed.

Additional quantitative tables with descriptions for numerous minerals and compounds found as secondary deposits in concrete are presented in the ASTM Method C856.

## B. Pleochroism

If one rotates the microscope stage during examination of transparent particles and fibers with one polarizer in the optical path, the crystallographic axes are variably aligned with the polarizer. If the material changes color during this rotation (pleochroism), the structure has absorbed different wavelengths for different light polarization directions. Neither colorless materials nor isotropic materials (containing only one refractive index: e.g., cubic crystals like sodium chloride) can show pleochroism; only colored anisotropic substances (containing two or three different refractive indices) exhibit this phenomenon. A few common materials that exhibit this effect are crocidolite (the only pleochroic

(a)

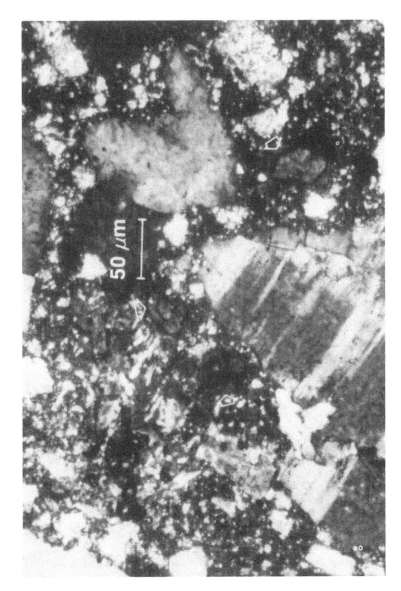

(b)

**Figure 10** Clinkers formed during cement hydration, viewed at 300× in (a) bright field and in (b) PLM.

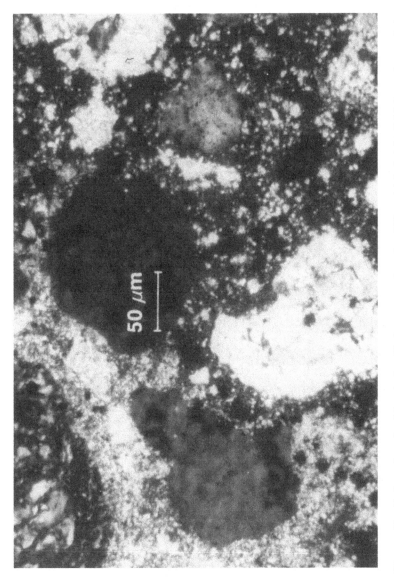

(b)

**Figure 11** Carbonation of calcium hydroxide in mortar, viewed at 300×. (a) In bright field the carbonate is invisible. (b) In PLM the carbonate appears as bright regions.

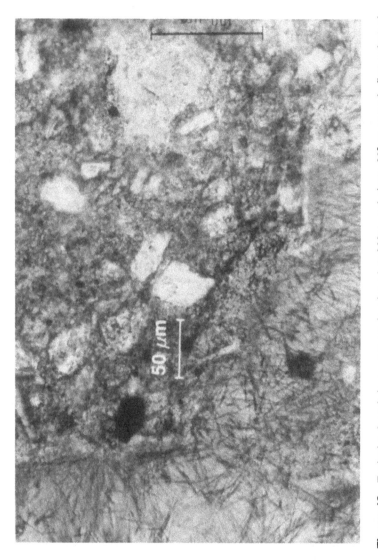

**Figure 12** Ettringite deposit in cement crack, viewed at 300×, polarizers 20° uncrossed, first-order red plate. The ettringite needles are seen at the left and in the lower half of the figure.

asbestos mineral), dyed fibers, tourmaline, rutile, azurite, and malachite. The intensity of pleochroism will depend on the orientation of the material and the magnitude of the difference between the highest and lowest refractive indices of the material (birefringence).

## C. Dispersion Staining

Another one-polarizer technique for identifying materials is dispersion staining. This technique requires a microscope equipped with an adjustable aperture diaphragm, the installation of a specially designed dispersion staining objective (either annular or central-stop design) on the microscope, and the use of high dispersion mounting media (available from Cargille). Since only the edges of the substance exhibit the effect, objectives with high numerical aperture, high magnification, or excellent resolution are not required. A $10\times$ objective magnification is normally used in this method. The sensitivity of the technique is best when using the central-stop objective with a stopped-down aperture diaphragm to provide a black background. Central-stop dispersion staining will be discussed, rather than the less sensitive annular-stop objective (white background) technique. Since the undeviated rays of the core of the particle are blocked by the design of the dispersion staining optics, the edge of the particle deviates the light path and acts like a prism in relation to the liquid in which it is immersed. The result is a colored edge on a black field, where at some wavelength of light $\lambda_0$, the refractive index of the particle and the mounting media match.

Dispersion staining colors and their associated wavelengths are summarized in Table 3.

If the mounting medium perfectly matches the refractive index of a particle, the edge will be magenta on a black background. Therefore, mounting a particle in a series of different media will generate a dispersion curve relating refractive index to a wavelength characteristic of the particle. To match the mounting media to particles with positive dispersion slopes, use the following rules for central-stop conditions:

1. If the edge color is yellow, increase the refractive index of the mounting media.
2. If the edge color is blue, decrease the refractive index of the mounting media.
3. Avoid pale colors (UV and IR). Use the deep solid colors.

As with pleochroism, the stage of the microscope is rotated to determine the color extremes of refractive indices in the orientation of the material. By iterating this exercise for several mounting media with a substance, the refractive index parallel of the polarizer orientation $n_{\parallel}$ and the refractive index perpendicular to the polarizer orientation $n_{\perp}$ can be plotted to determine the dispersion

**Table 3**  Dispersion Staining Colors

| Wavelength $\lambda_0$ (nm) | Color | |
|---|---|---|
| | Central stop | Annular stop |
| 300 | White | Black |
| 350 | Light yellow | Black-violet |
| 400 | Pale yellow | Dark violet |
| 440 | Yellow | Violet |
| 450 | Golden-yellow | Blue |
| 480 | Orange | Blue-green |
| 520 | Red-purple | Green |
| 560 | Purple | Green-yellow |
| 580 | Blue | Yellow |
| 610 | Blue-green | Orange |
| 650 | Light blue-green | Red-brown |
| 720 | Pale blue-green | Red-brown |
| 1500 | White | Black |

staining curve of a material. The dispersion staining curves for a mounting liquid and a solid are illustrated in Fig. 13.

In forensic applications, glass fragments from victims, the crime scene, or standards can be rapidly compared and distinguished by this technique, because identical materials have the same dispersion staining curves. An even more rapid comparison consists of exactly matching the refractive index of the mounting media to the standard particle in question.

For materials with identical $n_{\parallel}$ and $n_{\perp}$, the colors should match on rotation and, if the materials are mounted in exactly matching refractive index liquids, magenta edges should be encountered at each liquid. Once again, nonidentical materials will generate different colors on rotation in each refractive index mounting liquid. Therefore, since many particles can be examined on the same slide in the same mounting medium, particle identification can be quite specific and rapid.

The procedure is similar for fiber identification in dust and other fiber mixtures. Table 4 lists the conditions for identification of five common synthetic fibers. By employing only two liquids, each of the five fibers can be uniquely identified. If an unknown fiber of one of these types is to be identified, one simply mounts the unknown fiber in each of the two liquids, rotates the stage, and observes the color matches. Of course, many fibers and other materials will give similar responses and could be misidentified in a sample of completely unknown origin. Practical experience and familiarization with dispersion staining

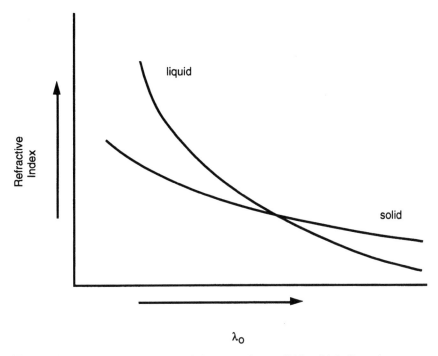

**Figure 13** Schematic dispersion staining curve for a solid in a high dispersion mounting liquid.

**Table 4** Dispersion Staining Data for Some Synthetic Fibers

| | | | | Dispersion Staining Colors | | | |
| | | | | in 1.550 HD | | in 1.520 HD | |
| Fiber | $n_{\parallel}$ | $n_{\perp}$ | $n_{\parallel} - n_{\perp}$ | $n_{\parallel}$ | $n_{\perp}$ | $n_{\parallel}$ | $n_{\perp}$ |
|---|---|---|---|---|---|---|---|
| Polyester | 1.71 | 1.54 | 0.16 | White | Pale blue | White | Pale yellow |
| Polyamide | 1.58 | 1.52 | 0.06 | Pale yellow | Pale blue | Pale yellow | Blue |
| Aramid | 2.37 | 1.65 | 0.72 | White | White | White | White |
| Polyethylene | 1.56 | 1.52 | 0.04 | Yellow | Pale blue | Pale yellow | Blue |
| Polypropylene | 1.52 | 1.49 | 0.03 | Pale blue | Pale blue | Blue | Pale blue |

data references is essential. For polymers and other materials, dispersion staining data may be found in the *Particle Atlas* [5] or other resources [38].

## D.  Crossed Polarized Light Microscopy

Examination of crystals, films, and fibers under crossed polarized filter optics in the microscope provides quantitative as well as qualitative identification parameters. Orthoscopic observation is aided through the use of the Michel–Levy chart, which correlates thickness, birefringence and retardation. The method is supplemented by application of a first-order red plate, quartz wedge, or Berek compensator. Normal observation provides one data point characteristic of the substance, whether the material is isotropic or birefringent. If the material is birefringent, the degree of birefringence may be an identifying parameter, defined by the formula

$$r = 1000tB$$

where $r$ = retardation (nm); $t$ = thickness ($\mu$m); and $B$ = birefringence, $n_2 - n_1$.

Retardation is the color hue of the sample that is defined by the wavelength of light generated by the optical path difference in the material. The eye can distinguish different color hues to about 10–20 nm, which is sufficient resolution for successful use of the technique. Black approaches zero thickness. In the thinnest areas after black, grays and yellows occur as thickness increases. Orders of red are marked successively by pastel reds at 550, 1100, 1650 nm, etc. Other colors are generated in each of the orders to correspond to a specific wavelength of light.

Specific orders may be determined from the hue by eye or by using quartz wedge or Berek compensator optics. These optics are placed in the optical train in such a way that orders of red are subtracted from the specimen as they are manipulated. The procedure is to count the number of red units subtracted until black replaces red, whereupon the color and order determine the wavelength used in the formula or found on the Michel–Levy chart. Counting must be done carefully, since the bands are close together near the edge. The thickness of the specimen in the observed orientation must also be accurately determined. For cylindrical fibers or accurately polished petrographic thin sections, this parameter is quite easily obtained.

For rapid identification, the Michel–Levy chart in the *Particle Atlas* (or those provided by microscope manufacturers) provides numerical birefringence values of several common minerals and fibers. A yellow (900 nm) fiber that is 45 $\mu$m in diameter would have a birefringence of 0.02 and is found to be viscose rayon. A similarly colored 15 $\mu$m diameter fiber would correspond to a birefringence of 0.06, which identifies it as nylon. A 24 $\mu$m sucrose crystal would have a sky

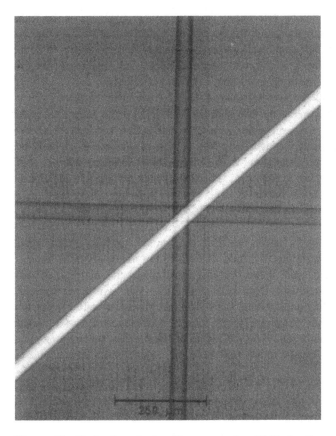

**Figure 14** Extinction of nylon fiber on rotation: PLM plus first-order red plate (10×).

blue (660 nm) color. Straw yellow (270 nm) crystal in a 30 μm petrographic thin section identifies quartz (birefringence of 0.009).

Normal observation of materials proceeds after the determination of interfacial angles and extinction (where the crystal orientation becomes dark). Extinction, illustrated in Fig. 14, is most easily determined under PLM conditions by observing a straight edge of a crystal as it is rotated with the microscope stage.

Every 90°, the crystal should become extinct, as the crystal's orientation allows light emerging from it to have the same polarization direction as the polarizer. Parallel extinction is observed if the material has extinction in the fixed axial direction of the polarizers in the microscope's optics (usually aligned north–south/east–west). Since many materials exhibit parallel extinction, those that have extinctions at some other orientation are more conveniently identified.

The extinction value can be determined from the degree markings on the rotating stage of a polarizing light microscope. For other determinations, crystals are usually observed at a 45° angle from extinction. At this position, they will be the brightest, and therefore, least aligned with the vibrational direction of the polarizer, and they will generate maximum interference effects.

Additional crystallographic information about birefringent materials can be obtained using conoscopic observation of the back focal plane generated by the specimen using higher magnification objectives and a Bertrand lens (or removing one ocular and using a magnifying lens or phase telescope to magnify the pattern). The best interference figures are detected by observing low order, preferably light gray, crystals in the field. Uniaxial crystals provide a Maltese cross interference pattern, while biaxial crystals produce split hyperbolic isogyres or brushes. These patterns are illustrated in Fig. 15 for sodium nitrate and an oriented polymer film.

Once the crystal classification has been determined, the sign of the crystal may be established using the first-order red plate. For uniaxially positive crystals, blue will be observed in the first and third quadrants and yellow in the second and fourth quadrants. Uniaxially negative crystals will generate an inverse pattern with blue in the second and fourth quadrants and yellow in the first and third. For biaxially positive crystals, blue will be observed on the outside of the isogyre arc, with yellow on the inside of the arc. For biaxially negative crystals, the inverse is again found, yellow on the outside of the arc and blue on the inside.

Additional quantitative data for determining optic axial angles for unknown biaxial materials can be calculated. A more rigorous discussion, procedures, and useful tables to calculate the values are found elsewhere [4]. Figures 16 and 17 and a brief discussion will summarize the quantitation parameters that identify materials from optical crystallography [39,40].

In Figures 16 and 17, $2V$ is the true optic axial angle that exists in the crystal; $2E$ is the apparent optic axial angle if air ($n = 1.0$) occupies the optical path between the object and the objective; $2H$ is the apparent optic axial angle if the optical path is filled with oil ($n = 1.515$). The following equations define the relationships between these parameters:

$$B \sin V = 1.00 \sin E = 1.515 \sin H$$

$$\sin E = \frac{d(\sin AA/2)}{D}$$

$$NA = 1.515 \left( \frac{\sin AA}{2} \right)$$

$$\sin H = \frac{d(NA/1.515)}{D}$$

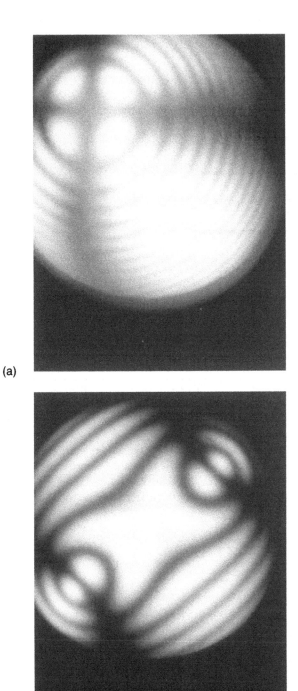

(a)

(b)

**Figure 15** Conoscopic interference figures from crystals. (a) Sodium nitrate, a uniaxial crystal, has a characteristic Maltese cross pattern. (b) An oriented biaxial polymer has characteristic isogyres.

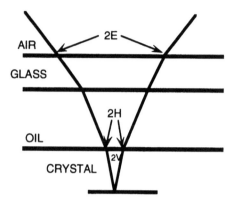

**Figure 16**   Relation between true and observed optical axis angles in a crystal. 2*E*, angle observed in air; 2*H*, angle observed in immersion oil; 2*V*, optical axis angle in crystal.

Where *B* is the refractive index for the vibration direction perpendicular to the optic axis, *d* is the distance between the centers of the brushes, *D* is the diameter of the field of view, *NA* is the numerical aperture of the objective, and *AA* is the angular aperture of the objective.

The true refractive index values ($\alpha$, $\beta$, and $\gamma$ for biaxial crystals; $\epsilon$ and $\omega$ for uniaxial crystals) may be determined for specific substance identification by establishing intermediate refractive indices, by using crystal rolling of the sample under the coverslip of a slide, or by fastening the material to the rotating needle

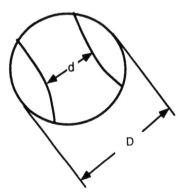

**Figure 17**   Definition of parameters in isogyre and field of view measurements: *d*, distance between center of brushes; *D*, diameter of field of view.

of the spindle stage. Crystallographic graphing and calculations provides the exact refractive indices [41].

Through observation and quantification of crystallographic parameters, materials may be identified. This is especially true when using PLM for evaluating newly developed research compounds. Pharmaceuticals, for instance, may possess different crystal forms, polymorphs, that have widely different activities, stabilities, and optical properties. The observation of pure materials and formulations for polymorphic forms and transformations aids in the determination of shelf life and effectiveness for these pharmaceuticals and can be used to measure the decay rate for each form. In adulteration and tampering cases, such as the recent Tylenol cyanide poisoning scares, PLM can rapidly screen capsule or tablet crystals for materials that have optical properties different from those of the correct ingredients. Some materials, such as the Maltese cross of starch shown in Fig. 18, provide unique identification markers.

Starch is found in pharmaceuticals, cosmetics, food products, and other substances, and its presence can be rapidly detected. In forensic studies of narcotics distribution, the rapid identification of white powders used for "cutting" or diluting street drugs, as well as the drug(s) itself, is important. The appearance of starch is different from that of sugar, and both differ from the drugs of interest. In the polymer area, spherulitic structure and material under strain affect the performance of a product. PLM often provides vital data in these determinations, where other analytical approaches may fail.

PLM plays an important role in the study of the structure of liquid crystals [42]. Usually liquid crystalline properties result from molecules that are quite elongated and flattened. Most contain parasubstituted aromatic rings and one or more polar group. Smectic forms (from the Greek word *smegma*, meaning soap) are phases in which chains of molecules lie parallel to one another in layers with long axes perpendicular to the layers. They are common to sodium and potassium salts of long chain fatty acids. Smectic phases are usually optically positive uniaxial materials. Nematic forms (from the Greek *nema*, meaning thread) are phases in which molecules also are parallel but are more random, since there is no layer arrangement. These phases are also optically positive uniaxial materials. Thin layers of these materials can be structurally ordered by using electrical potential, which is the basis for their use as digital display components. In cholesteric mesophasic forms, a nematic phase is twisted about an axis perpendicular to the long axes of the molecules. These structures are optically active and possess a negative optical sign. The changes from one form to another may be thermotropic (from a change in temperature) or lyotropic (from a change in the amount of solvent). Except for low angle resolved x-ray diffraction techniques, PLM is the only way to determine the difference between the hexagonal and lamellar forms, as shown in Fig. 19.

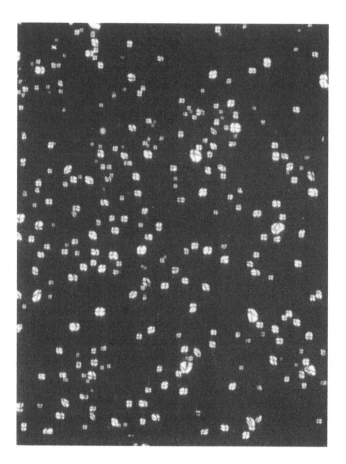

**Figure 18** Characteristic Maltese cross pattern of potato starch grains under PLM (100×).

## IV. INDIVIDUAL MICROCHEMICAL TESTS

With the advent of atomic absorption (AA) spectroscopy, inductively coupled plasma (ICP) emission spectroscopy, and x-ray fluorescence, elemental micro-chemical identification by the classical techniques of Mason and Chamot [1,2] is no longer common in analytical chemistry. These methods are still useful, however, where the element/reactant crystalline product form provides unequiv-ocal visual identification of limited amounts of material. We cite a few examples of specific analytical reagents [43]. Uranyl acetate provides faint yellow, regular tetrahedra or triangular plates with sodium, but long slender needles and rods with potassium. Zinc forms feathery crosses with potassium mercuric thiocy-

(a)

**Figure 19** (a) Hexagonal and (b) lamellar forms of liquid crystal standards viewed under PLM (100×).

anate. Squaric acid is a unique reagent that forms many distinct crystal forms with a wide variety of elements and is especially colorful in transition metal ion complexes [44–46]. Squaric acid is an ideal reagent for iron because of the habit and intense pleochroism of its ferric ion complex. The red–violet complex has a detection limit of 5 ng with a wavelength maximum at 545 nm. This reagent provides a test for iron almost two orders of magnitude more sensitive than other microchemical tests and distinguishes between the two oxidation states of iron in solution.

In the forensic world, specific chemical tests for illicit drugs (such as cocaine, heroin, new "designer" drugs), flammables, high explosives (nitrates,

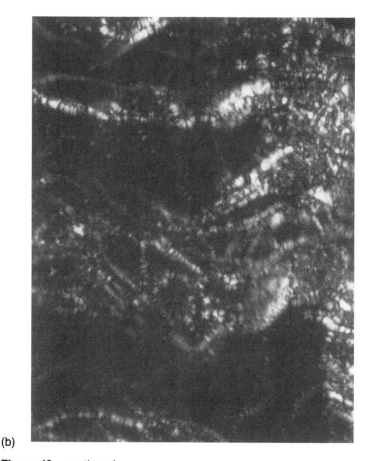

(b)

**Figure 19** *continued*

TNT, etc.), body fluids (blood, blood type), and gunshot residues afford rapid evaluation, classification, and identification of materials. Much of the information elicited by microchemical analysis provides an alternative to time-consuming, expensively maintained instrumental techniques. In addition, some of the microchemical results could not be obtained any other way from the trace quantities of material available. In courtroom trials photographs showing unique crystal structures may provide more convincing, conclusive evidence to lay jurors than spectral or numerical data.

## V. PHASE CONTRAST

Transmitted phase contrast microscopy distinguishes materials that have similar refractive indices, that is, $\Delta n < 0.1$. Normal bright field illumination would not differentiate these materials. In applications to plastics, this technique reveals the distribution of rubber particles, lattices, and different polymer blend phases, or the composition of gel phases. Particle size distribution and the extent of particle agglomeration in lotions, slurries, and reaction solutions may be characterized by phase contrast. However, the halo that surrounds individual particles and the matrix phase interface is a drawback of the procedure. Still, determination of a substance's refractive index to ±0.0001 refractive unit is possible.

## VI. HOT STAGE APPLICATIONS

Hot stage methods allow the microscopist to study the purity of a substance, determine phase diagrams, observe polymorphism, measure physical properties of the material, discover the kinetics of crystal growth, investigate boundary migration, and analyze mixtures. Information is available during both heating and cooling processes [3,47]. During heating, the investigator may observe the temperature of occurrence, if present, and occurrence of sublimation and the nature of the sublimate, polymorphism, loss of water or solvent, decomposition, and melting point. During cooling, supercooling, crystal growth, glass formation, and polymorphism, crystal anomalies (gas bubbles, shrink cracks, twinning, and unstable polymorphs) may be traced. Reheating, or alternately heating and cooling, at a particular transformation event allows more careful study of dynamic crystal alterations. These processes can be quite complicated. For example, ammonium nitrate and veronal have 5 and 11 polymorphic forms, respectively. Figure 20 illustrates some of the polymorphs of ammonium nitrate.

The study of pharmaceuticals also employs the hot stage technique [48]. With the appropriate chemical reagents, mixed fusions can reveal uniquely identifiable eutectics for many substances, solid solution type, miscibility, addition compounds, and zone of mixing events. Dinitrophenol, for example, forms identifiable eutectics with many polymers. Identification tables of organic compounds with mixed fusion provide useful reference data for the method [49].

Modern hot stage controllers, such as the Mettler FP80, also allow differential scanning calorimetry (DSC) under the microscope. An interchangeable DSC hot stage attachment (Mettler FP84) replaces the normal hot stage (Mettler FP82). Hot stage observation of events can be accomplished on any microscope equipped with long working distance objectives (LWDO). A normal lens is corrected for use with only a coverslip (size 1 1/2) that is 0.13 mm thick or for no coverslip at all (metallographic lens). LWDO objectives are necessary to protect the lens from heat and sample sublimation and to provide room for the specimen when it is surrounded by resistive heating elements.

Reaction progress and control of hot stage devices mounted on a PLM can proceed with a photomonitor to trace the loss or appearance of birefringence in materials as they are heated or cooled. With a video camera, dynamic events can be recorded and kinetically documented. The complete package allows not only visual observation of events, but quantitative photometric, DSC, and thermal

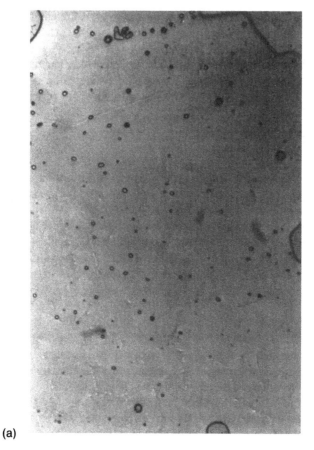

(a)

**Figure 20**  Selected polymorphic forms of ammonium nitrate: (a) isotropic melt, (b) isotropic melt with stress cracking, (c) a birefringent form, and (d) another birefringent form.

documentation, as well as accumulation of qualitative crystallographic data and kinetic information for identification of materials.

Gels in materials often are more highly crosslinked, than their matrices, hence more birefringent. Observation and temperature profile documentation of the melting behavior of a thin cross section of bulk polymer with embedded gel on the hot stage provide simultaneous, direct, quantitative comparison data for both materials. Since increased crosslinking is frequently related to increased melting temperature, the extent of crosslinking can be estimated. Anomalies in DSC data for other materials may be reconciled by observation of crystal changes with PLM.

(b)

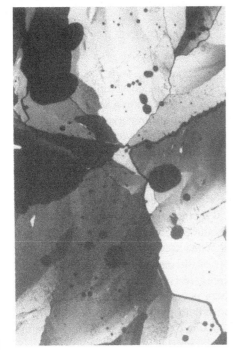

(c)

(d)

**Figure 20**  *continued*

An extension of the hot stage technique is the observation of materials during and after ignition on a platinum strip support using a small butane burner and stereo microscope. Many speck contaminations in polymers may be classified at relatively low magnification with a stereo microscope. Further confirmation may be necessary. In these cases, the concentration of specks is often extremely low, at ultratrace levels compared to the bulk sample. The surrounding organic matrix material usually interferes with the analysis. Even after exhumation of specks from the matrix, qualitative elemental identification by energy dispersive spectroscopy is often compromised by the organic continuum.

Concentration and separation of the contamination from the continuum may be accomplished through ignition of the specks. During ignition, the charring process and burning characteristics of the material are revealed. At red-heat of platinum, some of the common inorganic materials, such as titanium dioxide, zinc oxide, and zinc stearate (which is converted to zinc oxide), exhibit characteristic yellow colors, and tetrasodium pyrophosphate (TSPP) flows into a glass. On cooling, the titanium and zinc oxide return to white, while the transparent TSPP glass usually shrinks and cracks.

Many other materials have unique identifying properties. For example, polymer degradation products behave differently from rust, and polymer fibers behave differently from mineral fibers. The carbon is burned away from organic material, leaving little or no residue, while rust and other inorganic materials remain essentially unchanged. From the relative volume of residue lost, the ratio of organic to inorganic material present in the contamination can be estimated.

Care in observation must be used, however, because some inorganic materials—for example, iron halides—will decompose, altering their chemical composition. Attention to environmental and health precautions is mandatory, because the gases evolved from materials during combustion could be toxic. Once the material has been separated from the matrix, neat and residue samples may be observed using normal microscopical procedures or transferred to the appropriate device for instrumental analysis, (e.g., to a carbon or beryllium stub for EDS analysis).

## VII. CONFOCAL LASER SCANNING MICROSCOPY

Confocal laser scanning microscopy (CLSM) bridges the gap between the depth-of-field loss with increasing magnification of light microscopy and the inapplicability of the scanning electron microscope (SEM) to the analysis of wet materials or to direct imaging of the internal structure of materials or samples under actual environmental conditions (room temperature at normal atmospheric pressure). Even with the development of the environmental scanning electron microscope, the specimen is subject to a vacuum level of approximately 50 torr. As a result, confocal microscopy, especially CLSM, has been extensively used in the biological laboratory to study, by optical sectioning, materials that are electron beam or vacuum sensitive under atmospheric conditions. The three-dimensional image that is generated by the technique provides internal structures of materials that are not obtainable by any other means without physically sectioning the specimen. In addition, laser excitation and computer image enhancement make possible greater sensitivities for specimen staining and tracking stain traces in tissue. In the industrial world, the instrumentation is being increasingly used in areas such as paper and paper/binder studies, food investigations (oil and fat globule distribution in products such as margarine and mayonnaise), consumer product formulations (air and particulate segregation in lotions and shampoos), and pharmaceutical preparations (crystal morphologies in slurries and formulations).

## VIII. INSTRUMENTAL ANALYSIS COORDINATION

Although the light microscope alone is a powerful tool for characterizing the structure and chemical composition of materials, it enables the spectroscopist to use other instruments more effectively. Simple examination and classification of contaminants, material surfaces, bulk fibers, and other samples with the stereo microscope allow substantial time savings and reduce unnecessary analyses. The microscope, in many cases, provides the solution to many industrial problems without further work and in nearly always yields useful preliminary information. Its use in isolating small amounts of material that normally would be below the detection limit or lost in the confusion of a matrix takes patience and a steady pair of hands. Great care is necessary in removing tiny fragments from a matrix (especially from electrostatic media such as plastics and rigid polymer foams), followed by mounting and securing particles less than 100 μm in diameter on metal pinhole foils (10–100 μm in aperture) for IR and/or Raman spectral determinations or on carbon or beryllium stubs for electron probe analysis (EPA) and elemental SEM work. Without careful isolation of materials and elimination of contamination by the matrix from which the material was removed, the spectrometer would not provide the lowest, most reliable detection limits possible.

The extraction of fragments is not the only example of light microscopical aid to the instrumentalist. The detection and identification of materials on the surface of a substance is one goal of x-ray photoelectron spectroscopy (XPS) and scanning Auger microscopy (SAM). As these instruments are refined, their reduced beam spot size allows the area of analysis to shrink. Analysis of areas smaller than 2500 μm$^2$ is now possible by XPS. For example, to avoid the "needle-in-a-haystack" problem, these instruments employ light microscopes (and recently video cameras) for observation. The proper angle and type of illumination at the analysis surface must be selected for formation of useful images. Oblique illumination is adequate for surfaces that have rough topography but fails for smooth surfaces such as silicon wafers and semiconductor substrates. For this type of sample surface, coaxial illumination provides an image that allows precise location of the analytical region of interest.

## IX. CONCLUSION

In summary, the light microscope is a truly powerful and elegant instrument for determining the chemical state of a wide variety of materials, while giving the microscopist an aesthetic view of materials unparalleled in splendor. With proper alignment and operation, it sometimes yields information that no other technique can provide. In other cases, it can extend instrumental techniques to allow them to achieve previously impossible detection limits.

Application of light microscopical techniques to solve problems is one of the most complicated tasks in the analytical laboratory. Light microscopy requires more than just looking through the eyepiece. Light microscopy is a process of constant learning and investigation with light—a probe that is limited yet, with the proper optics, nearly unlimited. The microscope is optically attached to the human eye—a versatile, but limited detector. Microscopical information is processed by the human brain—the best computer available.

Unfortunately, colleges and universities do not teach chemical microscopy anymore. The light microscopist is truly "made, not born" in our scientific community. We have been caught up in an electronic revolution that abhors the manual and worships the electronic. In determination of the full chemical state of materials, the electronic and the manual merge to provide the best of both worlds. The light microscope takes a backseat to no instrument in its value and utility. Spectroscopists have much to gain by taking the time to learn the proper use of the light microscope or just by becoming acquainted with the light microscopists and their techniques.

## REFERENCES

1. Mason, C. W. (1958). *Handbook of Chemical Microscopy, Vol. I, 4th ed.* Wiley, New York.
2. Chamot, E. M. (1931). *Handbook of Chemical Microscopy, Vol. II*, Wiley, New York.
3. Behrends, T., and Kley, P. (1980). *Organische mikrochemische Analyse*, R. E. Stevens, transl., Microscope Publications, Chicago (original work published by Leopold Voss, Leipzig, 1922).
4. McCrone, W. C., McCrone, L. B., and Delly, J. G. (1984). *Polarized Light Microscopy*, McCrone Research Institute, Chicago.
5. McCrone, W. C., and Delly, J. G. (1973). *The Particle Atlas, Vols. I–VI*, Ann Arbor Science Publishers, Ann Arbor, MI.
6. Hemsley, D. A. (1989). *Applied Polymer Light Microscopy*, Elsevier Applied Science, New York.
7. Wilson, T. (1990). *Confocal Microscopy*, Academic Press, San Diego, CA.
8. Simpson, D., and Simpson, W. G. (1988). *An Introduction to Applications of Light Microscopy in Analysis*, Royal Society of Chemistry, Burlington House, London.
9. Tomer, A., *Structure of Metals Through Optical Microscopy*, ASM International, The Materials Information Society, USA.
10. McCrone, W. C. (1980). *The Asbestos Particle Atlas*, Ann Arbor Science Publishers, Ann Arbor, MI.
11. Hemsley, D. A. (1984). *The Light Microscopy of Synthetic Polymers*, Oxford University Press, Royal Microscopy Society, Oxford.
12. Pleom, J. S., and Tanke, H. J. (1987). *Introduction of Fluorescence Microscopy*, Oxford University Press, Royal Microscopy Society, Oxford.
13. Pawley, J. (1989). *The Handbook of Biological Confocal Microscopy*, IMR Press, Madison, WI.
14. Cocks, G. G. (1976). Chemical microscopy, *Anal. Chem.*, *48* (5):333R–340R.
15. Cocks, G. G. (1978). Chemical microscopy, *Anal. Chem. 50* (5):205R–213R.
16. Kentgen, G. A. (1982). Chemical microscopy, *Anal. Chem. 54* (5):244R–264R.
17. Kentgen, G. A. (1984). Chemical microscopy, *Anal. Chem. 56* (5):69R–83R.
18. Cooke, P. M. (1986). Review: Chemical microscopy, *Anal. Chem. 58* (5): 1926–1937.
19. Cooke, P. M. (1988). Chemical microscopy, *Anal. Chem. 60* (5):212R–226R.

20. Cooke, P. M. (1990). Chemical microscopy, *Anal. Chem. 62* (5):423R–441R.
21. *Microscope*, McCrone Research Institute, Chicago.
22. *Micron and Microscopica Acta*, Pergamon Press, Elmsford, NY.
23. *Journal of Microscopy*, Royal Microscopical Society, Blackwell Scientific Publications, Boston.
24. McCrone, W. C., and Delly, J. G. (1973). *The Particle Atlas*, Vol. IV, Ann Arbor Science Publishers, Ann Arbor, MI, pp. 857–871.
25. Brule, B., and Druon, M. (1975). Fluorescence microscopy applied to the study of thermoplastic bitumens, *Bull. Liasion Lab. Ponts Chaussées*, *79*:11.
26. Garrett, H. (1985). Use of fluorescent dye to reveal spray pattern on plant leaves, *Microscope*, *33*:115–117.
27. Cheever, G. D. (1987). Development of a rapid method for measuring solvent barrier properties of primers on SMC, *Finishing Automotive Plastics*, Federation of Societies for Coating Technology, Nov. 18–20.
28. Roberdeau, L., Bosch-Figueora, J., and Font-Altaba, M. (1975). Qualitative study of fluorescence of brilliant cut diamonds, *Gem Minerals and Gemmology, Papers*, 9th IMA Meeting, December, pp. 521–529.
29. Needham, G. (1958). *The Practical use of the Microscope*, Charles C. Thomas, Springfield, IL, pp. 101–120.
30. Krueger, D. A. (1990). Industrial use of the fluorescence microscope, *Microscope*, 38:51–59.
31. Billingham, N. C., and Calvert, P. D. (1982). Applications of ultraviolet microscopy to polymers, *Dev. Polym. Charact.*, 3:
32. Harris, J. M., and Youngman, R. A. (1990). Time-resolved luminescence of oxygen-related defects in aluminum nitride, *Mater. Res. Soc. Symp. Proc. 167*:253–256.
33. Youngman, R. A., Harris, J. H., Chernoff, D. A., and Slack, G. A. (1988). Cathodo-luminescence and photoluminescence in aluminum nitride, *Proceedings of MRS/ACS Meeting on Advanced Characterization Techniques in Ceramics*, San Francisco.
34. Delly, J. G. (1988). *Photography Through the Microscope*, 9th ed, Eastman Kodak, Rochester, NY.
35. Inoue, S. (1986). *Video Microscopy*, Plenum Press, New York.
36. Palache, C., Berman, H., and Frondel, C. (1944). *The System of Mineralogy of S. D. Dana and E. S. Dana*, 7th ed. Vol. 1, Wiley, New York.
37. Skinner, H. C. W., Ross, M., and Frondel, C. (1988). *Asbestos and Other Fibrous Materials: Mineralogy, Crystal Chemistry and Health Effects*, Oxford University Press, New York.
38. Skirius, S. A. (1986). Polymer identification by microscopical dispersion staining, *Microscope*, *34*:28–43.
39. Winchell, A. N. (1954). *Optical Properties of Organic Compounds*, 2nd ed., New York.
40. Winchell, A. N., and Winchell, H. (1964). *The Microscopical Characters of Artificial Inorganic Solid Substances: Optical Properties of Artificial Minerals*, 3rd ed., Academic Press, New York.
41. Bloss, D. F. (1981). *The Spindle Stage: Principles and Practice*, Cambridge University Press, New York.

42. Hartshorne, N. H. (1974). *The Microscopy of Liquid Crystals*, Microscopy Publications, Chicago.
43. Delly, J. G. (1989). Microchemical tests for selected cations, *Microscope*, *37*:139–166.
44. Stevens, R. E. (1974). Squaric acid: A novel reagent in chemical microscopy, *Microscope*, *22*:163–168.
45. Whitman, V. L., and Willis, W. F., Jr. (1977). Extended use of squaric acid as a reagent in chemical microscopy, *Microscope*, *25*:1–13.
46. Willis, W. F., Jr. (1990). Squaric acid revisited, *Microscope*, *38*:169–185.
47. Kofler, L., and Kofler, A. (1952). *Thermal Micromethods*, Trans. W. C. McCrone, McCrone Research Institute, Chicago (originally published as *Thermo-Mikro-Methoden zur Kennzeichnung organischer Stoffe und Stoffgemiche*, Innsbruck, Austria).
48. Haleblain, J., and McCrone, W. C. (1969). Pharmaceutical applications of polymorphism, *J. Pharm. Sci.*, *58*(8):911–929.
49. McCrone, W. C. (1957). *Fusion Methods in Chemical Microscopy*, Wiley, Interscience, New York.
50. Cook, P. M. (1992). Chemical microscopy, *Anal. Chem.*, *64*(12):219R–243R.

# 3

# Infrared and Raman Spectroscopic Imaging

**Patrick J. Treado**  *University of Pittsburgh, Pittsburgh, Pennsylvania*

**Michael D. Morris**  *University of Michigan, Ann Arbor, Michigan*

## I.  INTRODUCTION

Microscopies based on the vibrational spectrum of a molecule provide perhaps
the closest approach to universally applicable, chemically selective contrast generation techniques. Infrared and Raman microscopic imaging both use the technology of light microscopy, although in modified form. The mid-infrared and
the near-infrared are both employed for microscopic observation. Raman imaging uses the visible region. Like the underlying spectroscopies, infrared microscopy is more widely practiced than Raman microscopy. And like the underlying
spectroscopies, each has its characteristic advantages and drawbacks.

In this chapter we review the principles and major applications of both infrared and Raman imaging. We emphasize the development of the last decade
and describe emerging techniques that may become more widely used in the
near future. We briefly discuss infrared and Raman microspectroscopy, the measurement of spectra of microscopic particles. Spectroscopic imaging and microspectroscopy are closely related techniques. The vibrational spectrum of a
molecule is both rich and redundant, so that characterization of the infrared
or Raman spectrum of an unknown material is usually needed before imaging
can begin.

## II.  INFRARED MICROSCOPY

Infrared microscopy is a dynamic field, which uses techniques familiar and un-
familiar to chemists. We review the established experimental techniques and recent technological innovations in infrared microscopy as well as imaging

applications in microscale materials research. Every infrared microscopic technique combines elements of microscopy and infrared spectroscopy. Individual techniques can be distinguished by how they balance the two components, with some methods emphasizing the spectroscopy and others emphasizing the imaging features. The balance is realized, in practice, as a compromise between the need for well-resolved spectral bands essential for molecular characterization and the demand for visual clarity at high spatial resolution in order to image microscopic heterogeneities. Ultimately, specific applications determine where the tradeoffs are made.

Most chemists associate infrared microscopy exclusively with mid-infrared microspectroscopy, the acquisition of absorbance spectra from spatially isolated, microscopic regions of heterogeneous materials. Attempts to couple an optical microscope to an infrared spectrometer date back to 1949 [1,2]. Reviews of recent developments in mid-infrared microspectroscopy and its applications have appeared in the past several years [3–5]. These treatments have stressed the ability of infrared microprobes coupled to interferometers to quickly and accurately characterize microscopic features in complex matrices. Near-infrared microspectroscopy and its applications have also been reviewed [6].

Infrared microspectroscopy is commonly used in a wide array of applications including fiber analysis in forensic chemistry [7], polymer characterization [8], and biological tissue studies [9]. Infrared microspectroscopy applications blossomed in the 1980s. A comprehensive bibliography has been compiled [10].

Infrared microscopic imaging is considered to be a specialized extension of infrared microspectroscopy. This review focuses primarily on developments in infrared microscopic imaging with consideration of microspectroscopy only where appropriate. Emphasis also is placed on emerging technological developments, which may provide the kind of renaissance in infrared microscopy that Raman microscopy has enjoyed recently.

The infrared imaging techniques considered include methods in the near-, mid-, and the far-infrared regions using transmission and diffuse reflectance spectroscopy. The wavelength region and imaging technique employed are determined to a large degree by the sample composition. While the spectroscopic experimental requirements vary widely, two primary methods are generally employed for image generation. One method involves scanning of the sample in a systematic two-dimensional pattern with a point infrared source coupled to single-channel detection. The second method involves wide field microscope illumination with an infrared source tuned to a specific infrared spectral transition. Detection is performed with a multichannel detector. The majority of infrared microscopies employ some form of these approaches. Several novel imaging techniques employing alternative methods also are considered.

## A.  Near-Infrared Microscopy

Most near-infrared absorbance spectra of organic materials are generated from vibrational overtones and combination bands of OH, NH, and CH fundamental transitions [11]. The near-infrared absorbance bands between 0.7 and 2.5 μm are one to three orders of magnitude weaker than their corresponding fundamental transitions. This places some stringent experimental requirements on near-infrared spectroscopic analysis and near-infrared imaging of organic materials, namely that detector sensitivity and dynamic range be adequate to resolve small near-infrared absorbance changes in large signal backgrounds.

One advantage of near-infrared microspectroscopy and microscopy is the ability to analyze samples that are totally absorbing in the mid-infrared region [12]. Commercially available Fourier transform infrared (FT-IR) microscopes can rapidly be modified for near-infrared microscopy [8]. Figure 1 shows a common infrared microscope design [13] used primarily for mid-infrared microspectroscopy and imaging but suitable for near-infrared work with some modification. The all-reflecting microscope is achromatic and is designed to allow easy mechanical transition between visible transmission microscopy for sample placement and focus, and microspectroscopic operation in transmission or reflectance. Modification for near-infrared work requires replacement of the mid-infrared Globar light source with a quartz halogen lamp, and the use of a quartz or Si-on-CaF$_2$ beam splitter and mercury–cadmium–telluride (MCT) or indium–gallium–arsenide (InGaAs) detector. For work between 0.8 and 1.1 μm, a silicon detector is adequate.

In operation, the infrared light is focused at the sample on the microscope stage by an off-axis paraboloidal mirror or Cassegrain condenser. A Cassegrain objective collects and magnifies the infrared light and directs it to the detector. Aperturing the image to select the region to be analyzed is performed at the intermediate focal plane of the Cassegrain objective with a variable spatial filter. Some implementations employ double or *redundant aperturing* [14], which involves placement of multiple variable spatial filters at conjugate sample image planes of the Cassegrain condenser and objective to reduce stray light contributions.

Spectral maps or images can be constructed with the microprobe by translation of the sample through the spatially filtered source. Complete spectra are obtained at every position. Spectroscopic features are extracted and manipulated in software to construct individual images. The signal intensity used to construct an image from absorbance or from normalized band intensities is a function of several variables [15].

$$I \propto I([\epsilon(\nu), C(a_x, a_y, b), r(R, L), D(s_x, s_y)]) \tag{1}$$

where $I$ is the signal intensity and $\epsilon$ is the absorptivity coefficient, which is a function of $\nu$, the frequency of absorption; $C$ is the concentration, which is a

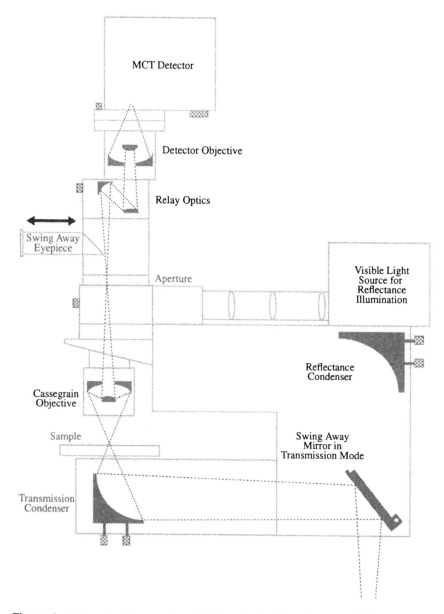

**Figure 1**   Schematic diagram of a mid-infrared reflective microscope. This instrument can be operated in either reflectance or transmission mode by operation of a swing-away mirror. (Adapted with permission from Ref. 13.)

function of the absorption cross section or sampling volume defined by $a_x$, $a_y$, and $b$. The display depth increment $r$ is defined as $R/L$, where $R$ is the range of intensity values and $L$ is the display depth. For contour plots $L$ is the total number of contour levels used to represent the image, and for bit-mapped displays $L$ is the number of gray or false color levels. $D$ represents the spatial resolution defined for scanned systems by the step sizes, $s_x$ and $s_y$.

Generation of spectral maps by sample scanning is slow. Even a crude image containing only a few hundred pixels takes several hours. Measurement time has limited the application of microspectroscopic mapping to low definition imaging in the past. While near-infrared microspectroscopy has been demonstrated [6], spectral mapping in the near-infrared is not widely used. Examples in the mid-infrared are more common.

Diffraction is the ultimate limit to spatial resolution in far field microscopies, while Seidel aberrations [16] often limit resolution in practice. Centrally obscured reflecting objectives exhibit poorer imaging performance than well-corrected refractive optics [17]. While immune to chromatic aberration, centrally obscured Cassegrain objectives are prone to spherical aberration, effectively limiting the flatness of fields for wide area imaging. In addition, the central obscuration reduces light throughput, effectively reducing image contrast. Centrally obscured objectives are limited to lower magnifications and numerical apertures (NA) than are possible with glass refracting objectives. For example, commercially available reflecting objectives typically provide magnification up to $50\times$ with NAs up to 0.65. Significantly higher magnifications and NAs are not practical with reflecting systems.

The spatial resolution of a mid-infrared microscope is limited to approximately 10–20 µm. Spatial resolution can be improved through the use of near-field or ultramicroscopy techniques [14]. Ultramicroscopy in the far-infrared has been investigated [18]. The scanned evanescent wave microscope employs an optically pumped methanol vapor laser emitting at 118.8 µm through small apertures. Spatial resolution down to 30 µm has been demonstrated.

Near-infrared microscopy can image materials at higher spatial resolution than mid- or far-infrared microscopy. For example, a reflecting microscope modified for near-infrared microscopy can image features with spatial resolution of approximately 2–5 µm, instead of the 10–20 µm mid-infrared limit. Further improvement is possible with the use of glass refractive optics. Refracting objectives corrected for the near-infrared at 0.8 and 1.3 µm can have higher NAs than reflecting objectives and can provide improved spatial resolution. For example, a high NA (0.9–1.0) refracting objective can provide spatial resolution of approximately $\lambda/2$.

Near-infrared refractive microscopes provide spatial resolution almost as great as that obtained with visible light microscopes [19]. In fact, near infrared refractive microscopes are usually constructed from visible microscopes with

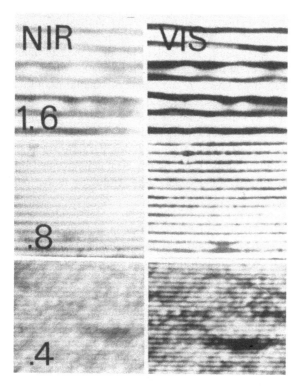

**Figure 2** Comparison of near-infrared (left) and visible (right) transmission microscope images of diffraction grating replicas. Grating lines spaced at intervals of 1.6, 0.8, and 0.4 μm are faithfully reproduced. (Adapted with permission from Ref. 19.)

only slight modification. Evidence of the imaging quality provided by a wide field illumination and detection near-infrared microscope is shown in Fig. 2, a direct comparison between a visible (0.4–0.7 μm) and near-infrared (0.8–1.1 μm) system. Figure 2 shows near-infrared and visible phase contrast images of diffraction grating replicas; the apparatus used to generate these images is comprised of a Zeiss standard RA microscope with phase contrast lenses and condenser. Near-infrared wavelengths are isolated from a quartz halogen lamp with a glass filter. The combination of the filter bandpass (0.8–4.4 μm) and a silicon vidicon camera response (0.4–1.1 μm) provides spectral response of the microscope between 0.8 and 1.1 μm. Notice that the 0.4 μm spaced grooves are still resolved in the near-infrared.

Broad band pass filters do not provide adequate selectivity for spectroscopic analysis of unknown materials. For spectroscopic imaging applications of spectrally well-characterized materials, however, this approach can be adequate.

For example, near-infrared dark field video microscopy is used for imaging band gap ($E_g$) defects in gallium arsenide (GaAs) wafers [20]. The major mid-band-gap donor defect, EL2, in semi-insulating GaAs absorbs at approximately 1 μm and can easily be visualized with a vidicon detector. Dark field microscopy is used to improve image contrast [21]. Figure 3 compares images made with bright field and dark field illumination: there is a central region of a 500 μm thick indium-doped GaAs wafer, focused at a depth of 150 μm below the surface. Features observed vary in size from 4 to 20 μm. In the bright field mode (top) the dark defects are distinguished from the light background with approximately 25% contrast. The dark field image (bottom) exhibits almost 100% contrast between the bright defects on a dark background. While the visible defect features have also been characterized as EL2 defects by complementary techniques [22,23], scattering from dust particles or indium microprecipitates tend to be visible as well. While EL2 is the major defect contributor, the spectral selectivity of broadband illumination is inadequate to distinguish between known and unknown defects with 100% confidence.

Buried layer indium–gallium–arsenide–phosphorus/indium–phosphorus (InGaAsP/InP) heterostructures have been imaged at visible and near-infrared wavelengths [24]. The microscope employs a variety of long pass, short pass, and band pass near-infrared filters in conjunction with multichannel detection. A silicon vidicon camera is employed for wavelengths below 1.1 μm, while a lead sulfide (PbS) infrared vidicon is used between 1.1 and 1.8 μm. Band gap absorptions of the InGaAsP quaternary structure at 1.3 and 1.5 μm have been used to visualize the buried structures within InP substrate during fabrication. The 1 μm spatial resolution and image contrast of the technique are adequate to show irregularities during etching or regrowth.

The point-focused laser scanning technique has been used for near-infrared imaging of semiconductor heterostructures and defects [25]. The laser scanning microscope consists of a 1 mW HeNe laser at 1.15 μm, which illuminates the GaAs wafer in transmission. At the laser wavelength, EL2 absorption can be distinguished readily. The beam is focused to a 2 μm spot at the specimen, which is mechanically scanned in an x-y raster pattern beneath the laser. Complete image frames of 512 lines are scanned in approximately 62 seconds. The transmitted beam is measured by a germanium (Ge) detector. A reference Ge detector is also employed in an analog noise reduction circuit to remove the 2% laser intensity ripple. Near-infrared laser scanned bright field images of undoped liquid encapsulated Czochralski (LEC) GaAs (001 surface) ingots are shown in Fig. 4. The broad, dark areas marked D in Fig. 4a are approximately 200–300 μm across and correspond to increased absorption at dislocation cell walls where EL2 absorption concentration is enhanced. The bright features correspond to material without EL2 defects. Figure 4b shows area $D_1$ (from Fig. 4a) at higher magnification. The dark spots of approximately 2 μm diameter are free arsenide

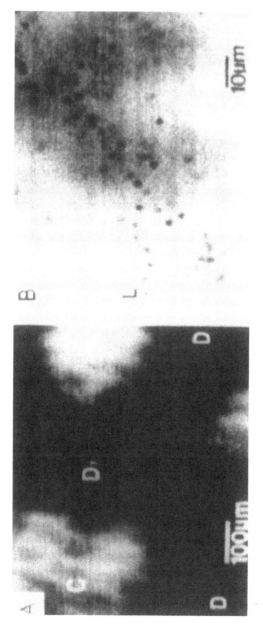

**Figure 3** Near-infrared bright field (A) and dark field (B) transmission microscope images of a 500 μm thick wafer of indium-doped GaAs. Visible features (dark regions in bright field, light regions in dark field) correspond to mid-band-gap donor (EL2) defects. (Adapted with permission from Ref. 20.)

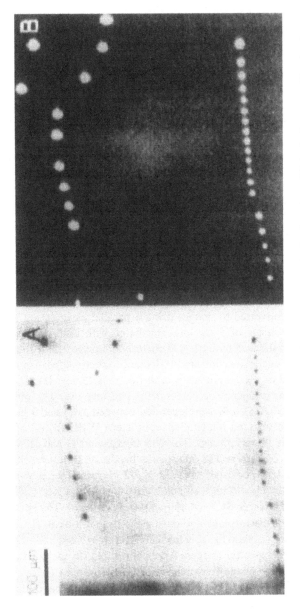

**Figure 4** Laser scanning microscope images of (001) undoped GaAs. Dark regions (D) in the low magnification image (top) correspond to EL2 absorption of the 1.15 μm He-Ne laser source. Individual EL2 defects are visible in a higher magnification image of area $D_1$ (bottom). (Adapted with permission from Ref. 25.)

particles. The spot size and density are in agreement with electron microscopy measurements [26].

The 1.15 μm wavelength is not useful for most other materials. Other near-infrared lasers can be used to image the absorptions of different substrates. For example, a 1.3 μm diode laser has been used to image oxide precipitate particles in Czochralski silicon wafers [27]. Near-infrared lasers including titanium: sapphire, f-center, and others would provide the broad tunability necessary to make laser scanned microscopy widely applicable to a variety of composite semiconductors.

Chemometrics has been applied to near-infrared reflectance scanned spectroscopic imaging. The technique combines a nonparametric multivariate algorithm and a cellular automation model to convert near-infrared spectra into digital images. Applications to the reconstruction of subsurface images [28] and imaging of low density lipoprotein (LDL) in rat arteries [29] have been reported. For arterial analysis, a fiber-optic probe that employed a compound parabolic concentrator (CPC) for efficient light compression of the near-infrared source was used. In operation, near infrared light from 1.1 to 2.5 μm was transmitted through the CPC, and diffusely reflected near-infrared light was detected by lead sulfide detectors. Spectra were acquired at scanned intervals within a 1 × 6 mm region and converted to images using a parallel assimilation algorithm. Figure 5 shows an image of a rat aorta incubated with LDL obtained by this technique. Changes induced by LDL that correspond to spectral features at 1.56 and 1.70–1.87 μm of the near-infrared reflectance spectra appear in the image.

At the other end of the length scale, airborne remote near-infrared imaging spectrometers perform macroscopic chemical state imaging. The airborne imaging spectrometers (AIS) [30] were operated between 1983 and 1987, and the airborne visible and infrared imaging spectrometer (AVIRIS) [31] began operation in 1988 and is in current use. The AIS instruments (I and II) viewed the spectral range 1.2–2.8 μm in 128 contiguous bands, at 10 nm resolution. The AIS I imaging system consisted of a 32 × 32 element mercury–cadmium–telluride array sandwiched with a silicon charge-coupled device (CCD), while AIS II consisted of a larger 64 × 64 element array detector. The spectrometers were mounted in a C130 airplane and during the time it took to fly 1 pixel (12 m), a stepping motor scanned the spectrometer gratings through the 128 spectral bands to build up the spectral images. Applications of AIS included estimation of the lignin content of whole forest canopies [32].

The AVIRIS (Fig. 6), is the next generation in airborne imaging spectrometers. The spectral imaging system is constructed of four spectrometers connected to the sensor scanning head by optical fibers. The detectors are linear arrays of silicon and indium antimonide (InSb) and consist of a total of 550 pixels across 224 contiguous spectral bands, providing 10 nm resolution. Spectral coverage is from 0.41 to 2.45 μm. The AVIRIS flies in a high altitude airplane, which

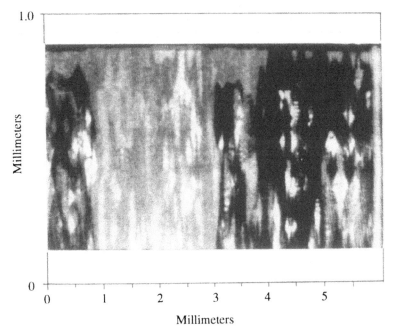

Millimeters

Millimeters

**Figure 5** Spectral image of low density lipoprotein (LDL) in a 1 × 6 mm section of rat artery. The image is reconstructed from diffusely reflected near-infrared spectral features at 1.56 and 1.70–1.87 mm using a parallel assimilation algorithm. LDL is visible as the dark region at the right. (Adapted with permission from Ref. 29.)

cruises at 65,000 feet or higher. Spatial resolution is approximately 20 m$^2$ with an 11 km$^2$ field of view. Initial studies have included detection of segregated regions of tree devastation and mineralogical variations in Nevada's cuprite mining area.

## B. Mid-Infrared Microscopy

The general principles of near-infrared microspectroscopy and imaging are applicable and widely implemented in the mid-infrared. We review imaging examples from several areas.

Line imaging is often adequate to define a sample. Figure 7 shows the chemical composition along the length of a microtomed corn kernel edge [5]. The spectral line image was acquired with a SpectraTech infrared microscope using a 32× Cassegrain reflecting objective. The 14 spectra were acquired with 8 cm$^{-1}$ resolution at 10 μm intervals with a 6 × 6 μm sampling area through the

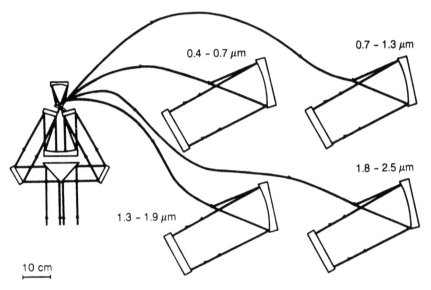

**Figure 6** Schematic diagram of the airborne visible and infrared imaging spectrometer (AVIRIS). The sensor consists of four spectrometers connected to the scanning head with optical fibers. Detection is provided by Si and InSb linear arrays for continuous wavelength coverage between 0.4 and 2.5 μm. (Adapted from Ref. 31 by permission of the American Chemical Society.)

four distinct regions of the kernel. The outermost edge of the kernel is the pericarp, which is followed by a layer of aleurone cells. Next is a band of subaleurone endosperm and a kernel core composed of starchy endosperm. The lipid carbonyl (1740 cm$^{-1}$), protein amide I (1650 cm$^{-1}$, and amide II (1520 cm$^{-1}$) bands are emphasized in Fig. 7. Amide II is the best marker for protein, because the HOH deformation band of water at 1640 cm$^{-1}$ overlaps the amide I band.

Within the endosperm, lipid carbonyl is absent and the protein level is low. The protein component of subaleurone endosperm increases, but the lipid component remains low. In aleurone, the lipid concentration increases dramatically and remains elevated into the pericarp. The high lipid content suggests that the pericarp consists, in part, of collapsed cell membrane walls.

An identical infrared microscope has been used to image in two dimensions an acrylic fiber contaminant on a microcircuit die [33]. Several methods for reconstruction of images from the mid-infrared reflectance spectrum have been compared. The results are shown in Fig. 8. The images were acquired with a 15× Cassegrain objective and a 25 × 25 μm aperture by averaging 64 scans at 8 cm$^{-1}$ spectral resolution. Image reconstruction using the aliphatic CH stretching peak height at 2968 cm$^{-1}$ yielded Fig. 8a. Integration of the CH stretching

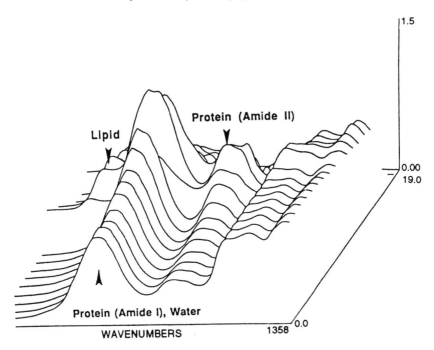

**Figure 7** Mid-infrared line scan image of a microtomed corn kernel edge at 10 μm steps. The 14 spectra, displaying between 1800 and 1358 cm$^{-1}$, were acquired with a 6 × 6 μm sampling aperture at 8 cm$^{-1}$ spectral resolution. (Adapted with permission from Ref. 5.)

peak from 2850 to 3000 cm$^{-1}$ was used for Fig. 8b, and full spectrum integration between 1000 and 4000 cm$^{-1}$ was used to generate Fig. 8c. All spectra were baseline subtracted before image reconstruction. Finally, the Gram–Schmidt orthogonalization method [34] was employed to reconstruct Fig. 8d. This chemometric approach utilized a 10-spectrum basis set and 100 data points of the interferogram for data analysis.

From visual inspection of Fig. 8 it is evident that peak height and peak integration are superior image reconstruction methods. Where isolated spectral markers for reconstructing images are not available, the Gram–Schmidt method is superior to whole spectrum integration. For completely unknown materials, where marker bands are not well established a priori, Gram–Schmidt reconstruction might be a generally suitable method.

Functional group imaging has been used to characterize heterogeneities in polymeric substrates [15]. Defects in polyethylene film are barely discernible when viewed by a visible transmission microscope. In contrast, a computer-controlled, scanning infrared microscope generates the functional group image

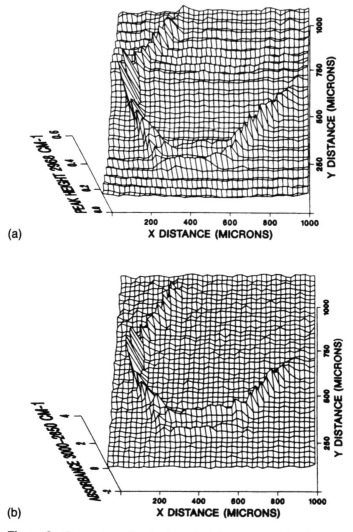

(a)

(b)

**Figure 8** Comparison of analysis methods for reconstructing images of an acrylic fiber contaminant on a microcircuit die. Image generation methods include the use of (a) absorbance intensity at 2968 cm$^{-1}$, (b) integrated absorbance between 2850 and 3000 cm$^{-1}$, (c) integrated absorbance from 1000 to 4000 cm$^{-1}$, and (d) the Gram–Schmidt method. (Adapted with permission from Ref. 33.)

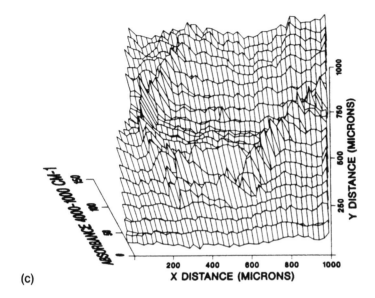

(c)

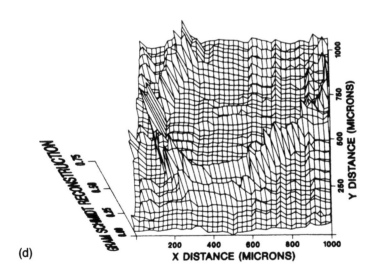

(d)

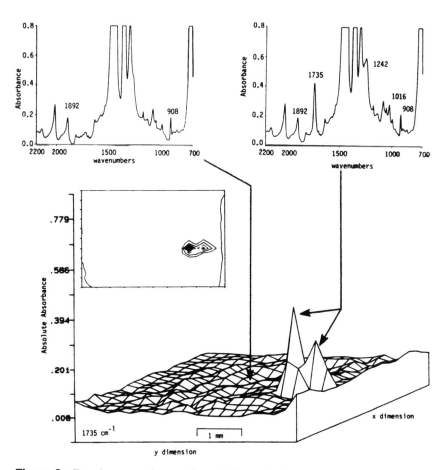

**Figure 9** Function group image of an ethylene–vinyl acetate heterogeneity in polyethylene film using the 1735 cm$^{-1}$ carbonyl absorption band. Images are displayed as a topographical map with a contour plot inset. Representative spectra between 2200 and 700 cm$^{-1}$ are included. (Adapted with permission from Ref. 15.)

of Fig. 9, and displays the defects clearly. The 20 × 20 pixel spectral map was generated using a 32× Cassegrain objective. A 250 × 250 μm aperture defined the pixel size, which is equal to the translation stage step size. At each pixel, 32 scans were averaged at 8 cm$^{-1}$ resolution.

The dimensions and orientation of the feature in the image agree with the dimensions of the defect barely seen under visual microscope observation. The axonometric and contour images are based on the 1735 cm$^{-1}$ ester carbonyl frequency. Other spectroscopic evidence suggests that the defect material probably is an ethylene–vinyl acetate copolymer.

Spectral images can be reconstructed from vibrational spectra by the methods discussed earlier. Another method is to use miniplots for display of spatially resolved spectra [35]. Miniplots are one-inch-square plots with relevant regions of the spectrum overlaid. The miniplots can be arranged in a two-dimensional array, corresponding to the spatial position from which the spectra were obtained. While the miniplot spatial "map" is not a conventional image, the display method does present complex information in an interpretable form. As the sample complexity increases, however, the miniplots become unwieldly unless the overlaid spectra are reduced in number. While not widely employed for spectral image display, the miniplot may find application as a graphic approach to spectral image archiving.

Mid-infrared macroscopic imaging combining Hadamard transform spatial multiplexing and Fourier transform infrared spectroscopy has been reported [36]. The imaging technique is similar to that developed for Raman microscopy (Section III). In the hybrid Hadamard–Fourier technique, a two-dimensional $15 \times 17$ element Hadamard mask is placed at the sample compartment focal plane of an interferometer. For each of the 255 independent mask positions, an FT-IR spectrum is collected. After the full complement of spectra has been obtained, specific vibrational frequencies corresponding to marker bands for the imaged sample can be extracted to vector data sets. Application of the Hadamard transform to these specific infrared frequency data sets is sufficient to generate coarse images of marker band absorbances.

## C. Emerging Technologies

Sensitive infrared multichannel detectors may soon have the impact on infrared imaging that diode array and CCD detectors have had on Raman imaging [38,39]. A variety of near- and mid-infrared, solid state, focal plane array detectors have the potential for analogous impact in infrared vibrational spectroscopy and imaging [40,41]. Several attractive infrared detector technologies for spectroscopy have been described [42]. Many of these devices are hybrid materials composed of epitaxial metallic layers on a semiconductor substrate, usually silicon. The photodetection mechanism is based on the generation of thermal carriers in the metallic layer upon exposure to infrared radiation with subsequent internal photemission into the silicon substrate.

A small selection of detector materials (and operating wavelengths) includes platinum silicide (PtSi, 1.0–5.7 µm), palladium silicide (Pd2Si, 1.0–3.2 µm), iridium silicide (IrSi, 1–8.2 µm), indium antimonide (InSb, 1.0–5.5 µm), indium gallium arsenide/indium phosphorus (InGaAs/InP, 1.0–1.7 µm), germanium (Ge, 1.0–1.6 µm), and mercury cadmium telluride (MCT, 2.5–12 µm). This list is by no means complete. Other exotic materials have been developed for infrared detection. A few of these materials have been incorporated into commercially available detectors. These include InGaAs, PtSi, Ge, and InSb focal plane arrays.

Historically these detectors have been used in defense and space imaging applications. They are only now migrating into spectroscopic laboratories as infrared cameras, enabling new spectroscopic and spectroscopic imaging applications. One potentially powerful imaging technique would integrate an infrared camera and an infrared microscope with a broadly tunable, high resolution optical filter. The spectral filtering method could be based on an interferometer design, a classical dispersion system, or some novel filtering technique. The resulting system would combine the selectivity of spectroscopy and the large format area of infrared array detectors to yield real time infrared video microscopy for strong infrared absorbers. For weak absorbers, particularly in the near-infrared, slow scan video detectors can be employed for higher sensitivity at much faster imaging times than are now possible.

Preliminary work in the author's laboratory (P.J.T.) has yielded promising results with a novel vibrational spectroscopic imaging approach. We couple a diffraction-limited refractive microscope to a near-infrared acousto-optic tunable filter [43] (AOTF) to provide broad source tunability over 0.6–1.1 μm with $25$ cm$^{-1}$ resolution. Visible/near-infrared detection is provided with a high dynamic range silicon CCD. The general approach is feasible and potentially would have the greatest impact if applied further into the infrared. Such a system would integrate an AOTF infrared source with a reflective infrared microscope and would employ an infrared focal plane array detector.

Alternative filtering approaches to the AOTF are plausible as well, including the use of tunable Fabry–Perot etalons in the near- and mid-infrared. These devices provide high throughput and high spectral resolution (1–5 cm$^{-1}$), making them ideal for imaging of spectrally well-characterized systems.

## III.  RAMAN MICROSCOPY

### A.  Instrumentation

Since development of the first practical systems in the early 1970s, the Raman microscope has emerged as a powerful sampling system, with applications to many different fields. Virtually all designs use the epi-illumination system developed by Delhaye and Dhamelincourt [44].

Most microscopes are attached to conventional dispersive Raman spectrographs with array detectors, or to scanning Raman spectrometers equipped with photomultipliers and photon counting. Triple spectrographs (i.e., a double subtractive Czerny–Turner stage operated as a filter, followed by a single Czerny–Turner stage for spectral dispersion) are commonly employed. Single-stage Czerny–Turner spectrographs equipped with prefilters are increasingly common. Scanning double monochromators are also used, although these are un-likely to be chosen for new installations. Fourier transform Raman microscopy is possible, using a microscope coupled to a Michelson interferometer.

The bulk of the applications have been to Raman microspectroscopy, that is, to the use of the device for obtaining a spectrum from a small, well-defined re-gion of a material. In this mode, the system is usually called a microprobe. Ra-man imaging, after some early demonstrations to strong scatterers, was dormant for some years. With advances in spectroscopic technology, imaging has re-cently been enjoying renewed popularity.

Rosasco has provided an excellent discussion of the optics of Raman micros-copy, as well as a review of early applications [45]. Much of the discussion of the optics remains pertinent, although hardware has changed in the last decade. Schrader has recently presented an integrated treatment of the optics of Raman and infrared microprobes, emphasizing throughput constraints [46]. The Fourier transform Raman microprobe has been discussed by Messerschmidt and Chase [47], Bergin [48], and Sommer and Katon [49].

A schematic diagram of a typical Raman microprobe is shown as Fig. 10. The microscope is usually a modified research grade fluorescence or metallo-graphic microscope, employing epi-illumination. Typically, there is also provi-sion for illuminating the sample with incoherent light, parallel to the incident laser beam, for visual inspection. Almost all commercial instruments (and many home-made ones) include a video camera and monitor for observation of the sample and positioning of the laser beam. In addition to the very real quantita-tive and convenience advantages of video microscopy [50], substitution of a camera for the traditional binocular eyepieces is an important safety feature. For safety reasons, the Raman microscopes in the senior author's laboratory are equipped for video only.

Well-corrected achromatic or apochromatic objectives are usually used on microprobes. Depending on the objective, the magnification will be $10\times-100\times$ and the numerical aperture will be 0.5–0.9. Some workers also employ water immersion objectives, which are typically $50\times/1.0$ NA or $100\times/1.2$ NA. In the ultraviolet and the near-infrared, Schwarzchild (reflecting objectives) have been used. Refractive objectives are available for the ultraviolet, down to 250 nm, and for the near-infrared (out to 2 $\mu$m). Refractive optics provide higher numer-ical aperture and no central obscuration; the objectives are limited production items, usually more expensive than reflective designs. Although visible light ob-jectives transmit in the near-infrared, most are not corrected for this region and can suffer from severe chromatic aberration.

The lateral resolution of a good microscope objective is nearly diffrac-tion limited. If the laser beam is expanded to fill the objective, the diameter $d$ of a laser focused through an objective onto an object can be approximated by Eq. (2).

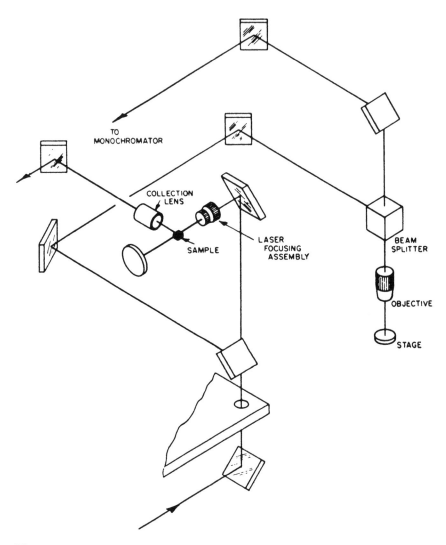

**Figure 10** Optical diagram of a Raman microprobe. A swing-away mirror is used to divert the input laser beam from the macro sample chamber to the microscope objective and stage. Another swing-away directs the Raman scatter from either sample system to the monochromator. (From Ref. 51).

$$d = 1.22 \frac{\lambda}{NA} \tag{2}$$

For green radiation ($\lambda = 0.5$ $\mu$m), the focused beam diameter can approach 1 $\mu$m.

Although the microprobe has excellent spatial sensitivity, it also generates high local power densities. If the incident laser bean of irradiance $I_0$ (W/cm$^2$) is absorbed by a particle of radius $r_p$ (cm) in contrast with a substrate or host of thermal conductivity $K$ (W/cm/K) with efficiency $Q_a$, then the steady state temperature rise, $\Delta T_s$ is given by Eq. (3) [45].

$$\Delta T_s = \frac{Q_a I_0 r_p}{4K} \tag{3}$$

For a typical transparent material, $Q_a$ is approximately $1 \times 10^{-4}$. If the sample is in contact with air ($K = 2.6 \times 10^{-4}$W/cm/K), the temperature rise can be quite high ($> 200°$C). However, for a particle in contact with a glass substrate, or other reasonably good thermal conductors ($K = 0.1$ W/cm /K), the temperature rise will be 10–20°C if the laser power is held to a few milliwatts.

Similar considerations hold for local illumination of an extended sample. A tightly focused beam can cause large local temperature rises if the laser power exceeds 2–10 mW. The problem can be alleviated by using line-focused or global illumination, as described below.

Resonance Raman microspectroscopy is possible but usually requires some form of external cooling. Cold stages, which are common microscope accessories, are a suitable approach. Alternatively, one can increase the thermal conductivity of a water-compatible sample by immersing it in water. This may be done by use of a water immersion objective, which will also provide higher numerical aperture than a conventional objective. In some cases resonance enhancement is great enough to allow reduction of sample power to microwatts. Where the resonance-enhanced scattering intensity increases by a greater factor than the heat generation caused by absorption, there is a net improvement in the heat dissipation problem. This gain can often be realized in practice.

Fluorescence interference is less troublesome with the Raman microprobe than with macro sample illumination systems [51]. Three factors contribute to this effect. First, with visible excitation, most fluorescence is caused by impurities, not by the sought component(s) of the sample. Impurities are typically heterogeneously distributed, so that a weakly fluorescent or nonfluorescent region can often be found. Second, the high laser flux from a tightly focused laser beam will burn out fluorescence more quickly than in a macro system. Third, because fluorescent lifetimes are nanoseconds long, excited state migration often causes emission to occur several micrometers away from the illumination point. Spatial filtering at any of several intermediate image planes in the optical system can block much of this radiation.

The possibility of thermal damage requires relatively low laser power densities. In turn, the intensity of the Raman scatter is much lower than is normally encountered with macro sampling systems. The microscope itself is usually the limiting optical conductance, or etendue, in the total system. Therefore, the total Raman signal is proportional to the microscope conductance $(NA)^2/M^2$, where NA is the numerical aperture and M the magnification of the microscope objective. For a given objective optical design, numerical aperture increases, but less than proportionately, with magnification. Consequently, high magnification, and with it high spatial resolution, will be obtained at the cost of a weaker signal.

A 50/50 beam splitter is commonly used in the Raman microprobe. The device is reasonably achromatic. It is inefficient because half the incident radiation and half the collected Raman scattered light are lost. For spectroscopy, some workers prefer to use a beam splitter that allows 90% transmittance and 10% reflectance. The system laser usually provides much more power than can be safely used, so attenuation of the incoming radiation by the beam splitter is less important than efficient transmission of the scattered light. A dichroic mirror can be used in place of the beam splitter, allowing 90% reflection of the incident laser and 90% transmission of the Raman scatter. However, a separate dichroic mirror is needed for each laser line, and the transmission efficiency of most dichroic mirrors falls off rapidly below about 800 cm$^{-1}$. Ultimately, holographic filters, now widely used as prefilters in Raman spectrographs, may provide sharp edge dichroic beam splitting in Raman microscope and microprobe applications as well.

Polarization ratio measurements under a microscope require special precautions [52,53]. Both the illumination and the gathering angles through microscope objectives are large. For example, NA 0.85 corresponds to a gathering angle of 39.1°. The standard formulas, which are based on the assumptions of illumination with collimated light and collection through a small angle, do not apply. Neglecting the polarizing effects of the microscope optics themselves, for isotropic samples, the polarization ratio $\rho$ is given by Eq. (4).

$$\rho = \frac{(A + B)\beta^2}{15A\alpha^2 + \left(\frac{4}{3}A + B\right)\beta^2} \tag{4}$$

where $\alpha$ and $\beta$ are the mean polarizability and anisotropy of the polarizability tensor, as conventionally defined [44], and A and B are defined by Eqs. (5) and (6), where $\Theta_m$ is the gathering half-angle of the objective.

$$A = \pi^2\left(\frac{4}{3} - \cos \Theta_m - \frac{1}{3}\cos^3\Theta_m\right) \qquad (5)$$

$$B = \left(\frac{2}{3} - \cos \Theta_m + \frac{1}{3}\cos^3\Theta_m\right) \qquad (6)$$

For $B \ll A$, Eq. (4) becomes the following familiar polarization equation.

$$\rho = \frac{3\beta^2}{45\alpha^2 + 4\beta^2} \qquad (7)$$

The apparent polarization ratio is a function, then, of the illumination angle. For totally polarized bands, Eq. (4) may not be adequate, because it neglects the depolarizing effects of the microscope objective itself. Similar effects are observed in single-crystal polarization measurements.

Most Raman microprobes have been based on wide field microscopes. Recently, confocal designs using an aperture placed at the back focal plane of the objective have been reported [4,54,55]. The Instruments SA Raman microprobe incorporates an aperture, although as noted above, fluorescence rejection, rather than rejection of out-of-focus Raman light, appears to be the reason for inclusion of a spatial filter. Puppels and coworkers [54,55] have incorporated a spatial filter in a Raman microprobe, explicitly for confocal operation. Their design is otherwise a fairly conventional epi-illumination system, coupled to a prefilter/single-stage spectrograph/array detector system. Puppels and coworkers point out that careful optimization of aperture diameter is necessary to optimize the weak Raman signal.

Imaging Raman microscopy, using epi-illumination, is instrumentally quite similar to fluorescence imaging. Inherently weak Raman signals make image acquisition slow compared to fluorescence microscopy. Most of the practical techniques were understood (and many were demonstrated) early in the development of Raman microscopy [44].

The sample can be illuminated globally, and the spectrograph operated as a filter [44]. In addition to requiring a highly stigmatic spectrograph, this approach requires wide entrance slits, causing loss of resolution and susceptibility to stray light. Consequently, it is satisfactory only for use with isolated lines from strong scatterers having little fluorescent background.

In principle, the spectrograph can be replaced by any narrow band filter, such as an interference filter, holographic filter, or interferometer, in exact analogy to fluorescence microscopy. If the filter is not tunable, a complete spectrum might be obtained with a tunable laser and fixed filters. This approach was attempted by Delhaye and Dhamelincourt [44] but failed for lack of adequate filters. Even now, unless three or four are stacked, existing filter systems cannot adequately reject laser light to make this approach practical for most real samples. However, rapid advances in technology may soon make microscopes and filters a viable Raman imaging approach, particular for quality control or other applications where the underlying Raman spectrum of the object is already known.

Recently, global illumination has been combined with one-dimensional spatial Hadamard multiplexing for Raman imaging [56,57]. The image is multiplexed in the spatial direction parallel to grating dispersion and is imaged directly parallel to the spectrograph entrance slit. An image is recovered by computer processing after a series of exposures with different multiplexing masks. With a two-dimensional detector, multispectral imaging is possible [58]. This approach allows high incident powers and correspondingly strong Raman signals, so that image acquisition times are short. We have called the incident power increase a power distribution advantage, in contrast to the conventional (detector noise) multiplex advantage normally associated with Hadamard transform or Fourier transform spectroscopy.

The Hadamard-multiplexed Raman microscope (Fig. 11) uses the conventional epi-illumination mode to generate Raman scatter, with fore-optics modified for global illumination. A magnified image is projected onto an encoding (Hadamard) mask, where the spatial dimension parallel to the dispersion direction of the spectrograph is encoded through a series of apertures. The encoded image is then compressed into a line image, with the unencoded dimension focused onto the spectrograph slits unchanged. This transformation is accomplished by a cylindrical focusing/compression system. A cylinder lens is used to collimate the image in the encoded axis only, and an achromat–meniscus pair focuses the image to the desired shape. The instrument uses a 255-element encoding mask and requires 255 exposures to generate a complete $255 \times 256$ pixel image. Compared to raster scanning of a focused laser beam, this approach reduces local power densities by a factor of more than 65,000. In practice, high total power (100–500 mW for nonabsorbing samples) can be used, so that high definition images can be built up in 3–10 minutes.

The laser can be focused into a line with cylindrical lenses. Raman scatter from the line is imaged on an array detector, oriented parallel to the spectrograph entrance slit. The image is built up by moving the sample under the line or by scanning the line across the sample. With a linear detector, a Raman image at a single band is obtained. With a two-dimensional (video) detector, multiband (multispectral) imaging is possible. Compared to point scanning, this approach reduces local power density by a factor of 100–500. Versions of this technique are in current use [59–61]. Although microscope objectives are most commonly used in Raman imaging, high quality camera lenses provide good images if a resolution of about 8–10 μm is acceptable [16,17].

The focused laser beam can be raster scanned across the sample, or the sample can be systematically moved under the laser beam. With an array detector,

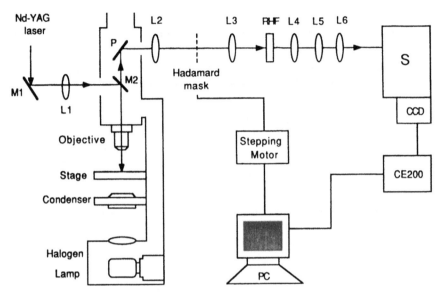

**Figure 11**  Schematic of a Hadamard Raman microscope: M1, input turning mirror; L1, 50 mm $f_1$ defocusing lens; M2, dichroic beam splitter; L2, 5× projection eyepiece; L3, 120 mm $f_1$ collimating lens; RHF, 532 nm holographic notch filter; L4, 160 mm $f_1$ projection lens; L5, −100 mm $f_1$ cylindrical lens; L6 +100 mm $f_1$ cylindrical lens; S, spectrograph. (From Ref. 58).

multispectral imaging is possible. This technique is slow and is most useful for coarse mapping or for relatively low resolution images. Raster scanning of the sample stage is readily automated, however [16,17,62,63].

## B.  Imaging Applications of the Raman Microscope and Microprobe

The Raman microprobe has several important areas of application. Although the principal use is microspectroscopy, the microprobe is practical for rough mapping, particularly when only linear or radial distributions are needed. In such cases, 10–20 spectra are used to define the spatial features, and the microscope state may be manually scanned or incremented. We briefly review applications and manual mapping.

### 1.  Raman Microspectroscopy and Line Imaging

Semiconductor structural features, at least for current-generation devices, are large enough for easy observation by visible light microscopy, including Raman microscopy. Raman scattering is a nondestructive microprobe for several properties of devices based on both silicon and gallium arsenide [64]. Phonon frequencies are a function of strain for both silicon and gallium arsenide structures. The silicon mode (ca. $520 \text{ cm}^{-1}$) decreases in frequency with increasing stress, so that its frequency reports crystal strain at boundaries. Since the frequency range is only $1-2 \text{ cm}^{-1}$, precise high resolution measurements are necessary. With a microprobe, it is possible to construct rough cross-sectional maps—for example, across a region of 10–20 μm—which reveal the extent of strain across a feature. In polar semiconductors, such as gallium arsenide, gallium–aluminum–arsenide, and zinc selenide, the free carrier plasma is a space charge with an electric field, which can interact with LO phonon modes. If the plasmon and phonon frequencies are similar and the plasmon damping is small, two well-defined bands are observed. Peak intensity monitors carrier concentration. Raman scattering is also widely used to monitor damage by ion implantation in silicon as the decrease in intensity of a silicon crystalline lattice vibration. Quantitative measurements are possible by normalization to scattering intensity from an undamaged region.

Thermometry and radial thermal profiling of silicon on silica and sapphire substrates is possible, using the $520 \text{ cm}^{-1}$ phonon band [65]. This band shifts to lower frequency and broadens as temperature is increased. Two-beam heat and probe thermometry with 1 μm resolution is possible, up to close to the melting point of silicon. Silicon is strongly absorbing in the midvisible, and small thin disks of silicon (4–20 μm diameter) can be melted by 20–100 mW of continuous wave power from an argon laser. Temperature calculations are complicated by the presence of lattice strain, which also shifts phonon frequencies and bandwidths, and by the large temperature gradients themselves.

Shifts in the phonon bands of graphitic carbon have been used for spatially resolved Raman thermometry of carbon fibers. Using a strong point-focused beam to heat the fiber and a weak line-focused probe, Ager and coworkers have mapped the temperature distribution along an 8 μm diameter fiber [66]. As shown in Fig. 12, the temperature varies from about 150°C at the fiber ends to about 750°C at the illumination point. The measurements were made in a line imaging mode and required 10 minutes.

The same group has provided an elegant demonstration of the utility of line imaging, mapping local changes in structure of a diamond film grown by chemical vapor deposition [67]. They have mapped the change in the frequency of the diamond $1331 \text{ cm}^{-1}$ phonon with $4 \times 15$ μm pixels. Because the frequency shift over a 2 mm length is less than $2.5 \text{ cm}^{-1}$, the usual tactic of sacrificing spectral resolution for signal intensity could not be used. At $0.5 \text{ cm}^{-1}$ spectral resolution, image acquisition required 1.5 hours with 400 mW laser power. The frequency shift, which is correlated with bandwidth and is attributed to lattice strain, caused the film to be thermally stressed during the cooling step of its fabrication.

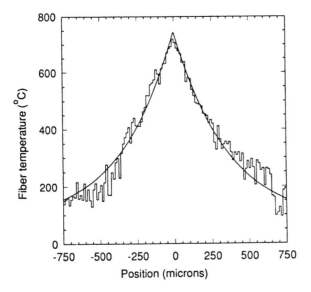

**Figure 12** Temperature distribution along a graphite fiber as measured by shift in G-band frequency (1582 cm$^{-1}$). The smooth curve is the theoretical distribution for uniform heating with adsorbed power 20 mW and convective cooling with fiber thermal conductivity 1.1 W/cm/K. The jagged line connects experimental points. (From Ref. 66).

The Raman microprobe is a nondestructive probe of fluid inclusions in minerals [68,69]. Except for $SO_4^{2-}$ and $HS^-$, most polyatomic anions are present at concentrations too low for Raman measurement, but it is sometimes possible to identify monatomic ions, including $Na^+$, $K^+$, $Ca^{2+}$, $Mg^{2+}$ and $Fe^{2+}$ from bands of their water complexes, normally in the range of 350–600 cm$^{-1}$. The major use is to measure the permanent gases, $CO_2$, $CO$, $CH_4$, $C_2H_6$, $H_2$, $H_2S$, $SO_2$, $N_2$, and $O_2$, which may be present at several atmospheres. Measurement of composition can be made with high precision, often approaching ±1%. Accuracy is limited by the accuracy of Raman cross sections and by the accuracy of the correction factors used to extrapolate to high pressure conditions. Pasteris and coworkers [68] have shown that the calculated mole fractions can vary by a factor of 2, depending on the extrapolations and corrections used.

The Raman microprobe has several applications in disease diagnosis [70], although it remains a research tool rather than a routine clinical instrument. The Raman spectrum can identify inclusions of foreign materials in tissues. Examples include mineral particles in the lung tissues of silicosis victims, fragments of prosthetic silicone implants that have broken loose from their implantation sites, and the compositions of gall stones and kidney stones. Infrared microscopy can provide similar information, although with lower spatial resolution.

Both techniques are used in gallstone and kidney stone research [71]. Little Raman imaging has yet been carried out.

The Nelson group [72] demonstrated that resonance Raman microprobe spectra can be obtained from the carotenoid pigments of individual freely moving bacteria. Using a microscope with extended-ultraviolet fluorite optics, Sureau and coworkers [73] recorded spectra from the DNA of a single living tumor cell. With a single-grating spectrograph and multichannel detection, spectra could be obtained in 100 seconds using 10 µW 257 nm from a frequency-doubled argon ion laser. The relative ease with which spectra can be obtained suggests that it is possible to use UV Raman spectroscopy as a probe of some dynamic events, such as metabolism, or of the effects of introduced materials, at the single-cell level.

## 2.   Two-Dimensional Raman Imaging

Raman imaging is potentially a powerful tool in materials science, with applications to technologically important classes of materials including semiconductors, graphitic materials and diamond films, and oxide superconductors. Lattice modes are a good indicator of crystal structure, and defects are easily mapped by imaging on phonon Raman bands. Gross composition can be mapped on intramolecular vibrations. Most workers have employed 5–30 µm pixels, rather than the 0.5–1 µm pixels of which the technique is capable.

Laboratories investigating new Raman imaging instrumentation have typically included inorganic crystals among their demonstration images. Many are strong Raman scatterers and they are inexpensive, commonly available, and undamaged by exposure to air, humidity, or moderate laser power. Examples include Hadamard Raman images [10] of potassium nitrate and raster-scanned multispectral images of potassium dihydrogen phosphate.

Using line imaging and mechanical sample scanning, Dauskardt and coworkers [74] have studied the monoclinic/tetragonal phase transformations around a crack in partially stabilized zirconia. Using the integrated intensity under the $181-192$ cm$^{-1}$ band pair as a measure of monoclinic zirconia and the intensity under the 264 cm$^{-1}$ band as a measure of tetragonal zirconia, these investigators were able to map fractional content of the monoclinic phase with a spatial resolution of about 27 µm. More recently, the experiment has been refined to a resolution of about 16 µm (7.84 µm steps) [75]. Zirconia morphology has also been mapped with 5 µm pixels by point–point scanning [76]. Medium definition (70 × 100 pixels) multispectral images are obtainable in about 10 hours.

A typical example [75] of transformation zone morphology is shown in Fig. 13, which indicates the changes in the local crack shape as different stresses are applied. The zirconia sample is a difficult one because it is translucent and scatters light, thus reducing contrast and resolution. Nonetheless, the apparent half-width obtained by this technique agrees well with that obtained by Nomarski interference microscopy.

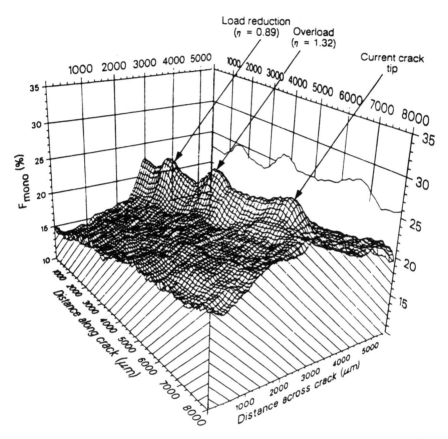

**Figure 13** Transformation zone morphology of a zirconia crystal, as measured by Raman microscopy. The crack growth is followed as the fraction of the monoclinic phase, seen as an increase in integrated Raman intensity at 181 and 192 cm$^{-1}$. The ratio of steady state stress to applied stress at a point is denoted by $\eta$. (From Ref. 75).

Treado and coworkers [56] have employed Hadamard Raman microscopy to image edge plane microstructures in laser-modified highly ordered pyrolytic graphite (HOPG) electrodes. They have shown that the defects are uniformly distributed. This work also illustrates one advantage of global illumination for imaging. Graphite has a large coefficient of expansion in the direction perpendicular to the sheet structures. Sufficient laser power for Raman imaging heats the graphite and changes its position. In this example, 250 mW causes the graphite to expand about 40 μm along the optical axis during image acquisition, blurring the Raman image almost beyond recognition. Preequilibration for about 15 minutes brings the sample to thermal equilibrium, so that sharp images are obtainable. With a raster scan or even with a line scan imaging system, it would be difficult to achieve thermal equilibrium at high enough power levels for rapid imaging.

The mixed semiconductor $AnS_xSe_{1-x}$ prepared by vapor phase deposition is spatially inhomogeneous, as shown by Raman mapping of the ZnS phonon band at ca. 300 $cm^{-1}$ [77]. Figure 14 shows spatial distribution of ZnS on the (100) plane of a crystal of the semiconductor as a function of mole-fraction sulfide. At 0.04 sulfide, the distribution is reasonably uniform. As the sulfide content is increased, stripes oriented along the (110) growth boundary begin to appear. These are not homogeneous, but are clustered at intervals of about 100 μm.

Less work has been done on organic molecules and polymers than on inorganic materials. Demonstration images of benzoic acid, naphthalene and N, N-dimethyl p-nitroaniline have been obtained with the Hadamard transform Raman microscope [10]. The resolution is 1.2–5 μm, and the images are 127 × 128 pixels. The substituted nitroaniline is interesting because it is a low melting (159°C) compound that absorbs green laser light. Preresonance Raman imaging is possible with 6 mW, 532 nm distributed over the crystal. The same power tightly focused causes local melting of the crystal.

The Hadamard microscope has been used to obtain multispectral images of a polystyrene–polyethylene laminate (Fig. 15), using the 1450 $cm^{-1}$ $CH_2$ wagging mode of both polymers and the polystyrene-specific phenyl stretch at 1590 $cm^{-1}$. In the aromatic ring mode–$CH_2$ region, there are few windows in which the spectra of these two polymers do not overlap. Although the authors image on only two bands and use simple subtraction to identify the two polymers, a multivariate approach may prove a more powerful general approach to polymer mapping.

## 3. Depth-Resolved Imaging

Confocal Raman images have not yet been reported. The computationally simple nearest-neighbors image restoration algorithm [78] has been used to sharpen Hadamard Raman images of benzoic acid [79].

## 4. Surface Raman Imaging

Very recent work has demonstrated that surface plasmon enhanced Raman spectroscopy may be a practical contrast generation technique for imaging monolayer or multilayer ultrathin films on metallic substrates. The lateral resolution is still determined by diffraction of visible light, however, and remains in the micrometer range. The technique may prove useful for imaging film defects or decomposition on certain metal surfaces, as well as for other specialized studies. Alternatively with a homogeneous film, surface Raman imaging can provide spatially localized information about the structure of the underlying metal surface.

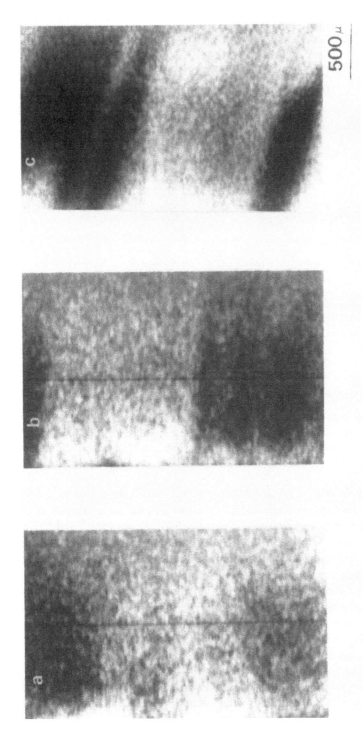

**Figure 14** Distribution of sulfur in $ZnS_xSe_{1-x}$ mapped as intensity of the ZnS 300 cm$^{-1}$ band: (a) $x = 0.04$, (b) $x = 0.07$, and (c) $x = 0.1$. (From Ref. 77.)

500 $\mu$

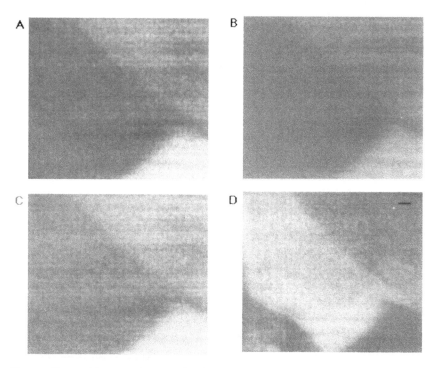

**Figure 15** Multispectral images of polystyrene–polyethylene laminate obtained with 532 nm excitation: (A) 1450 cm$^{-1}$ (polystyrene + polyethylene); (B) 1525 cm$^{-1}$ (fluorescence background); (C) 1600 cm$^{-1}$ (polystyrene); and (D) bright field transmission. (From Ref. 58.)

Laser damage to roughened silver electrodes has been imaged as surface-enhanced Raman spectroscopy of pyridine [80]. Green (514.5 and 532 nm) radiation focused to about 6 μm radius gives an apparent radius, as mapped by pyridine surface-enhanced Raman scattering (SERS), which is greater than the focused radius of the Gaussian laser beam, as measured by diffuse reflectance. Even with 15 mW focused on the sample, the SERS radius is about 10 μm, compared to an actual radius of about 6μm (Figure 6). As the power is increased, the discrepancy increases. This behavior is attributed to greater surface damage in the middle of the beam, with radially decreasing damage. Because the SERS enhancement is greatest on the undamaged surface, the apparent SERS radius is greater than the true beam radius.

Surface plasmon field-enhanced Raman imaging on silver layer gratings on glass substrates has been shown for cadmium arachidate films [81]. High con-

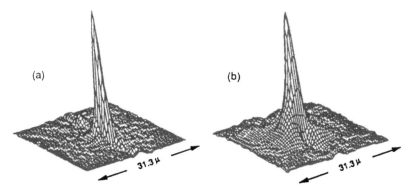

**Figure 16** SERS image of laser-illuminated roughened silver electrode: (a) reflectance, 514.5 nm; and (b) pyridine SERS, 1005 cm$^{-1}$. The apparent beam waist ($1/e^2$ intensity decrease) is approximately 4 μm greater in the SERS image than reflectance (From Ref. 80.)

trast C—H stretch images have been obtained for 8-molecule thicknesses. Plasmon luminescence can also be imaged. SERS, the most common version of plasmon field-enhanced Raman scattering, usually refers to enhancement on roughened electrodes, silver island films, colloids, and other irregular surfaces. Enhancement on gratings or other periodic surfaces shows stronger angle dependence than is observed on the irregular surfaces commonly used for SERS.

## 5. Coherent Raman Imaging

A scanning coherent anti-Stokes Raman scattering (CARS) microscope has been constructed and used to image biological samples. Imaging on O—D of tissue water and C—D stretches of lipids in liposomes has been demonstrated [82]. Micrometer resolution has been demonstrated. CARS is a difficult experimental technique, which has fallen from favor among biological Raman spectroscopists. However, the signal is generated at the common focus of two laser beams. For this reason, the coherent Raman microscope, like the closely related two-photon fluorescence microscope [83], has inherently good depth resolution.

## IV. CONCLUSION

Together, infrared and Raman imaging offer the chemist a powerful set of tools for characterization of the spatial distribution of subtle molecular properties. Between them, the techniques can be used to image virtually any condensed phase material. That a light microscope is, or can easily be, part of every imaging system is an important strength. Except for microscopies based on electronic spectroscopy, no other imaging techniques offer such direct connections to the visible world.

## REFERENCES

1. Barber, R., Cole, A. R. H., and Thomas, H. W. (1949). Intra-red spectroscopy with the reflecting microscope in physics, chemistry and biology, *Nature*, *163*:198.
2. Gore, R. C. (1949). Infrared spectrometry of small samples with a reflecting microscope, *Science*, *110*:710.
3. Messerschmidt, R. G., and Harthcock, M. A., (1988). Infrared microspectroscopy: Theory and applications in *Practical Spectroscopy Series*, Vol. 6, (Messerschmidt, R. G., and Harthcock, M. A., Eds.), Dekker, New York.
4. Roush, P. B., Ed. (1987). The design, sample handling, and applications of infrared microscopes, *ASTM Spec. Tech. Publ. 949*. American Society for Testing and Materials, Philadelphia.
5. Reffner, J. A. (1990). Molecular microspectral mapping with the FT-IR microscope, *Inst. Phys. Conf. Ser. 98*:559.
6. Smith, M. J., and Carl, R. T. (1989). Applications of microspectroscopy in the near-infrared region, *Appl. Spectrosc. 43*:865.
7. Grieve, M. C., and Kotowski, T. M. (1977). The identification of polyester fibers in forensic science, *J. Forensic, 21*:390.
8. Benedetti, E., Galleschi, F., D'Alessio, A., Ruggeri, G., Aglietto, M., Pracell, M., and Ciardelli, F. (1989). Microscopic FT-IR analysis of blends from functionalized polyolefins and polyvinyl chloride or polystyrene. *Chem. Macromol. Symp.* 23:265.
9. O'Leary, T. J., Engler, W. F., and Ventre, K. M. (1989). Infrared microspectroscopy of human tissue, *Appl. Spectrosc. 43*:1095.
10. Nyquist, R. A., Leugers, M. A., McKelvy, M. L., Papenfuss, R. R., Putzig, C. L., and Yurga, L. (1990). Infrared spectrometry, *Anal. Chem. 62*:223R.
11. Wetzel, D. C. (1983). Near-infrared reflectance analysis, *Anal. Chem. 55*:1165A.
12. Hill, S. L., and Krishnan, K. (1988). Some applications of the polarized FT-IR microsampling technique, in *Practical Spectroscopy Series*, Vol. 6 (Messerschmidt, R. G., and Harthcock, M. A., Eds.), Dekker, New York, p. 115.
13. Shearer, J. C., and Peters, D. C. (1987). Fourier transform infrared microspectrophotometry as a failure analysis tool, *ASTM Spec. Tech. Publ. 949*: (Roush, P. B., Ed.), American Society for Testing and Materials, Philadelphia, Ann Arbor, p. 27.
14. Messerschmidt, R. G. (1987). Photometric considerations in the design and use of infrared microscope accessories, in *ASTM Spec. Tech. Publ. 949* (Roush, P. B., Ed.), American Society for Testing and Materials, Philadelphia, Ann Arbor, p. 12.
15. Harthcock, M. A., and Atkin, S. C. (1988). Imaging with functional group maps using infrared microspectroscopy, *Appl. Spectrosc. 42*:449.
16. Born, M., and Wolf, E. (1980). *Principles of Optics*, 6th ed., Pergamon Press, Elmsford, NY.

17. Heimann, P. A., and Urstadt, R. (1990). Deep ultraviolet microscope, *Appl. Opt.* 29:495.

18. Massey, G. A., Davis, J. A., Katnik, S. M., and Omon, E. (1985). Subwavelength resolution far-infrared microscopy, *Appl. Opt.* 24:1498.

19. Albrect-Beuhler, G. (1984). Movement of nucleus and centrosphere in 3T3 cells, in *Cancer Cells*, Vol. 1, *The Transformed Phenotype* (Levine, A. J., Van de Woude, G. F., Topp, W. C., and Watson, J. D., Eds.), Cold Spring Harbor Laboratory, Cold Spring Harbor, NY, p. 87.

20. Montgomery, P. C., and Fillard, J. P. (1990). Near-infrared dark-field microscopy with video for studying defects in III–V compound materials, *Meas. Sci. Technol.* 1:120.

21. Loveland, R. P. (1981). *Photomicrography, A Comprehensive Treatise*, 2nd ed., Wiley, New York.

22. Weyher, J. L., and Van de Ven, J. (1987). Selective etching and photoetching of gallium arsenide in chromium trioxide–hydrofluric acid aqueous solutions. III. Interpretation of defect-related etch figures, *J. Cryst. Growth*, 78:191.

23. Moriya, K., and Ogawa, T. (1983). Observation of lattice defects in GaAs crystals by infrared light scattering tomography, *Jpn. J. Appl. Phys.* 22:L207.

24. Sartorious, B., and Rosenzweig, M. (1986). Wavelength-selective infrared microscopy: A nondestructive and quick method for the investigation of buried structures in optoelectronic devices, *J. Appl. Phys.* 60:3401.

25. Kidd, P., Booker, G. R., and Stirland, D. J. (1987). Infrared laser scanning microscopy in transmission: A new high-resolution technique for the study of inhomogeneities in bulk GaAs, *Appl. Phys. Lett.* 51:1331.

26. Stirland, D. J. (1986). *Semi-Insulating III–V Materials*, Hakoni, Ohmsha, p. 81.

27. Laczik, Z., Booker, G. R., Bergholtz, W., and Falster, R. (1989). Investigation of oxide particles in Czochralski silicon heat treated for intrinsic gettering using scanning infrared microscopy, *Appl. Phys. Lett.* 55:2625.

28. Lodder, R. L., and Hieftje, G. M. (1988). Surface image reconstruction by near-infrared reflectance analysis, *Appl. Spectrosc.* 42:309.

29. Lodder, R. L., Cassis, L., and Ciurczak, E. W. (1990). Arterial analysis with a novel near-IR fiber-optic probe, *Spectroscopy*, 5:13.

30. Geotz, A. F. H., Vane, G., Solomon, J. E., and Rock, B. N. (1985). Imaging spectrometry for Earth remote sensing, *Science*, 228:1147.

31. Amato, I. (1988). Remote sensing: A distant view of chemistry, *Anal. Chem.* 60:1339A.

32. Wessman, C. A., Aber, J. D., Peterson, D. L., and Melillo, J. M. (1988). Remote sensing of canopy chemistry and nitrogen cycling in temperate forest ecosystems, *Nature*, 335:154.

33. Ward, K. J. (1989). Applications of image analysis for infrared microspectroscopic detection of contaminants on microelectronic devices, *Proc. SPIE*, 1145:212.

34. Hanna, D. A., Hangac, G., Hohne, B. A., Small, G. W., Niebolt, R. C., and Isenhour, T. L. (1979). A comparison of methods used for the reconstruction of GC-FT/IR chromatograms, *J. Chromatogr. Sci.* 17:423.

35. Milledge, H. J., and Mendelssohn, M. J. (1988). Infrared microspectroscopy with special reference to computer controlled mapping of inhomogenous specimens, in

*Analytical Applied Spectroscopy* (Creaser, C. S., and Davies, A. M. C., Eds.), Royal Society of Chemistry, London, p. 217.

36. Zhang, F., and Gu, T. (1990). Hadamard transform FT-IR spectrometry for two-dimensional surface imaging spectroscopy, *Proc. SPIE, 1205*:150.

37. Treado, P. J., and Morris, M. D. (1989). Hadamard transform Raman microscopy, *Appl. Spectros. 43*:190.

38. Bowden, M., Gardiner, D. J., Rice, G., and Gerrard, D. L. (1990). Line-scanned micro-Raman spectroscopy using a cooled CCD imaging detector, *J. Raman Spectrosc. 21*:37.

39. Treado, P. J., Govil, A., Morris, M. D., Sternitzke, K. D., and McCreery, R. L. (1990). Hadamard transform Raman microscopy of laser modified graphite electrodes, *Appl. Spectrosc. 44*:1270.

40. Chase, B. (1991). Array detector technology beyond 1 micron, 18th Annual FACSS, Anaheim, CA.

41. Chase, B., and Talmi, Y. (1991). The use of a near-infrared array detector for Raman spectroscopy beyond 1 micron, *Appl. Spectrosc. 45*:929.

42. Dereniak, E. L., and Sampson, R. E., Eds. (1990). Infrared detectors and focal plane arrays, *Proc. SPIE*, and *1308*.

43. Bennett, J. M. (1976). Tunable acousto-optic filters, *Appl. Opt. 15*:2705.

44. Delhaye, M. and Dhamelincourt, P. (1973). Raman microprobe and microscope with laser excitation. *J. Raman Spectrosc. 3*:33.

45. Rosasco, G. J. (1980). Raman microprobe spectroscopy, in *Advances in Infrared and Raman Spectroscopy*, Vol. 7 (Clark, R., and Hester, R. E., Eds.), Heyden and Son, London, p. 223.

46. Schrader, B. (1990). Micro methods in infrared and Raman spectroscopy, *Fresenius J. Anal. Chem. 337*:824.

47. Messerschmidt, R. G., and Chase, D. B. (1989) FT-Raman microscopy: Discussion and preliminary results, *Appl. Spectrosc. 43*:11.

48. Bergin, F. J. (1990). A microscope for Fourier transform Raman spectroscopy, *Spectrochim. Acta, 46A*:153.

49. Sommer, A. J., and Katon, J. E. (1991). The development of a Fourier transform Raman microprobe, *Appl. Spectrosc. 45*:527.

50. Inoue, S. (1986). *Video Microscopy*, Plenum Press, New York.

51. Adar, F. (1988). Developments of the Raman microprobe—Instrumentation and applications, *Microchem. J. 38*:50.

52. Turrell, G. (1984). Analysis of polarization measurements by Raman spectroscopy, *J. Raman Spectrosc. 15*:103.

53. Bremard, C., Dhamelincourt, P., Laureyns, J., and J. Turrell, G. (1985). The effect of high numerical aperture objectives on polarization measurements in micro-Raman spectrometry, *Appl. Spectrosc. 39*:1036.

54. Puppels, G. J., de Mul, F. F. M., Otto, C., Greve, J., Robert-Nicoud, M., Arndt-Jovin, D. J., and Jovin, T. (1990). Studying single living cells and chromosomes by confocal Raman microspectroscopy, *Nature, 347*:301.

55. Puppels, G. J., Colier, W., Olmikhof, J. H. F., Otto, C., de Mul, F. F. M., and Greve, J. (1991). Description and performance of a highly sensitive confocal Raman microspectrometer, *J. Raman Spectrosc 22*:217.

56. Treado, P. J., and Morris, M. D. (1990). Multichannel Hadamard transform Raman microscopy. *Appl. Spectrosc 44*:1.

57. Treado, P. J., A. Govil, Morris, M. D., Sternitzke, K. D., and McCreery, R. L. (1990). Hadamard transform Raman microscopy of laser-modified graphite electrodes, *Appl. Spectrosc. 44*:1270.

58. Liu, K.-L. K., Cheng, L.-H., Sheng, R.-S., and Morris, M. D. (1991). Multispectral Hadamard transform Raman microscopy, *Appl. Spectrosc. 45*:000.

59. Viers, D. K., Rosenblatt, G. M., Dauskardt, R. H., and Ritchie, R. O. (1988). Two-dimensional spatially resolved Raman spectroscopy of solid materials, *Microbeam Analysis*—1988 (Newbury, D. J., Ed.), p. 179.

60. Viers, D. K., Ager, J. W., Loucks, E. T., and Rosenblatt, G. M. (1990). Mapping materials properties with Raman spectroscopy using a 2-D detector, *Appl. Opt. 29*:4969.

61. Bowden, M., Gardiner, D. J., Rice, G., and Gerrard, D. L. (1990). Line-scanned micro-Raman spectroscopy using a cooled CCD imaging detector, *J. Raman Spectrosc. 21*:37.

62. Gardiner, D. J., Littleton, C. J., and Bowden, M. (1988). Automated mapping of high-temperature corrosion products on iron chromium alloy using Raman microscopy, *Appl. Spectrosc. 42*:15.

63. Watanabe, M., and Ogawa, T. (1988). Raman scattering and photoluminescence tomography, *Jpn. J. Appl. Phys. 27*:1066.

64. Nakashima, S., and Hangyo, M. (1989). Characterization of semiconductor materials by Raman microprobe, *IEEE J. Quantum Electron. QE-25*:965.

65. Pazionis, G. D., Tang, H., and Herman, I. P. (1989). Raman microprobe analysis of temperature profiles in CW laser heated silicon microstructures, *IEEE J. Quantum Electron. QE—25*:976.

66. Ager, J. W., Viers, D. D., Shamir, J. and Rosenblatt, G. M. (1990). Laser heating effects in the characterization of carbon fibers by Raman spectroscopy, *J. Appl. Phys. 68*:3598.

67. Ager, J. W., Viers, D. K. J., and Rosenblatt, G. M. (1991). Spatially resolved Raman studies of diamond films grown by chemical vapor deposition, *Phys. Rev. B, 43*:6491.

68. Pasteris, J. D., Wopenka, B., and Seitz, J. C. (1988). Practical aspects of quantitative laser Raman microprobe spectroscopy for the study of fluid inclusions, *Geochim. Cosmochim. Acta, 52*:979.

69. Dubessy, J., Poty, B., and Ramboz, C. (1989). Advances in C-O-H-N-S fluid geochemistry based on micro-Raman spectrometric analysis of fluid inclusions, *Eur. J. Mineral. 1*:517.

70. Ozaki, Y. (1988). Medical application of Raman spectroscopy, *Appl. Spectros. Rev. 24*:259.

71. Ishida, H., Kamoto, R., Uchida, S., Ishitani, A., Iriyama, K., Tsukie, E., Shibata, F., Ishihara, K., and Kameda, H. (1987). Raman microprobe and Fourier transform infrared microsampling studies of the microstructure of gallstones, *Appl. Spectrosc. 41*:407.

72. Dalterio, R. A., Nelson, W. H., Britt, D., Sperry, J., and Purcell, F. J. (1986). A resonance Raman microprobe study of chromobacteria in water, *Appl. Spectrosc. 40*:271.

73. Sureau, F., Chinsky, L., Amirand, C., Ballini, J. P., Duquesne, M., Laigle, A., Turpin, P. Y., and Vigny, P. (1990). Ultraviolet micro-Raman spectrometer: Resonance Raman spectroscopy within living cells, *Appl. Spectrosc. 44*:1047.

74. Duaskardt, R. H., Viers, D. K., and Ritchie, R. O. (1989). Spatially resolved Raman spectroscopy study of transformed zones in magnesia–partially-stabilized zirconia, *J. Am. Ceram. Soc. 72*:1124.

75. Dauskardt, R. H., Carter, W. C., Viers, D. K., and Ritchie, R. O. (1990). Transient subcritical crack-growth behavior in transformation toughened ceramics, *Acta Metall. Mater. 38*:2327.

76. Bowden, M., Dickson, G. D., Gardiner, D. J., and Wood, D. J. (1990). Automated micro-Raman mapping and imaging applied to silicon devices and zirconia ceramic stress and grain boundary morphology, *Appl. Spectrosc. 44*:1679.

77. Sakai, K., Sawahata, K., and Ogawa, T. (1990). A study on defects in a $ZnS_xSe_{1-x}$ crystal by Raman scattering tomography, *J. Cryst. Growth, 103*:61.

78. Agard, D. A., Hiraoka, Y., Shaw, P., and Sedat, J. W. (1989). Fluorescence microscopy in three dimensions, in *Methods in Cell Biology*, Vol. 30. *Fluorescence Microscopy of Living Cells in Culture*: Part B, *Quantitative Fluorescence Microscopy—Imaging and Spectroscopy* (Taylor, D. L., and Wang, Y.-L., Eds.), Academic Press, New York, P. 353.

79. Govil, A., Pallister, D. J., Chen, L.-H., and Morris, M. D. (1991). Optical sectioning Raman microscopy, *Appl Spectrosc. 45*:0000.

80. McGlashen, M. L., Guhathakurta, U., Davis, K. L., and Morris, M. D. (1991). SERS microscopy: Laser illumination effects, *Appl. Spectrosc. 45*:543.

81. Knobloch, H., and Knoll, W. (1991). Raman imaging and spectroscopy with surface plasmon light, *94*:835.

82. Duncan, M. D., Reintje, J., and Manuccia, T. J., (1985). Imaging biological compounds using the coherent anti-Stokes Raman microscope, *Opt. Eng. 24*:352.

83. Denk, W., Strickler, J. H., and Webb, W. W. (1990). Two-photon laser scanning fluorescence microscopy, *Science, 248*:73.

# 4
# Fundamentals of Computer Image Processing

**Robert A. Morris**  *University of Massachusetts at Boston, Boston, Massachusetts*

## I. INTRODUCTION

Image processing systems include familiar optical systems, complex laboratory apparatus, robot vision, human vision, and more. For a system to be labeled "image processing," an informal but necessary condition is that it transform images in a way that permits the extraction of information more easily than is represented in the original. We will make a more precise definition which happens to admit of the destruction of information also, but the usual point to doing this is to reveal some otherwise obscured information. To analyze such a system we need a definition: an *image* is a continuous function $I(x,y)$ on the real plane taking nonnegative real values. We should think of an image as a picture, with $I(x,y)$ representing the intensity of the picture at $(x,y)$.

In this chapter, we will restrict our attention to monochromatic images. Often, color images are modeled by combinations of several (typically three) monochrome images of different colors, together with functions which describe the spectral distribution in the image. For details see Refs. 1 and 2.

Images arising from the real world have further restrictions, which usually play a role in their analysis. They are typically nonnegative and bounded, because neither negative nor unbounded intensity is physically meaningful. Furthermore, although we will have important occasions to talk about periodic images, those that arise directly from pictures of the world will also be of finite extent (or more precisely, will be zero outside some finite region of the plane).

Images arising from scenes in the natural world typically arise from the reflection of incident light on the scene. Such images can conveniently be regarded as the product of the illumination $i(x,y)$ and the reflectance $r(x,y)$:

$$I(x,y) = i(x,y)\, r(x,y) \tag{1}$$

where $0 \le r(x,y) \le 1$. Since $I(x,y)$ is positive and bounded, we will have some bounding intensities $I_{min}$ and $I_{max}$, and it is common to normalize $I$ by shifting the origin to $I_{min}$. In this case, we often speak of $I_{min} = 0$ as *black* and $I_{max}$ as *white*. (Often we further normalize by dividing $I(x,y)$ by $I_{max}$, but in digital image processing we also find it convenient to have $I_{max} = 2^N - 1$ for suitable $N$.)

Use of the words "black" and "white" arises from our visual perception of scenes in the world. Many image processing techniques originate in the desire to overcome limitations imposed by the visual system or by whatever may have recorded or transmitted a representation of the scene. From certain common experiences, we can deduce that our visual perceptions cannot be determined on a pointwise basis solely from the intensities in the image. For example, the black ink on this page has a reflectance much less than 0.1, but the outdoor illumination on a bright day is more than 100 times that in a typical office. Thus, Eq. (1) shows that outdoors, the text on this page might have 10 times the intensity the white part has when viewed indoors. Nevertheless, after a suitable period of adaptation, we perceive both cases to be black text on a white page. Contemporary explanations of visual perception often rest on models arising from image processing theory [3]. This is not surprising, since, as we remarked above, the historical purpose of image processing techniques was to enhance perception. Conversely, in recent years, substantial attention has been given to modeling machine vision on the neural models underlying visual image processing [4,5].

## II. TRANSFORMING IMAGES

We can give a simple definition of the fundamental mechanism of image processing: an *image transform* $T$ is a function from the set of images to itself. $T$ is *linear* if

$$T(I_1 + I_2)(x,y) = T(I_1)(x,y) + T(I_2)(x,y)$$

and

$$T(\lambda I_1)(x,y) = \lambda T(I_1)(x,y) \tag{2}$$

for any images $I_1$ and $I_2$ and scalar $\lambda$.

Linear transforms have advantages familiar to all physical scientists, but many image transforms have a further important property: $T$ is called *shift invariant* if

$$T(I(x - x_0, y - y_0)) = T(I)(x - x_0, y - y_0) \tag{3}$$

for every image $I$ and points $(x_0, y_0)$. This means that shifting an image before transforming it has the same result as shifting it afterward.

We will shortly see that linear shift-invariant transforms are quite easily modeled. But note that the human visual system is neither linear nor shift invariant. (For example, acuity is greater at the center of vision than at the periphery. This cannot happen for a shift-invariant system. Also, we suggested above that multiplying image intensity by a factor of 100 might have little effect on the perceptions of that image.)

## III. CONVOLUTION AND THE FOURIER TRANSFORM

The *Fourier transform* $F(f) = F(u,v)$ of a function $f(x,y)$ is

$$F(u,v) = \int_{-\infty}^{\infty} \int_{-\infty}^{\infty} f(x,y) \exp(-2\pi j(ux + vy))dx\ dy$$

where $j = \sqrt{-1}$.

The *inverse Fourier transform* $F^{-1}(F)$ is given by

$$f(x,y) = \int_{-\infty}^{\infty} \int_{-\infty}^{\infty} F(u,v) \exp(2\pi j(ux + vy))du\ dv$$

In an obvious way, these generalize the well-known definitions for functions of one variable, and we refer to standard works for the properties and conditions of existence of the Fourier transform and its inverse [1: Chap. 3]. Among the most important are those justifying the term "inverse":

$$F^{-1}(F(f)) = f \quad \text{and} \quad F(F^{-1}(F)) = F$$

We will call $f$ and $F$ a *Fourier transform pair*. $F$ and $F^{-1}$ are both linear operators. In one dimension, where the function variable is often interpreted as time and the functions are interpreted as signals, the Fourier transform has an interpretation as frequency. This interpretation extends to the two-dimensional case. It is simplest to see this in the case of a periodic image. (Even though they do not meet the criterion of finite extent arising from real images, periodic images play an important role in higher dimensions similar to that in one dimension). Figure 1 shows a *cosine grating*, a rotation of a periodic image of the form

$$I(x,y) = \cos(2\pi\omega x)$$

The intensity of $I(x,y)$ is varying at the rate of $\omega$ cycles per unit distance. It is not difficult to show that the Fourier transform of such a function is a pair of impulses. For a discussion of impulse functions in two dimensions, see below and Ref. 6.

**Figure 1**   A cosine grating given by rotating $I(x,y) = \cos(2\pi\omega x)$ through 45°.

As a result of standard rotational properties of the Fourier transform (see Section 3.3.4 in Ref. 1), if we rotate the grating in direction $(a,b)$, the Fourier transform would be a pair of impulses at points $(a,b)$ and $(-a,-b)$, in the $(u,v)$ plane, and we would say that the grating is varying $\omega$ cycles per unit distance in the direction of $(a,b)$. For these reasons, the value of the amplitude $|F(u,v)|$ is sometimes called the *spatial frequency* in the direction $(a,b)$.

Somewhat more precisely, $F(u,v)$ is a complex-valued function and so can be written

$$F(u,v) = |F(u,v)| \exp(j\phi(u,v))$$

where $\phi(u,v)$ is the *phase* at $(u,v)$ given by

$$\tan(\phi(u,v)) = \frac{\text{Im}(F(u,v))}{\text{Re}(F(u,v))}$$

Here, $\text{Re}(z)$ and $\text{Im}(z)$ denote the real and imaginary parts of the complex number $z$. The function $|F(u,v)|$ is called the *amplitude spectrum* and $\phi(u,v)$ the *phase spectrum*. Figure 2 shows representations of two spectra of an image consisting of the letter "p."

It is not difficult to show that $|F(u,v)| < F(0,0)$. In the amplitude spectrum (Fig. 2a), the intensity at $(u,v)$ is proportional to $|F(u,v)|/|F(0,0)|$. In the phase picture (Fig. 2b), intensity is proportional to $\phi(u,v)/2\pi$.

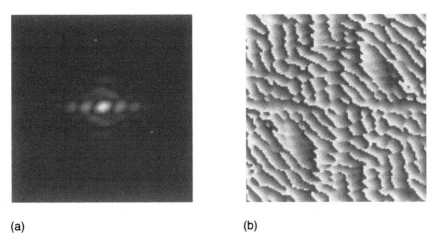

(a)                                                          (b)

**Figure 2**   (a) Amplitude Spectram of the letter "p." (b) Phase spectrum.

Actually, some further processing has been done to make the amplitude spectrum more comprehensible. It can be shown that the amplitudes may be expected to have a Rayleigh distribution (Ref. 7, p. 49), and most of them would be too small to be distinguishable from one another in the printed page. Indeed, to enhance the image we have used a technique known as histogram equalization, which we discuss later. Logarithmic transformations are also commonly used (Ref. 1, p. 72).

If $f(x,y)$ and $g(x,y)$ are two functions, their *convolution* $f*g$ is the function $h(x,y)$ given by

$$h(x,y) = \int_{-\infty}^{\infty} \int_{-\infty}^{\infty} f(u,v)g(x - u, y - v)du\,dv$$

From the corresponding properties for integrals, it is easy to verify that $f*g = g*f$, $f*(g_1 + g_2) = f*g_1 + f*g_2$, and $f*(\lambda g) = \lambda f*g$.

Convolutions play a central role in the analysis of linear, shift-invariant transforms. Let $\delta$ be the two-dimensional Dirac delta function. (Ref. 6, p. 6). If **H** is a linear shift-invariant operator, then one can show that $\mathbf{H}(f) = f * \mathbf{H}(\delta)$. The image $\mathbf{H}(\delta)$ is called the *impulse responses* of **H**. In optics, the impulse function $\delta$ represents a point light source and $\mathbf{H}(\delta)$ is known as the *point spread function* of **H**.

The application of **H** to an image $f$ is often called *filtering*, and **H** is referred to as a *filter*. Linear shift-invariant filters always take complex harmonic functions to other complex harmonic functions of the same frequency but possibly different amplitude. (Ref. 8, p. 142). Precisely,

$$\mathbf{H}(\exp(j(ux + vy))) = H(u,v)\exp(j(ux + vy))$$

$H$ is called the *transfer function* (in optical systems, the *optical transfer function*) and in fact is simply the Fourier transform of the impulse response of $\mathcal{H}$. $|H(u,v)|$ is sometimes called the *modulation transfer function*. Some authors use slightly different definitions of the transfer functions. All the Fourier transforms are normalized by dividing by their values at the origin in the frequency domain. Transfer functions then vary between 0 and 1 (see Ref. 2, p. 351 for this convention).

The results and definitions above are especially useful because of the relationship between convolution and the Fourier transform:

*Convolution Theorem. Let f and g be images, and h = f∗g their convolution. If F, G, and H and are their respective Fourier transforms, then H(u,v) = F(u,v) G(u,v).*

This theorem means that we can compute the effects of filtering simply by multiplication in the frequency domain followed by application of an inverse Fourier transform. In subsequent sections, we express all these ideas for discrete images. There we will see that there is sometimes an advantage to working in the frequency domain because of the existence of fast algorithms for computing discrete Fourier transforms.

## IV. DISCRETE IMAGES

For computer processing of images we must represent discretely both the image intensities and the points in the plane. The former representation is called *gray level quantization* and the latter is called *sampling*. Typically, sampling is done by selecting points in the plane on a uniform grid, although stochastic sampling models have been used to explain the absence in human vision of an expected artifact due to inadequate sampling [9]. This artifact, called *aliasing*, is discussed below for arbitrary sampled systems.

Image samples are usually called *pixels*. Rectangular sample grids are most convenient for analysis and for the construction of processing algorithms arising from functions expressed in Cartesian coordinates. They also correspond to many real input and output devices (CRTs, digital scanners, laser printers, etc.) in which linear motions are clocked by uniform pulses.

Hexagonal sample mosaics occasionally arise when the sampling elements are modeled by close-packed uniform circles. This includes the central field of the human visual system, where we may consider the close-packed, circularly shaped photoreceptors to be such a system. Hexagonal sampling grids also find some application in digital halftoning, the process by which gray scale pictures are rendered on black-and-white devices by using varying size clusters of dots to give the impression of varying gray levels [10].

Some forms of nonuniform sampling are used for image compression techniques (Ref. 2, Section 21.6), but such sampling is rare in simple image processing systems. However, the convenience of uniform gray level quantization often is abandoned. The reason for this is that intensities are rarely uniformly distributed across the range of intensities present in a real image. In such a case, if very few pixels fall in a certain range of intensities, it is not effective to use the same number of bits to represent that range as a range in which many pixels fall and in which there is much fine variation. Pratt [2, Chap. 6] discusses choosing a gray level quantization that minimizes a useful error measure.

For output devices, nonlinearities in the display system often complicate the matter and require careful measurement and the use of software table lookup to recover uniform linear quantization. For example, many image processing systems for personal computers use 8 bits of gray level quantization. This means that the intensities from black to white are represented by numbers between 0 and 255. One might wish that assigning a value of 128 would produce an intensity that is about half the maximum, and more generally that each of the 256 possible choices of gray level would lead to 256 equally spaced intensities on the screen. In fact, this is rarely the case. Where this kind of linearity is critical, extensive measurement is used to select an appropriate subset of possible values [11].

Using too few bits to quantize the gray levels of an image leads to "false contouring," the appearance of sharp edges that would have looked smooth had there been more gray level resolution to represent the transitions. However, taking too few *samples* of the image leads to altogether different problems, to which we now turn.

The two-dimensional *Dirac delta function* or *impulse function* $\delta(x,y)$ is defined by

$$\delta(x,y) = \lim_{a \to 0} \left(\frac{1}{a^2}\right) \prod\left(\frac{x}{a}\right) \prod\left(\frac{y}{a}\right) \tag{4}$$

where

$$\prod(x) = \begin{array}{l} 1 \text{ if } |x| < \frac{1}{2} \\ \frac{1}{2} \text{ if } x = \pm\frac{1}{2} \\ 0 \text{ elsewhere} \end{array} \tag{5}$$

The delta function thus represents a limit of a square column of height $1/a$ and width $a$. Equivalent definitions, including the representation as the limit of a circular column, are discussed by Goodman ([6], Appendix A). The impulse function is a reasonable model of a point light source, which can provide some intuition for the properties below:

$$\int_{-\infty}^{\infty} \int_{-\infty}^{\infty} f(u,v) \, \delta(x - u, y - v) du \, dv = f(x,y) \tag{6}$$

(Strictly speaking, this applies only at points where $f$ is continuous, but we will apply it only to images in any case.) Since the convolution $f*\delta$ is exactly represented above, we may formulate it as

$$f*\delta = f \tag{7}$$

The integral form of this relationship is sometimes called the *sifting property* of the impulse function. Of substantial usefulness in manipulation of the impulse function is

$$\delta(ax,by) = \frac{1}{|ab|} \Delta(x,y) \tag{8}$$

Applying the convolution theorem to Eq. (6) shows that the Fourier transform of the impulse function is identically 1.

We have seen that by convolving with an impulse, we sample a function at a single point. If we wish to build samples of the function of a uniform rectangular grid, of spacing $\Delta$, we can multiply by the function

$$S(x,y) = \Delta^2 \sum_{m=-\infty}^{\infty} \sum_{n=-\infty}^{\infty} \delta(x - m\Delta, y - n\Delta)$$

Now using Eq. (8), one can show (Ref. 8, p. 227) that Fourier transform of $S$ is given by

$$F(S) = \sum_{m=0}^{\infty} \sum_{n=0}^{\infty} \delta\left(u\,\frac{m}{\Delta},\, v\,\frac{n}{\Delta}\right)$$

So by the convolution theorem

$$F(fS) = \sum_{m=0}^{\infty} \sum_{n=0}^{\infty} F*\delta\left(u\,\frac{m}{\Delta},\, v\,\frac{n}{\Delta}\right) \tag{9}$$

But since $F*\delta(u,v) = F$, Eq. (9) says that the spectrum of an image sampled at intervals of width $\Delta$ is simply an infinite sum of copies of the unsampled spectrum, replicated on a grid of width $1/\Delta$.

The situation is illustrated in Fig. 3. By considering this figure, it is clear that if the original spectrum is nonzero only on some finite region (called the *support* of the spectrum) and if the replicas do not overlap, we can recover the original spectrum—hence the original function (by applying the inverse Fourier transform)—from the spectrum of the sampled function, hence from the sampled function. From Fig. 3, we can see that a sufficient condition for this is that

$$F(u,v) = 0 \text{ whenever } |u| > 1/2\Delta \text{ and } |v| > 1/2\Delta$$

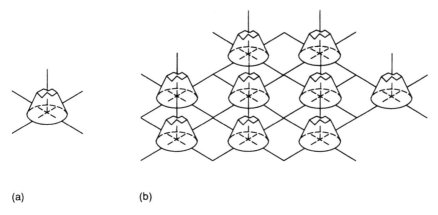

(a)                          (b)

**Figure 3**   (a) Spectrum of a band-limited function. (b) Spectrum of the sampled function.

This result is explicitly formulated as the:

*Nyquist Sampling Theorem: If f(x,y) is a function whose spectrum F(u,v) vanishes outside some square |u| > D₀ |v| > D₀ and fₛ(x,y), the discrete function obtained by sampling f(x,y) on a uniform grid of width Δ, then f is completely determined by fₛ provided 1/Δ > 2D₀.*

Actually, the support of $F(u,v)$ need not be square, but there is no loss of generality in assuming this, and the notation becomes slightly simpler.

Recovering an image from its samples becomes a matter of convolving the sampled function with a *reconstruction function*, the inverse Fourier transform of some function that will eliminate all the spectral replicas but one. The simplest case of this is a function that is 0 outside the rectangular region beyond which the spectrum vanishes, and 1 inside it. The corresponding reconstruction function is

$$\frac{\sin(\pi\Delta_0 x)}{\pi x} \frac{\sin(\pi\Delta_0 y)}{\pi y}$$

An image whose spectrum has finite support is said to be *band limited*. An image of finite extent is never band limited because the abrupt truncation at its boundaries gives rise to arbitrarily high frequencies, albeit of diminishing amplitude. But it is often the case that an image processing system (e.g., a lens) will attenuate all high frequencies sufficiently to permit the original image to be satisfactorily modeled as band limited. When the spectral replicas do overlap (either because the image is not band limited or because the sampling rate is too low), the summations in the overlapping portions will lead to high frequency components of one replica artificially added to lower frequency portions of the adjacent unit. If reconstruction is done by isolating this "polluted" replica, the resulting image suffers from "aliases." In addition, when the aliases are periodic features in a different direction from some periodic feature in the image, the familiar moiré pattern can result.

## V. DISCRETE CONVOLUTION AND FAST TRANSFORMS

Section IV established some of the utility of discrete images. Now we describe how convolution and Fourier (and other) transforms have discrete representations suitable for digital image processing.

We can think of a discrete image (e.g., one formed by sampling, as described above) as a function $f(i,j)$ on pairs of integers. Typically the values of the image are also quantized, so may be some discrete set of numbers. In a digital image processing context, these numbers are typically integers in some range $0 \le f(i,j) < 2^n$. However, the mathematics below does not depend on the values taken by images, and for some purposes it is convenient to permit them to be arbitrary complex numbers.

If $f(i,j)$ defined for integers $0 \le i < M$, $0 \le j < N$ is a discrete image, we define its *discrete Fourier transform* (DFT) $F(u,v)$ as the complex-valued function of integer pairs $(u,v)$ given by

$$F(r,s) = \frac{1}{MN} \sum_{\substack{m<M \\ n<N}} f(m,n)\exp\left[-2\pi j\left(\frac{mr}{M} + \frac{ns}{N}\right)\right]$$

This definition depends on $M$ and $N$, which should properly accompany the notation; that is, we should write $F_{M,N}(r,s)$. However, we usually leave this to the context. Typically $M = N$ and its value is a 2-power. Note that $F(0,0)$ is just the average value of all pixels in the image.

As with the continuous Fourier transform, the value of the discrete transform is called the *spatial frequency* at $(r,s)$. If $f(m,n)$ are the samples of a continuous function at uniform sample points $(m\Delta x, n\Delta y)$, then $F$ are the values of the continuous Fourier transform sampled at the points $(r/M\Delta x, s/N\Delta x)$.

The *inverse discrete Fourier transform* of a sequence $F(r,s)$ of complex numbers is given by

$$f(m,n) = \sum_{\substack{r<M \\ s<N}} F(r,s)\exp\left[2\pi j\left(\frac{mr}{M} + \frac{ns}{N}\right)\right]$$

and it is an easy exercise in algebra to demonstrate that this function is indeed inverse to the discrete Fourier transform.

If we let $g(r)$, $0 \le r < M$ be a sequence of numbers, the one-dimensional discrete Fourier transform is the special case of the foregoing definition of the DFT given by

$$G(r) = \frac{1}{M} \sum_{m<M} g(m)\exp\left[-2\pi j\left(\frac{mr}{M}\right)\right]$$

Similarly, the one-dimensional inverse transform can be written

$$f(x) = \sum_{u<m} G(u)\exp\left[2\pi j\left(\frac{ux}{M}\right)\right]$$

and it is straightforward to prove that both the Fourier and inverse Fourier transforms are *separable*, in the sense given below. Separable transforms, which include the Hadamard transform, are computationally convenient because they permit a two-dimensional transform to be reduced to the one-dimensional transform of a sequence of one-dimensional transforms.

If a transform $T(r,s)$ has the form

$$T(r,s) = \sum f(m,n)g(m,n,r,s)$$

we say that $g(m,n,r,s)$ is the *transform kernel* of $T$. The transform is *separable* if its kernel can be written

$$g(m,n,r,s) = g_1(m,r)g_2(n,s)$$

The transform is said to be *symmetric* if $g_1 = g_2$. The two-dimensional DFT of a square image is separable, since its kernel can be written as

$$\frac{1}{N}\exp\left(\frac{-2\pi rm}{N}\right)\frac{1}{N}\exp\left(\frac{-2\pi sn}{N}\right)$$

A separable transform can be computed by first transforming along each row

$$T(m,s) = \sum_n f(m,n)g_2(n,s)$$

and then transforming the result along the columns

$$T(r,s) = \sum_m T(m,s)g_1(m,r)$$

Separable symmetric transforms have particularly simple representations by symmetric matrices, which permits them to yield fast algorithms, the most well known of which is the fast Fourier transform (FFT). For more details, see Gonzalez and Wintz [1, Section 3.5].

Next we describe discrete convolution, a construction that permits us to apply image processing operations in the digital image domain. As with continuous convolution, the (discrete) Fourier transform will convert convolution into multiplication, making it sometimes easier to apply image processing in the frequency domain and sometimes easier in the image domain.

It is easy to see that the discrete Fourier transform and its inverse are periodic functions, namely (in the notation of the definition) $F(r + aM, s + bN) = F(r,s)$ for every pair of integers $(a,b)$. To formulate a definition of convolution that is sympathetic to this periodicity, assume that we have images $f$ and $g$ of sizes $A \times B$ and $C \times D$. We wish to extend these images with zero entries to be periodic of period $M \times N$, where

$$M \geq A + C - 1 \quad \text{and} \quad N \geq B + D - 1$$

The *discrete convolution* is the $M \times N$ image $h(i,j) = f*g$ given by

$$h(i,j) = \sum_{m=0}^{M-1} \sum_{n=0}^{N-1} f(l,m)g(i - l, j - m)$$

With this definition we obtain the discrete form of the convolution theorem:

*Discrete Convolution Theorem.* If $f$ and $g$ are discrete images, $h$ their convolution, and $F$, $G$, $H$ their respective discrete Fourier transforms, then $H = FG$. Conversely, if $F$ and $G$ are (complex-valued) "images" and $H$ their convolution and $f,g,h$ their inverse discrete Fourier transforms, then $h = fg$.

In other words, the Fourier transform and its inverse each exchange convolution for multiplication when the underlying images are suitably extended.

As with the continuous case, convolution can be regarded as *filtering* the image. In many applications the filter can be conveniently taken as a relatively small image, often as small as $3 \times 3$, and rarely more than $9 \times 9$. Indeed, if the filter $g$ has some size $N \times N$, there is a straightforward way to produce an $n \times n$ filter $\hat{g}$ which minimizes a reasonable error quantity. See Gonzalez and Wintz [1, Section 4.6] for details. In some examples below we show the effects of several simple small filters.

Because of the existence of fast Fourier transforms, this relationship can make it more convenient to transform, compute a product in the transform domain, and transform back. For example, consider an *ideal filter*, which is defined as one that is, in the frequency domain, the identity on some region and zero off that region. In the frequency domain this can be computed merely by setting the spectrum to zero at the appropriate points. No multiplication is involved at all in the application of the filter.

Figure 4 shows a simple but realistic application of this kind of image processing. Figure 4a is distorted by periodic sinusoidal noise, which manifests itself in the spectrum as a pair of pulses at the corresponding frequency and its negative (Fig. 4b). To remove the noise, it suffices to set the spectral values to zero at those points (or, sometimes more effectively, to the average value of the neighboring points). The result is shown in Fig. 4c.

## VI.  IMAGE PROCESSING

## A.  Filtering Techniques

In this section we briefly sketch several representative imaging processing techniques in terms of the ideas mentioned above. Our intention is only to give a sample of these techniques. The interested reader may consult Refs. 1, 2, 8, and 12, which contain details and further examples, as well as references to the literature, which is extensive.

Small "images" used as filters are often called *masks*. Convolution by the mask may be thought of as an average of the pixel values in the filtered image, weighted by the mask values. For example, if the mask is a $3 \times 3$ matrix

$$m_1 \quad m_2 \quad m_3$$
$$m_4 \quad m_5 \quad m_6$$
$$m_7 \quad m_8 \quad m_9$$

and we consider the values of an image just in the neighborhood of a pixel at position $(i,j)$, the definition of convolution will reduce to overlaying the mask on the neighborhood and summing the pixel values, each multiplied by the corresponding mask value. The case of $m_i = 1/9$ actually gives the average pixel value, giving the effect of smoothing edges in the processed image.

Unfortunately, simple averaging blurs edges, so it is not suitable for removing isolated spikes of noise, which often are introduced in imaging systems by interference during transmission or other inherent problems in the data gathering. In this case, replacing a pixel intensity by the *median* of the intensities of its neighbors (including itself) will have the effect of making points be similar in intensity to their neighbors only when they are very distinct from *most* of their neighbors. For example, consider the simple case of a black point on a vertical boundary between a black and a white region in an image. Median filtering taken over a $3 \times 3$ neighborhood of the point would find six adjacent points of intensity 0 and three of intensity 1, resulting in no change in the intensity of the selected point. Note that median filtering is not linear, since the median of a sum of sequences is not the sum of the medians. Consequently, we do not expect to represent median filtering by the application of convolution (i.e., by a mask). But its simplicity and superiority over averaging for removing impulse noise makes it widely used.

Often the task of image processing is not to remove unwanted features so much as to accentuate features of particular interest. Among the most important cases of this is *edge detection*, in which we seek to make a new image that highlights the edge pixels in the original image. Because edges correspond to points at which the image is changing most rapidly, the simplest edge detectors are based on approximations to the gradient of the image. Pixels at which the

(a)

**Figure 4**   (a) Image distorted by sinusoidal noise. (b) Amplitude spectrum of noisy image. (c) After removal of noise by suppression of impulse in amplitude spectrum.

gradient magnitude is largest will correspond to edges in the image, especially if isolated pixels have been removed by techniques such as described above.

The masks

$$
\begin{array}{ccc}
-1 & -2 & -1 \\
0 & 0 & 0 \\
1 & 2 & 1
\end{array}
\quad \text{and} \quad
\begin{array}{ccc}
-1 & 0 & 1 \\
-2 & 0 & 2 \\
-1 & 0 & 1
\end{array}
$$

when applied to a pixel, yield approximations to the gradients $G_y$ and $G_x$, respectively. This provides the basis for edge detection: if we construct an image $g(i,j)$ whose value at each point is the magnitude of the gradient vector (in practice, $|G_y| + |G_x|$ is a suitable approximation), this new image will have high intensity pixels where the gradient is large (i.e., at edges) and low intensity pixels where the gradient is small (i.e., at interior points). Figure 5 shows an example. The matrices described above are called the *Sobel* operators.

(b)

(c)

**Figure 5**   The image from Fig. 4 after applying Sobel operator gradient approximations for edge detection.

## B.   Histogram Modification

Sometimes features in an image, especially edges, are difficult to see because the contrast is too low in the image; that is, the distribution of pixel intensities is confined to a range that is too narrow. This difficulty arises in part because of properties of the human visual system, which cannot detect intensity differences less than about 0.025% under the best circumstances. Furthermore, this so-called *contrast sensitivity* of the visual system falls off rapidly as spatial frequency increases (and, perhaps surprisingly, as spatial frequency *decreases below an optimum value*). Thus, even if we had an imaging system with the ability to present very subtle intensity differences (e.g., some photographic film), we still might not be able to detect those differences visually. Consequently, we can accommodate our visual system by keeping the ordering of pixel intensities the same, but stretching out their dynamic range so that the differences between pixels become more apparent to us. One important technique for this is *histogram equalization*.

The histogram of an image is the graph of the number of pixels at each of the intensities representable by the digital imaging system. Histogram equalization modifies the pixel intensities in a way that makes the distribution of intensities approximately uniform. This stretches the dynamic range of the intensities to the full range accommodated by our imaging system. The algorithms arise from elementary statistical theory and are simple enough to produce with quite straightforward computer programs. In fact, the production of images with uniformly distributed intensities is not the only application of these algorithms, which can produce an approximation to any reasonable distribution. This makes it feasible to accentuate features in an image based on known properties of their "true" intensity distributions, or even in an interactive fashion to enhance the contrast of particular regions of an image until the user recognizes the objects in the image.

In what follows, we will assume that gray levels have been normalized so that they lie in the interval $0 \leq I(x,y) \leq 1$. A gray *level transformation* is a function

$$v = T(u)$$

which is monotonically increasing and maps the unit interval to itself. The monotonicity guarantees that the ordering of gray level is maintained.

Now let $P_u$ be the probability that a pixel has gray level less than $u$. In discrete terms $P_u(u)$ is approximately $n(u)/N$, where $n(u)$ is the number of pixels whose intensity is less than $u$ and $N$ is the total number of pixels in the image. From elementary probability theory, recall that the *probability density $p(u)$* is given by

$$p_u(u) = \frac{dP_u}{du}$$

Now if $u = T^{-1}(v)$ denotes the inverse function to $T$ (which is also a gray level transform), then

$$p_v(v) = \left[ p_u(u)\frac{du}{dv} \right]_{u \,=\, T^{-1}(v)}$$

Now set

$$v = T(u) = \int_0^u p_u(w)dw$$

the cumulative distribution function of $u$. From elementary calculus these two equations yield

$$p_v(v) = 1$$

that is, $v$ is uniformly distributed.

In practice, if $n_k$ is the number of pixels of the $k$th gray level and $N$ the total number of pixels, then as above we can approximate $P_u(n_k)$ by $n_k/N$ and the cumulative distribution of $u$ by

$$P_u(u_j) \approx \sum_{k=0}^{j} \frac{n_k}{N} \tag{10}$$

where $u_j$ is the $j$th gray level, and the left-hand side approximates the probability that a pixel has intensity at most $u_j$. If we transform every pixel of intensity $u_j$ to one of intensity $P_u(u_j)$ given by Eq. (10), the development above shows that we will have an image whose pixels have approximately equally distributed gray values. This technique, known as *histogram equalization*, is illustrated in Fig. 6, where a selected portion of the image has been equalized to show detail.

Processing images to have approximately uniformly distributed gray values is only one useful histogram modification. For example, images with both a very dark and a very bright region, one of which is not of interest, can benefit more

**Figure 6**   Image from Fig. 4 showing a portion with histogram equalization to enhance region of inadequate contrast.

by specifying an output histogram that expands the dynamic range mostly of the region of interest at the expense of the remainder of the image. A detailed example is given by Gonzalez and Wintz [1, p. 154]. The transformations necessary to obtain other useful output distributions, including exponential, Rayleigh, and hyperbolic distributions, can be found in Pratt ([2: Table 12.2.1]).

## C. Feature Extraction

Often it is desired to find the location of particular features in an image. For example, locating the boundaries of a tumor in a medical image will be of great importance for the surgeon. We discussed some edge detection methods earlier, but such techniques alone cannot distinguish one feature from another—for example, healthy from diseased tissue in medical images, a river from a winding road or one sort of crop from another in satellite images. To do this, some relations between the pixels in the features sought out must be found which distinguishes them from pixels in the features to be ignored. There are many techniques for this, we mention only two as illustrative.

*Pseudo-color* processing assigns colors to portions of monochrome images that have been manipulated in some information-enhancing way. For example, an image might be passed through multiple filters of varying pass-bands, with the output of each displayed in a different color. Instead of being separated by their spatial frequencies, features may be separated by their intensities. In *density slicing*, the range of gray levels in the image is divided into some (usually small) subset, with a different color assigned to pixels whose intensity falls in each division. This makes features especially visible, to the extent that particular features are characterized by intensities different from their surroundings, as might happen in a medical x-ray image, where pixel intensity is a function of x-ray opacity. *False color* maps original color images into some color-distorted version. This may be to convert an image (e.g., infrared) to the human-visible spectrum in some way reasonably faithful to the original range, or it may be a gross distortion of colors to reveal features camouflaged by similar colors in the original. The recent availability of low cost color hardware has made interactive color manipulation a particularly popular technique for searching for features.

*Segmentation* describes the process of dividing an image into regions of similar properties. Segmentation often starts with edge detection methods such as described earlier, but since many objects in a scene may have edges, it is often necessary to determine which edge pixels are in the edges of which object. This large collection of techniques, called *edge linking*, attempts to decide which pixels in the image belong to the boundary of a particular object. For example, given two nearby pixels both determined to be edge pixels, if the differences of their gradient vectors is below some particular threshold, we might conclude that these pixels were part of the edge of the same object and assign them both the same gray level. When successful, such a technique will show the outlines of different objects in different gray levels (or pseudocolors). The choice of thresholds for such techniques is itself a topic of some breadth [1, Section 7.3]. Besides these essentially local techniques based on neighboring pixels, one may use global techniques such as the Hough transform [1, p. 130, 345] to decide whether a collection of pixels lies on a particular curve.

## VII. SUMMARY

Many of the techniques applied to processing images are extensions of one-dimensional techniques familiar to scientists for the smoothing, enhancement, and interpretation of time-series, spectroscopic, chromatic, and other data subject to noise, uncertainties in the data gathering, or the influence of phenomena ancillary to the ones under study. For many real-world laboratory problems, the techniques described here and in the references are well within the capability of the current generation of low cost microcomputers such as the IBM PC-compatible or Apple Macintosh systems commonly used in laboratories. A Macintosh II Ci or a 25 MHz 80386 PC-compatible machine can compute a 256 × 256 fast Fourier transform in about a minute, and at a small extra cost, special-purpose add-in hardware is available which accelerates these computations by one or two orders of magnitude. For these computing environments, commercial and public domain packages are available that perform a wide variety of image manipulation. Three such packages for the Macintosh are reviewed by Morris [13], including the excellent and widely used Image program of the National Institutes of Health (NIH). The NIH program is available without charge on many computer networks and bulletin boards and for a nominal fee from the National Technical Information Service. The image manipulation of the figures in this chapter was done with IPLab, a product of Signal Analytics.

## REFERENCES

1.  Gonzalez, R. C., and Wintz, P. (1987). *Digital Image Processing*, 2nd ed., Addison-Wesley, Reading, MA.
2.  Pratt, W. (1971). *Digital Image Processing*, Wiley-Interscience, New York.
3.  Wilson, H. R., et al. (1990). The perception of form: Retina to striate cortex, in *Visual Perception: The Neurophysiological Foundations* (Spillmann, L., and Werner, J. S., Eds.), Academic Press, San Diego, CA, Chap. 10, pp. 231–272.
4.  Schwartz, E. L., Ed. (1990). *Computational Neuroscience*, MIT Press, Cambridge, MA.
5.  Richards, W. (1988). *Natural Computation*, MIT Press, Cambridge, MA.
6.  Goodman, J. W. (1968). *Introduction to Fourier Optics*, McGraw-Hill, New York.
7.  Goodman, J. W. (1985). *Statistical Optics*, Wiley-Interscience, New York.

8. Castleman, W. (1979). *Digital Image Processing*, Prentice-Hall, Englewood Cliffs, NJ.

9. Yellott, J. I., Jr. (1982). Spectral analysis of spatial sampling by photoreceptors: Topological disorder prevents aliasing, *Vision Res.* 22:1205–1210.

10. Ulichney, R. (1987). *Digital Halftoning*, MIT Press, Cambridge, MA.

11. Naiman, A. C. (1991)., CRT spatial non-linearities and luminance linearization, in *Raster Imaging and Digital Typography*, Vol. II (Morris, R. A., and Andre, J., Eds.), Cambridge University Press, Cambridge, pp. 43–53.

12. Chellappa, R., and Sawchuk, A. A., Eds. *Digital Image Processing and Analysis*, Vols. 1 and 2, IEEE Computer Society Press, New York.

13. Morris, R. (1990). Image processing on the Macintosh [in Product Reviews column], *IEEE Comput.*, August, pp. 103–107.

# 5
# Fundamentals of Scanning Probe Microscopies

**Bradford G. Orr**   *University of Michigan, Ann Arbor, Michigan*

## I. INTRODUCTION

This chapter is intended to serve as an introduction to scanning probe microscopies. Led by scanning tunneling microscopy (STM), these techniques have given researchers the ability to examine phenomena with extraordinary resolution. The field has evolved rapidly over the past decade. In its maturation phase, STM is no longer studied in and of itself, but is used as a standard tool. Other less well-developed techniques will require additional effort before they become part of a standard repertoire.

## II. HISTORY

Present-day scanning probe microscopies (SPM) were developed out of the merging of two areas: near-field microscopy and stylus profilometry. Microscopies relying on propagating modes are limited in resolution by diffraction. The Abbe criterion states that the lateral resolution limit of a microscope is determined by the wavelength of the illuminating radiation. To extend beyond this limit, it was proposed in 1956 that evanescent radiation produced by a small aperture be used to image small particles [1]. By positioning the particle near a subwavelength diameter hole and scanning their relative positions, an image could be formed with a resolution limited by the size of the aperture and the distance between the aperture and particle. This approach used the evanescent wave emitted from the hole, thereby circumventing the problem of diffraction.

There were obvious difficulties with this approach. The intensity of the radiation is strongly reduced after transmission through the imaging aperture. To increase the resolution of the microscope, the diameter of the hole must be decreased to less than that of the wavelength of the illuminating radiation. Concomitant with this, the transmitted signal decreases exponentially. The viability of the technique then becomes a question of aperture size and signal-to-noise level. Of more immediate concern at the time was the ability to position and scan the aperture with the necessary subwavelength precision and control. To test the near-field imaging concept, measurements using microwave radiation were performed. Experimental demonstration of the technique was achieved in 1972 using radiation with a wavelength of 30 mm [2]. The aperture diameter used for imaging was 1.5 mm, and a resolution of $\lambda/15$ was clearly achieved. In addition, image contrast due to variations in the dielectric constant of the sample was shown. Even after this demonstration of the near-field scanning technique, the construction of a superresolution optical microscope appeared to be problematic because of the submicrometer positioning and control necessary.

Development of superesolution optical microscopy proceeded in parallel with the development of stylus profilometry. Scanning profilometers typically employ a diamond stylus, which contacts the sample surface and produces a three-dimensional map. Lateral resolution is limited by the size of the stylus and the minimum force necessary for a detectable deflection. A major problem associated with this technique is the contact interaction between the stylus and sample, which can lead to contamination and damage of the sample surface. In 1972 a group at the National Bureau of Standards led by R. Young, developed a non-contacting profilometer [3,4]. This instrument used field emission to pass a current between the probe and the sample. By injecting a constant current of electrons into the sample and monitoring the voltage, the distance between probe and sample could be determined from Fowler–Nordheim theory [5]. The last step necessary to produce a microscope involved feeding the junction voltage into a servo loop and raster scanning the field emission probe. By recording the correction signal from the servo loop, a map of the surface was produced. Young named this instrument the "topografiner" from the Greek "to describe a place." Typical resolution of the instrument was 4 nm vertically and 400 nm laterally. The topografiner is the forerunner of the scanning tunneling microscope.

A major advance toward present-day scanning probe microscopies involved combining near-field superresolution techniques with that of the topografiner. If instead of using field-emitted electrons from a probe one used the evanescent wave associated with quantum mechanical tunneling, the resolution of the technique could be increased substantially. Young had explored the possibility of tunneling with the topografiner [6]. However, instead of examining the increased spatial resolution of the instrument in this mode, he concentrated on its spectroscopic capabilities [7].

The first near-field microscopy to combine these ideas was the scanning tunneling microscope invented by G. Binnig, H. Röhrer, C. Gerber, and E. Weibel in 1981 [8]. The lateral resolution of the instrument was estimated theoretically to be 4.5 nm; almost immediately after the first images were taken, however, atomic scale features were observed [9]. These images were initially met with polite skepticism, but after observations of the surface reconstruction on Si(111), the scientific community was ready to embrace a near-field scanning probe with atomic resolution [10].

Researchers soon realized that with the invention of STM, many of the problems associated with generic near-field scanning microscopies had been solved. It was then straightforward to generalize the techniques and principles to a host of alternate tip–sample interactions. Examples include thermal profilometry [11], near-field optical microscopy [12,13], and force microscopy [14]. The rest of this chapter will serve as an introduction to these techniques.

## III. PRINCIPLES

The principles of near-field microscopies were first demonstrated with the scanning tunneling microscope. A probe is positioned close enough to the sample surface to permit some form of interaction. If the probe–sample separation is reduced, so that the interaction is through evanescent radiation, the resolution of the instrument will not be diffraction limited. For the cases of near-field scanning optical microscopy (NSOM) and scanning tunneling microscopy, the interaction is mediated by such evanescent waves, photon and electron, respectively. This is the classical definition of a near-field probe. However, present-day use has generalized the concept of near-field probes to focus on probes for which the lateral extent of the interaction with the sample is determined by the local geometry in the region of minimum tip–sample separation.

With this generalization, new types of interaction can now be thought of within the same paradigm. It is possible to choose a specific probe–sample interaction that is appropriate for a particular problem. A near-field microscope can be designed to examine specific features of a sample by choosing the correct contrast mechanism. Herein lies much of the strength, and promise, of scanning probe microscopy techniques. To date, interactions such as tunneling, van der Waals, magnetic, electrostatic, thermal convection, and hard core repulsion have been used to produce near field scanning microscopes.

Although probe shape and probe–sample separation strongly influence the observed resolution of an instrument, the ultimate resolution is determined by the length scale over which the chosen interaction varies. The STM utilizes electron tunneling, for which this scale is fractions of an angstrom, whereas for NSOM the evanescent radiation sets the interaction range at approximately 20 nm. Each type of near-field microscopy has a different ultimate sensitivity set by the appropriate interaction range.

So far, differences between types of near-field microscopy have been emphasized. When the discussion leaves the details of the probe–sample interaction, however, the differences become few. The operating principle for all near-field microscopies is exactly that of Young's topografiner. First, interaction strength is quantified by measuring photon intensity, tunneling or field emission current, deflection of a vibrating cantilever, and so on. The signal is fed into a servo feedback loop, which attempts to maintain the signal at a constant value. Interaction strength is controlled by altering the vertical position of the probe. If the probe is now raster scanned over the sample surface, the feedback loop will guarantee that the probe traces out contours of constant interaction strength. To the extent that the interaction depends solely on the distance between probe and sample, these images reflect the sample surface topography. Figure 1 shows an atomic resolution STM topograph of a stepped Si(001) surface. The dimer chains of the reconstructed surface are visible as white lines. These rows alternate by 90° on adjacent terrace levels.

One of the difficulties associated with scanning probe microscopies lies in the assumption that the interaction strength varies only with probe–sample distance. In fact, seldom can a topograph be interpreted as a true representation of the physical sample surface. Usually the interaction depends on several parameters and, while the feedback loop is maintaining a fixed interaction strength during a scan, the individual parameters may be changing. Therefore, topographic images are actually derived from a convolution of these parameters. This makes the definition of resolution problematic. To give a concrete example, imagine an NSOM scanning a sample comprised of material with differing dielectric constants. The surface is physically flat, and so should a true topograph be flat. However, as the surface is scanned, whenever the probe passes over a boundary between the two materials, the probe height is altered to maintain a constant photon intensity. The measured topograph is actually a map of the spatial variation in dielectric constant, not sample height.

While this might seem to complicate the interpretation of an image, and it does, it also allows for the exciting possibility of spectroscopy. By intentional varying of one of the parameters during a scan, a spatially resolved spectroscopic map can be produced. This technique is most well developed in STM, where it is relatively easy to vary the voltage between tip and sample while recording the apparent change in height of the sample. Such STM spectroscopic images can give information about variations in the local density of states with atomic scale spatial resolution [15]. Spectroscopic techniques are also rapidly being developed for NSOM.

## IV. INSTRUMENTATION

Although general design goals for scanning probe microscopies have not changed since the first topografiner, the manner in which they have been implemented has changed tremendously. The first STM was designed with

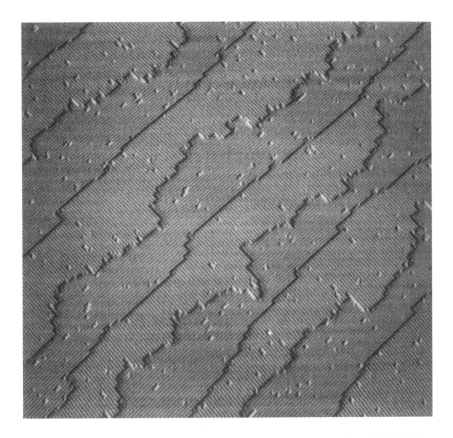

**Figure 1** STM image of a stepped Si(001) surface. The sample miscut is 0.5°. Dimer rows, surface steps, and adatom groups are clearly seen. The topograph was taken in the constant current mode. Light regions of the image indicate positions at which the tip was retracted to maintain a constant current. (Courtesy of M. Lagally.)

superconducting magnetic levitation as the means for vibration isolation. Present-day STMs will operate adequately on a table on an upper floor of a tall laboratory building. The development of standard scanning probe microscopy designs has resulted in the creation of approximately 10 companies that sell instrumentation.

The microscope design can be divided into three parts: scanning head, feedback electronics, and computer display. Figure 2 shows a block diagram of the complete system and an example of a typical scan head design. The following is a description of the instrumentation, operation, and probe fabrication techniques.

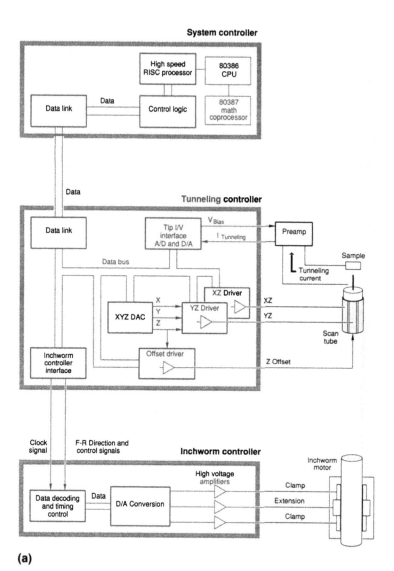

**(a)**

**Figure 2** (a) Block diagram of a scanning tunneling microscope system. This design uses a digital feedback loop. (b) Scan head for a vacuum-compatible STM. The vibration isolation is performed by soft springs, and the coarse approach uses a piezoelectric linear motor (Inchworm). (Courtesy of Burleigh Instruments).

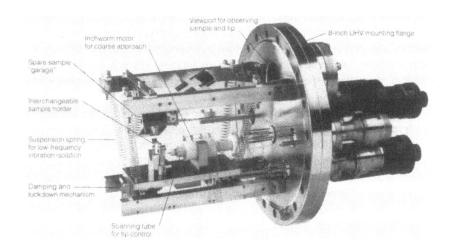

Viewport for observing sample and tip
8-inch UHV mounting flange
Inchworm motor for coarse approach
Spare sample "garage"
Interchangeable sample holder
Suspension spring for low-frequency vibration isolation
Damping and lockdown mechanism
Scanning tube for tip control

**(b)**

## A. Scanning Microscope Head

One of the virtues of near-field microscopies is the ability to operate in a variety of environments. Operation in vacuum, liquid, and gas atmospheres is possible. Clearly, there will be differences in the design of scan heads for different operating conditions. However, the general schemes are similar.

The microscope requires a transducer that monitors the interaction strength between probe and sample. This transducer may take many forms: in STM, the probe is connected to a current amplifier; in force microscopy, a cantilever beam is coupled to a fiber optic interferometer; and in NSOM, a micropipette and a photomultiplier tube are employed. The signal output from the transducer is fed to the servo electronics.

The remainder of the scan head is used for positioning and vibration isolation. Two types of probe motion are needed, coarse approach and raster scanning. The coarse approach motion must be precise enough to permit the probe to be brought within the interaction range without crashing into the sample. Since the interaction range for tunneling is so small, the STM is the instrument presenting the most severe test for the coarse approach mechanism. A number of methods have been successfully demonstrated. The first was the "louse" [16]. This three-legged platform, constructed from piezoelectric elements, is moved by electrostatically clamping two ends while extending the third. Repeating this for alternate legs allows the probe to be shuffled about in small steps.

In a desire to simplify the coarse approach mechanism, many mechanical methods have also been used. Stepper motors with gear reduction [17], fine-pitched screws with lever reduction [18], and differential spring systems [19] have all been operated successfully. More recently, electromagnetic impulse walkers and piezoelectric linear motors have proven useful [20–22]; each of these methods has its advantages and disadvantages. There is no generally correct implementation; specifics such as environment, size, and precision all bear on the choice.

When considering the active element for raster scanning, the choices are much reduced. All near-field microscopes use piezoelectric elements for raster scanning. Most microscopes use piezoelectric tubes that are metalized on the inner and outer diameters [23]. The outer electrode is divided into four longitudinal quadrants. By grounding the inner electrode and applying a bipolar voltage to opposite quadrants of the outer electrode, the tube can be warped. This motion is used to sweep the probe along the sample surface. The other two quadrants allow orthogonal motion.

To provide motion perpendicular to the sample surface, a common signal is summed into all four quadrants of the piezoelectric tube. This causes the tube to extend or contract along its axis. If reduced sensitivity is desired in the axial direction, the outer electrodes can be radially scribed to provide segments of unequal length. Typical sensitivities for a tube constructed from the piezoelectric crystal PZT [$PB(Ti,Zr)O_3$] with dimensions 5.0 cm long, 0.625 cm diameter, and 0.05 cm wall thickness are 0.4 $\mu$m/V for motion parallel to the sample surface and 0.026 $\mu$m/V normal to the sample surface. In addition to the tube scanner, there are other less commonly used drivers for the raster scan. Piezoelectric sticks and bimorphs can be used. The bimorphs act as benders, providing great lateral range.

Microscopes should be designed with the goal of minimizing vibration and thermal drift. Vibrations can be reduced by making a small, rigid scan head with the fundamental resonance frequency of the instrument much higher than typical environmental noise frequencies. Modern tube scanners have their first resonance at several kilohertz. In addition, external vibration isolation is usually needed. Instruments suspended from springs and elastic cords are often successful. Although these instruments can be made to work in less than ideal environments, it is best to start out in a quiet location.

Thermal drift can pose a severe problem for the stability of a microscope. Temperature changes caused by air conditioning, heating systems, or microscope illumination can bring about a noticeable drift in an image. To reduce thermal effects, materials should be matched for small differential expansion, and the instrument head should be symmetrical. Thermal drift problems can be conquered. The most dramatic example of this is the recent developments in variable temperature STM. Operation of the microscope has proven possible even when the sample temperature is above 200°C.

## B. Feedback System

Feedback circuitry receives the signal from the probe transducer and outputs a correction signal to the z-piezo element. Typical analog systems use proportional–integral control. With this scheme, the correction signal is proportional to the integral of the difference (error) between the actual transducer signal and the desired set point. The circuitry will have external controls for the time constant of the integrator, gain of the proportional amplifier, and set point level.

It is often useful to modulate the signal coming from the transducer and demodulate it before the input of the feedback loop. The demodulated signal can then be fed into the servo system or recorded to produce an image. The technique might be employed to reduce low frequency noise and improve sensitivity, or to perform spectroscopic measurements. A typical example would be to place an ac dither signal on the z-piezo element and demodulate the response using a lock-in amplifier. If this is done when performing STM, the demodulated signal is related to the local work function of the material. A similar technique can be used with force microscopy.

More recently, a new approach has been used for the feedback system. With the advent of fast digital system processing (DSP) boards, it has become possible to digitize the transducer signal and perform all the servo functions digitally. The analog signal is typically sampled at a frequency greater than 50 kHz. Then the data stream is digitally filtered and compared to the set point level. The value of the output signal is then passed to a digital-to-analog converter to drive the z-piezo. With DSP many different types of servo algorithm can implemented. This allows the system to dynamically change the feedback characteristics over different regions, rough or smooth, of the sample. DSP feedback systems are now being implemented on commercial instruments. However, some researchers feel that the proprietary nature of commercial DSP systems limits their accessibility and therefore the variety of approaches that can be taken with such instruments.

## C. Computer Control and Display

Scanning probe microscopies have benefited from advances in laboratory microcomputers. The first STM images were recorded with a chart recorder. Today, computer display is universal. Data acquisition rates have also increased dramatically. Whereas early STMs required several minutes to complete a scan, image rates presently are on the order of one frame per second. These advances in the state of the art have pushed computer control and display from the role of a convenience to that of a necessity.

The computer performs two functions: real-time control of the instrument and data display and storage. The x-y raster scan of the probe over the surface is synchronized by the computer. Typical implementations involve the use of two digital-to-analog (D/A) lines and a pair of high voltage amplifiers to drive the piezoelectric elements. The feedback loop will drive a third high voltage supply controlling the probe–sample separation. To form a topograph, the $z$-piezo output of the feedback circuit is recorded by the computer and displayed on a monitor. If an analog feedback circuit is utilized, an analog-to-digital (A/D) converter must be used to digitize the feedback output, whereas with a digital feedback loop the values are immediately available. Within the approximations just mentioned, the data represent the height of the surface as a function of lateral position. If a second channel is available for recording data, two images may be acquired simultaneously. By multiplexing the input data, a series of simultaneous images is produced. This is useful for spectroscopic measurements. The accuracy of the A/D and D/A converters should be at least 12 bits.

Display of recorded data is a very important aspect of the microscope system. The first image display methods mimicked the format produced by a chart recorder: a line scan much like the tracing on an oscilloscope is produced, and each successive line is displayed downward to give the impression of a surface. The technique has largely been replaced by three-dimensional surface graphics routines. A complementary method of data display involves mapping each height on the surface to a different color. If only shadings between black and white are used, a gray scale representation of the surface is produced. These two image formats remain the most popular. When scanning highly symmetric surfaces, the gray scale format is very useful. Symmetry is highlighted and defects can be easily spotted. For rough surfaces, however, gray scale results can be difficult to interpret. Three-dimensional surface representation is very useful for rough surfaces. Figure 3 shows the STM image of a rough surface of gold islands on germanium in both gray scale and surface formats.

Once the data have been acquired and displayed, it may be desirable to save them for future reference. Data storage can become a serious problem. A modest image of $256 \times 256$ pixels recorded in both the forward and reverse directions (they do not always appear the same) will require 256K bytes. In a relatively short time, 20 such images can be taken and recorded. In other words, 5M bytes of storage will have been required. If spectroscopic data are being recorded, the storage requirements may be 10 times larger. The solution is to use data compression, have good-sized hard disk capacity, and a means of data backup.

Many commercial systems distribute the two functions of the computer, data acquisition and data display, among separate microprocessors. DSP used for feedback control started this trend. One microprocessor controls the real-time aspects of the instrument operation, positioning the probe and data acquisition, while the second handles data display and storage. The two CPUs are connected by a high speed data link and operate asynchronously. The natural division allows each processor to be optimized for its respective tasks and leads to increased performance.

Finally, it may be necessary to produce hard copy output of the images. Several techniques can be used, including photographic reproduction of the screen, conversion of files to a format that can drive standard laser printers, or use of specialized color and gray scale printers. Local graphic arts shops can use these files to prepare publication quality prints. In general, it is difficult to reproduce the same image quality on paper as on the monitor screen. The technology exists, but at present it is quite expensive. As the hardware evolves, equipment will become available to solve the problem at a reasonable cost.

## D. Operation

There are two basic modes of operation for scanning probe microscopies. The first uses the feedback circuit to hold interaction strength constant as the probe is scanned over the surface as described earlier. The second allows the probe to scan in a uniform trajectory over the sample. In this mode the interaction strength will vary as the distance changes between the probe and sample. By recording the output of the transducer as the probe is scanned, a map of the surface is produced. The second mode is termed "constant height." There is an obvious disadvantage to the constant height mode: if the sample roughness is greater than the starting distance between sample and tip, there is a good chance of suffering a tip crash. To avoid crashes, a hybrid mode can be used in which the time constant of the feedback circuit is set so that rapid variations in the surface are not tracked, but long wavelength changes are. Then if the sample is relatively flat, the tip will track the average slope of the sample while imaging the more rapid variations on the surface. Since only electronic signals, not mechanical parts, vary rapidly in this technique, the scan speed can be greatly increased.

The constant height mode was developed for STM. Because of its very short interaction distance, only in special circumstances can one find suitably flat samples to scan in this mode. Graphite is a natural candidate. Beyond that, the constant height mode has had limited applicability for STM. However, in other scanning probe microscopies, where the interaction range is much longer, the constant height mode may find much greater application. Currently, most NSOMs are run in the constant height mode. With an interaction range of several hundred angstroms, tip crashes are a much less serious problem than for STM.

## E. Probes

So far in our discussion of instrumentation, the components and techniques have been common to all scanning microscopies. Now we come to the key difference

**(a)**

**Figure 3**  Comparison between top-down gray scale (a) and three-dimensional surface (b) views for a rough sample. The image is of a gold film 50 nm thick, deposited on a germanium substrate. The scan size is 80 nm × 80 nm. A topography with deep features is more clearly represented in a surface format. (Courtesy of RHK Electronics).

between the techniques: the probe. When Binnig calculated the resolution of the STM, he estimated 45Å [9]. Only later, when atomic scale features were observed, did the true nature of the probe–surface interaction reveal itself. Naturally occurring asperities on the probe acted as minitips, which increased the resolution beyond the expected value. The central reason for this effect was the extremely rapid variation of the electron tunneling probability with tip–sample separation. The probability of tunneling decreases by approximately an order of

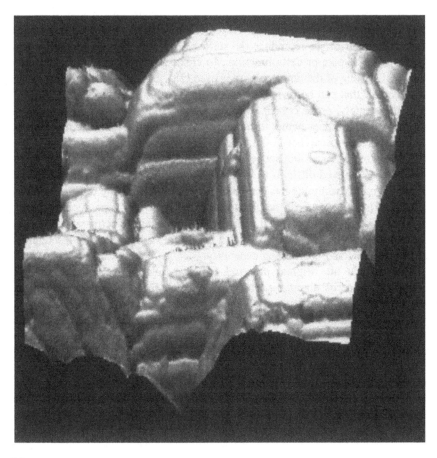

**(b)**

magnitude with a change in separation of 0.1 nm. This leads to a very strong selection of tunneling between the points of closest approach. Tunneling current is narrowly confined (~0.5 nm) to the region around the active asperity [24]. This is the explanation for the STM's high resolution [25].

At first glance, it might appear that to achieve such resolution one must fabricate tips with atomic precision. In practice, this has been done, and the tip size is related to the observed image resolution [26,27]. But because of the strong distance dependence of tunneling, if a tip has an end, it will probably work. This simple observation has been verified by experiment. One of the now-standard procedures for forming tips is to cut a tungsten wire with scissors. Tips so formed work well on reasonably flat surfaces.

If the surface that is being scanned has deep trenches, the overall profile of the tip is important. The observed image is a convolution of both tip and sample surface geometries. To reproduce the surface accurately, the tip geometry must approximate a spike or delta function. To create a sharply peaked tip, electrochemical etching of the tungsten wire is used [28]. Typically, the smallest radius of curvature that can be formed in this manner is $0.1$ $\mu$m.

Unfortunately, probe construction for other scanning microscopies is not so simple. No longer does nature form the tip for us; we must fabricate it ourselves. To paraphrase a researcher in the field, ''Tips are the name of the game'' [29]. The first atomic force microscope tips were simply STM tips with a bend in the shank. These proved difficult to form and mount. More recently, the technique of silicon micromachining has been used to form free-standing cantilever beams [30,31]. These devices can be batch fabricated with very reproducible characteristics. The Si micromachined probe in Fig. 4 contains electrostatic drive and sense plates and an integrated stylus. In the future such probes will also contain active circuitry and positioning capability.

O-Keefe, in his original proposal for near-field optical microscopy [1], expressed concerns about the ability to position the small aperture with respect to the sample. That problem has been surmounted. The problem that remains is formation of the subwavelength source. During the past few years, significant progress has been made. Recent work has focused on two approaches: the drawing of micropipettes, and optical fibers [32,33]. By heating a section of the glass and pulling, apertures down to several tens of nanometers can be formed. A metal film is evaporated onto the shank to complete the tip. These tips have produced images with resolution comparable to the aperture diameter.

One of the exciting aspects of scanning probe microscopy is the continuing development of new classes of probes. One such device is the thermal profilometer [11]. The thermal probe consists of an ultrasmall thermocouple (bimetallic junction) formed around an insulated STM tip. The device has a temperature sensitivity of less than $1 \times 10^{-3}$ K. With such a tip, the image contrast mechanism depends on the thermal properties of the sample. New probes expand our means to study and visualize surfaces. Much of the promise of scanning probe microscopy depends on the development and utilization of these techniques.

## V. EXAMPLES OF NEAR-FIELD SCANNING MICROSCOPIES

The following sections offer a brief introduction to the state of the art for several scanning probe microscopies. The examples illustrate the current techniques and the kinds of problems that can be studied with the various probes. More detailed descriptions can be found in the chapters that follow.

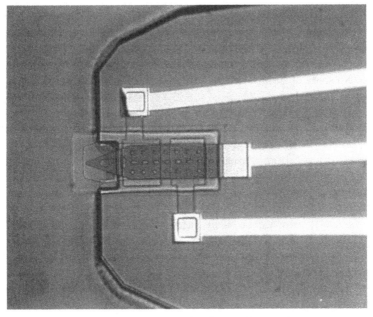

**(a)**

**(b)**

**Figure 4** An Si-micromachined sensor for force microscopy. (a) The probe is 200 μm long and 50 μm wide. The cantilever can be electrostatically driven for ac detection. (b) High magnification view of the apex of the probe. The stylus is made of poly-Si and is 5 μm high.

## A. Scanning Tunneling Microscopy

Scanning tunneling microscopy revolutionized the art of surface examination. Its inventors were awarded the Nobel prize for physics in 1986. STM has had major impact on a number of fields, however none so large as in the area of surface science. No study of surface structure can be considered to be complete without inclusion of STM data.

The ultimate goal of these studies is to determine the position and elemental nature of the surface atoms. STM data give atomic resolution information on the local density of states of the sample. Although the density of states is derived from the ensemble of surface atoms, there is no general way to invert this information to determine the surface structure. To proceed further, a model of the surface is needed. Predictions of the model may then be tested against the data taken by STM. Figure 5 shows two images of the GaAs(110) surface taken at different bias voltages, namely 1.9 and −1.9 V. With a positive sample bias, electrons flow from the filled states of the tip to the empty states of the sample. Conversely, with a negative sample bias, electrons flow from the filled states of the sample to the empty states of the tip. In GaAs the filled states are concentrated on the As atoms and the empty states on the Ga atoms. With this model and the STM images, it is possible to determine the locations of the Ga and As in the surface unit cell [34]. Similar studies have been done for other materials. However, as the system becomes more complex a unique interpretation of the data becomes more difficult.

Recently, STM has been used in conjunction with laser illumination to examine the nonequilibrium distribution of photoexcited charge carriers on a semiconductor surface [35]. Figure 6 shows images of the Si(111) surface topography and photovoltage acquired simultaneously. It is clear that the surface photovoltage (Fig 6b) is greatly reduced in regions near defects and grain boundaries. This can be explained by a larger recombination rate of the excited carriers near these imperfections. Atomic resolution images such as these are leading to much better understanding of surfaces.

An interesting variation of the tunneling microscope was developed by Kaiser and Bell [36], who constructed an instrument capable of imaging the properties of a buried interface. Ordinary STM is of little use in the study of such structures because of its extreme surface sensitivity. However, when electrons are injected by a tunneling tip they travel approximately 10 nm before undergoing an inelastic collision. If a buried interface is less than this distance from the surface, the ballistic electrons can probe the interface. This technique, called BEEM (ballistic electron emission microscopy), allows both imaging and spectroscopy of electronic state within the sample.

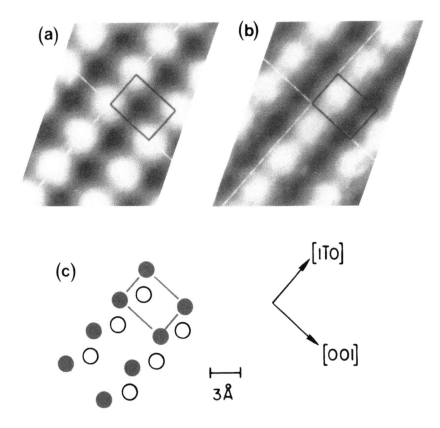

**Figure 5** Image of GaAs(110) surface by STM. (a) Topograph of the surface when scanned with a bias of 1.9 V on the sample. (b) Topograph of the surface when scanned with a bias of −1.9 V. (c) Top view of the surface atoms. Arsenic atoms are represented by open circles and gallium atoms by solid circles. The rectangle indicates a unit cell, whose position is the same in all three figures. (Courtesy of R. Feenstra.)

## B. Scanning Force Microscopy

Dc tunneling techniques require transport of a current through the sample. Scanning force microscopy was born out of the need to examine insulating surfaces. To study insulators, Binnig and coworkers [14] replaced the tunneling tip in an STM by a cantilever, creating the atomic force microscope (AFM). The interaction between the probe and surface is a repulsive contact force. To detect the sample surface, deflections of the cantilever are measured, and no tunneling current is necessary.

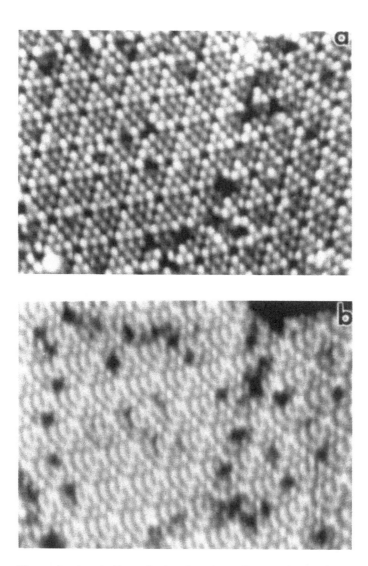

**Figure 6**   Atomically resolved surface photovoltage on Si(111) (7 × 7). (a) Topographic image including a domain boundary and several defects. (b) Spatially resolved photovoltage image showing reduction in the surface photovoltage around defects and at domain boundaries. (Courtesy of R. Hamers.)

Cantilever deflection was originally measured by a secondary tunnel junction, but optical techniques have largely replaced this method. There are presently two commonly used schemes in force microscopy. They differ in cantilever–sample interaction and in detection method. The first uses the repulsive interaction between two surfaces and is called AFM. This contacting method is very similar to that of a conventional stylus profilometer, but the interaction force is much lower and resolution much higher. The cantilever is brought into weak contact with the surface. Substrate–probe forces are typically in the range of $10^{-9}$ N. A laser beam incident at approximately 45° is reflected off the cantilever to a position-sensitive photodiode detector [37]. As the sample is scanned, the cantilever rides up and down with surface contours. The reflected laser spot moves across the position-sensitive diode. The photodiode output is recorded and produces a topograph of the surface. Atomic scale features have been observed with this method. However, one of the disadvantages of the contacting technique is the potential for altering the sample surface.

The second standard method of force microscopy uses van der Waals, electrostatic, or magnetic noncontacting interactions. These longer range interactions allow the probe to remain several hundred angstroms away from the surface as it is being scanned. The tradeoff is lower resolution and a more complicated detection method. An ac interferometric technique is generally used [38,39]. A common technique is to drive the cantilever probe near its resonant frequency and use the dynamic properties (amplitude, phase, etc.) as control parameters for the servo loop. Extremely high sensitivities can be achieved with these techniques. Changes in the ac dither amplitude of $10^{-5}$ nm can be measured. Consequently, force gradients of $10^{-4}$ N/m may be detected. The high sensitivity is required to measure small variations in the long-range interactions. With the small signals comes a reduction in lateral resolution, caused by the slower decay of the interaction. Larger areas of the probe and sample contribute to the total force. This nonlocal interaction reduces the sensitivity to lateral features and reduces the resolution.

As an example of a long-range interaction, consider Fig. 7. In this image of a thin film disk used for magnetic recording, the true surface is planar. However when scanned with a ferromagnetic tip, regions of varying magnetization can be observed. This image was taken with a derivative mode, where contrast is obtained over the magnetic domain walls. The ridges correspond to a series of "1's" written on the film. Such magnetic force microscopy is a good example of tailoring the tip–surface interaction to suit a particular purpose.

## C. Near-Field Optical Microscopy

NSOM represents an important advance in the field of optical microscopy. Optical microscopy has many advantages over alternative methods. It is well

**Figure 7**  Magnetic force microscopy image of a CoCr alloy thin film disk. The features recorded are the boundaries between magnetic bits. The track is 17 μm wide, and each bit is 2 μm wide. The detail outside the track is real and shows random magnetization. (Courtesy of D. Abraham.)

understood, it usually provides a nondestructive interaction, and it can operate in biological environments (air and water). Its major limitation is the resolution, which is about one-fourth of the observation wavelength. NSOM represents a means to increase the resolution while preserving the strengths of conventional optical microscopy.

The key to NSOM lies in the creation of the subwavelength-sized aperture. Two methods are presently being used. The first uses technology borrowed from neurology [40]. A glass pipette is heated and pulled until it separates. During the process the glass is drawn to a taper and a very small aperture is formed. The inner diameter of the pipette can be as small as 20 nm. A metal film is deposited onto the tip to complete the fabrication and form the waveguide. The second method is very similar except the pipette is replaced by an optical fiber [33]. The core is reduced in size as the fiber is drawn, forming a very small aperture. Once again, to complete fabrication the probe is coated with metal.

Recently, much effort has been devoted to fabrication of tips with a high luminosity for a given aperture size. One of the successful methods entails the use of a molecular crystal grown in the tip of a pipette [41]. As the crystal is illuminated on the inside of the pipette, excitons are created. A concentration gradient is produced, and the excitons diffuse toward the aperture, where they recombine and produce radiation. The decay of excitons varies as a power of the distance traveled, whereas the intensity of evanescent radiations decays exponentially. For appropriately sized crystals, a relative gain in intensity can be realized by pipettes with crystals over that of hollow pipettes. Although much work remains to be done in this area, the crafting of smaller and brighter probes appears quite promising.

## D.   Thermal Profilometry

Photothermal imaging has a long history. In the far field, resolution is limited by the wavelength of the excitation or detection radiation. Williams and Wickramasinghe developed near-field photothermal scanning to increase the resolution of the technique [11]. They fabricated an ultrasmall thermocouple to replace the tunneling tip. This probe is placed near the sample surface so that the two are thermally coupled. The electromotive force measured from the thermocouple is used as the feedback parameter. AC techniques provide improved sensitivity and allow simultaneous topographic and thermal imaging to be performed. Figure 8 shows topographic and temperature maps of a grating imaged by this technique. The temperature variation over the surface is approximately 0.01°C.

Fabrication of the thermal probe is accomplished by insulating an STM tip except at the apex. A second metal is then deposited over the apex to form an ultrasmall bimetallic junction. This process is difficult and has a low yield. Recently, Si micromachining has been used to form the basis for an alternative design of the probe. Figure 9 shows the shank and shuttle mechanism of the new probe. AC modulation of the signal is accomplished by electrostatically driving the probe. The thermocouple will be formed at the tip from a semiconductor–metal junction.

Near-field thermal profilometry is still in its infancy. In addition to topographic imaging, this technique allows the possibility of high resolution imaging of biological and chemical reactions. Exothermic and endothermic activity can be recorded with better than 100 nm resolution. When probe fabrication becomes routine, near-field thermal profilometry should reach its full potential.

## VI.   LIMITATIONS

Scanning probe microscopies make up a powerful class of techniques. Their strength lies in the variety of interactions that can be used to examine the sample

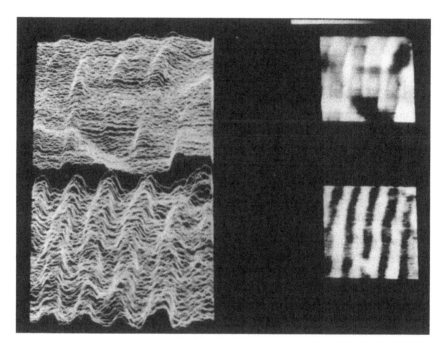

**Figure 8**  The top two images are thermal profilometer results, line scan and gray scale representation, of a photoresist grating. The scan area is 900 nm × 900 nm. The lower images are a simultaneously acquired thermal map of the same area. (Courtesy of C. Williams.)

at high resolution. There are, however, limitations to the techniques. The foremost is the very real difficulty in the interpretation of an image. One must be aware of the many different causes for an apparent feature: it may be very difficult to determine the actual change in the probe–sample interaction that has led to an apparent height change in the topograph.

An example of this difficulty, which is fundamental to all near-field probes, involves the probe–sample interaction. Interaction implies a coupling between tip and sample. Two difficulties can arise if this interaction is too strong. First, the probe can induce changes in the sample as it is scanned over the surface. Both structural changes and field-induced polarization can occur. Second, the

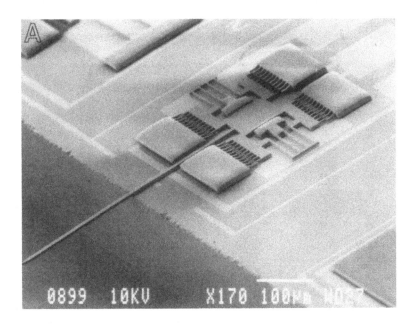

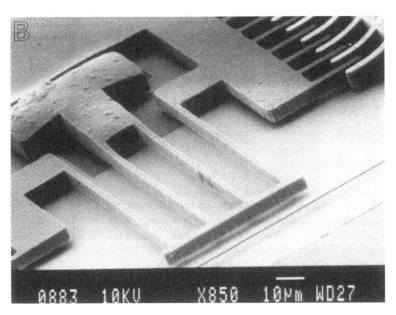

**Figure 9** (A) Si micromachined oscillator to be used as a scanning thermal profilometer. (B) The comb structure used to electrostatically drive the probe normal to the surface. This motion provides the high frequency thermal signal detected by a thermocouple at the apex of the probe. (Courtesy of K. Najafi.)

**(a)**

**Figure 10**  Two examples of common image artifacts caused by the tip–sample convolution. (a) Image of gold on germanium. The true topography is shown in Fig. 3. The repeated hillock structure is due to the rough sample imaging the tip. (b) Topograph of cobalt on silicon. The multiple ghost images are caused by the active tunneling region of the tip shifting as the scanning is taken. This leads to a lateral displacement of the surface features.

surface may induce temporary or permanent changes in the probe. The radiation pattern produced by a NSOM, for example, depends on both the geometry of the sample and that of the aperture. Therefore, interaction strength depends on the surface features in a complicated manner. Because the radiation field changes as the probe is scanned over the sample, interpretation of the images can be difficult.

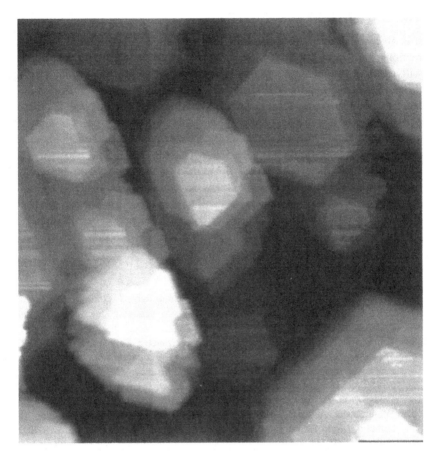

**(b)**

Reciprocity among these effects should be appreciated. There is nothing fundamentally different between the probe and the sample. Interaction strength is determined self-consistently, and induced changes in tip or sample will be measured. A common example is observed during STM examination of a rough surface. A topograph will occasionally be generated in which the same shape is repeated again and again across the surface. The cause is a reversal of the roles of tip and sample. If the sample is rough and the features on the surface are "sharper" than those of the tip; the sample will scan the tip repeatedly. The periodic shapes observed in the topograph are images of the tip! Figure 10 shows typical examples of tip artifacts. Such aberrations can be observed in all scanning probe microscopies. Proper tip design helps reduce these effects, but they can be a source of confusion even to the experienced investigator.

## VII.  FUTURE APPLICATIONS

It is always dangerous to extrapolate into the future. However, some future directions and applications are clear. Scan speeds of near-field microscopies will continue to increase. As the drive mechanics are reduced in size and the probes become more sophisticated, the goal of video rate imaging will be met in some techniques. Scanning instruments will be combined with other experimental methods to allow not only high spatial resolution but also high temporal resolution. As an example, ultrafast laser pump–probe methods can be combined with near-field optical scanning. There is clearly a push to apply these methods to soft materials, especially biological systems. When the contrast mechanisms are better understood, many advances will be made in biological studies. Finally, even more new probes will be developed. No one can predict the future utility of such devices, but their potential for new areas of application remains great.

## REFERENCES

1.  O-Keefe, J. A. (1956). Resolving power of visible light, *J. Opt. Soc. Am. 46*:359.
2.  Ash, E. A. (1972). Super-resolution aperture scanning microscope, *Nature, 237*:510.
3.  Young, R. D. (1966). Field emission ultramicrometer, *Rev. Sci. Instrum. 37*:275.
4.  Young, R. D., Ward, J., and Scire, F. (1972). The topografiner: An instrument for measuring surface microtopography, *Rev. Sci. Instrum. 43*:999.
5.  Fowler, R. H., and Nordheim, L. (1928). Electron emission in intense electric fields, *Proc. R. Soc. London, A119*:922.
6.  Young, R. D., Ward, J., and Scire, F. (1971). Observation of metal–vacuum–metal tunneling, field emission, and the transition region, *Phys. Rev. Lett. 27*:922.
7.  Young, R. (1971). Surface microtopography, *Phys. Today*, November, p. 42.
8.  Binnig, G., Röhrer, H., Gerber, C., and Weibel, E. (1982). Surface studies by scanning tunneling microscopy, *Phys. Rev. Lett. 49*:57.
9.  Quate, C. F. (1986). Vacuum tunneling: A new technique for microscopy, *Phys. Today*, August, p. 26.
10. Binnig, G., Röhrer, H., Gerber, C., and Weibel, E. (1983). $7 \times 7$ Reconstruction of Si(111) resolved in real space, *Phys. Rev. Lett. 50*:120.
11. Williams, C. C., and Wickramasinghe, H. K. (1986). Scanning thermal profiler, *Appl. Phys. Lett. 49*:1587.
12. Pohl, D. W., Denk, W., and Lanz, M. (1984). Optical stethoscopy: Image recording with resolution λ/20, *Appl. Phys. Lett. 44*:651.
13. Lewis, A., Isaacson, M., Muray, A., and Harootunian, A. (1984). Development of a 500 Å spatial resolution light microscope, *Ultramicroscopy, 13*:227.
14. Binnig, G., Quate, C. F., and Gerber, C. (1986). Atomic force microscope, *Phys. Rev. Lett. 56*:930.
15. Hamers, R. J., Tromp, R. M., and Demuth, J. E. (1986). Surface electronic structure of Si(111)-$7 \times 7$ resolved in real space, *Phys. Rev. Lett. 56*:1972.

16. Binning, G., and Röhrer, H. (1986). Scanning tunneling microscopy, *IBM J. Res. Devl. 30*:355.

17. Park, S. I., and Quate, C. F. (1987). Scanning tunneling microscope, *Rev. Sci. Instrum. 58*:2010.

18. Demuth, J. E., Hamers, R. J., Tromp, R. M., and Welland, M. E. (1986). A simplified STM for surface science studies, *J. Vac. Sci. Technol. A*, 4:1320.

19. Fein, A. P., Kritley, J. R., and Feenstra, R. M. (1987). Scanning tunneling microscope for low temperature, high magnetic field and spatially resolved spectroscopy, *Rev. Sci. Instrum. 58*:1806.

20. Cole, B. W., Ringger, M., and Guntherodt, H.-J. (1985). An electromagnetic microscopic positioning device for the scanning tunneling microscope, *J. Appl. Phys. 58*:3947.

21. Pohl, D. W. Sawtooth nanometer slider: A versatile low voltage piezoelectric translation device, *Surf. Sci. 181*:174.

22. Inchworm is commercially available from Burleigh Instruments, Inc., Fisher, NY.

23. Binnig, G., and Smith D. P. E. (1986). Single-tube three-dimensional scanner for scanning tunneling microscopy, *Rev. Sci. Instrum. 57*:1688.

24. Lang, N. (1986). Theory of single-atom imaging in the STM, *Phy. Rev. Lett. 56*:1164.

25. This selection does not preclude multiple tip effects. Ghost images and imaging of the tip are often observed when scanning rough surfaces. See Fig. 10.

26. Fink, H. W. (1986). Mono-atomic tips for scanning tunneling microscopy, *IBM J. Res. Devl. 30*:460.

27. Kuk, Y., and Silverman, P. J. (1986). Role of tip structure in scanning tunneling microscopy, *Appl. Phys. Lett. 48*:1597.

28. Melmed, A. J. (1991). The art and science and other aspects of making sharp tips, *J. Vac. Sci. Technol. B*, 9:601.

29. Wickramasinghe, H. K., private communication.

30. Kong, L. C., Orr, B. G., and Wise, K. D. (1990). A micromachined silicon scan tip for an atomic force microscope, *Sensors Actuators Tech. Digest*, June, p. 28.

31. Albrecht, T. R., and Quate, C. F. (1988). Atomic resolution with the atomic force microscope on conductors and nonconductors, *J. Vac. Sci. Technol. A*, 6:271.

32. Betzig, E., Isaacson, M., Barshatzky, H., Lewis, A., and Lin, K. (1988). Near-field scanning optical microscopy (NSOM), *Proc. SPIE*, 897:91.

33. Betzig, E., Trautman, J. K., Harris, T. D., Weiner, J. S., and Kostelak, R. L. (1991). Breaking the diffraction barrier: Optical microscopy on a nanometric scale, *Science, 251*:1468.

34. Feenstra, R. M., Stroscio, J.A., Tersoff, J., and Fein, A. P. (1987). Atom-selective imaging of the GaAs(110) surface, *Phys. Rev. Lett. 58*:1192.

35. Hamers, R. J., and Markert, K. (1990). Atomically resolved carrier recombination at Si(111)-(7 × 7) surfaces, *Phys. Rev. Lett. 64*:1051.

36. Kasier, W. J., and Bell, L. D. (1988). Direct observation of subsurface interface electronic structure by ballistic-electron-emission-microscopy, *Phys. Rev. Lett. 60*:1406.

37. Meyer, G., and Amer, N. M. (1988). Novel optical approach to atomic force microscopy, *Appl. Phys. Lett. 53*:1045.

38. Martin, Y., and Wickramasinghe, H. K. (1987). Magnetic imaging by "force microscopy" with 1000 Å resolution, *Appl. Phys. Lett. 50*:1455.

39. Rugar, D., Mamin, H. J., Erlandsson, R., Stern, J. E., and Terris, B. D. (1988). Force microscope using fiber-optic displacement sensor, *Surf. Sci. 59*:2337.

40. Brown, T., and Flamming, D. G. (1977). New microelectrode techniques for intracellular work in small cells, *Neuroscience*, *2*.

41. Lieberman, K., Harush, S., Lewis, A., and Kopelman, R. (1989). A light source smaller than the optical wavelength, *Science*, *247*:59.

# 6

# Scanning Tunneling and Atomic Force Microscopy: Tools for Imaging the Chemical State

**David L. Patrick and Thomas P. Beebe, Jr.** *University of Utah, Salt Lake City, Utah*

## I. INTRODUCTION

Surface chemistry is one of the most rapidly expanding and productive areas of scientific research today. Its success is due in part to a strong symbiotic connection with a diverse range of technologies and scientific disciplines that it both relies on and supports. Even the most fundamental surface chemistry is likely to have technological consequences. Disciplines as diverse as catalytic studies, integrated circuit design, interfacial biochemistry, and corrosion science all depend on surface chemistry. Likewise, many of the advances that surface chemistry has made in the last 30 years have been possible only through the advent of new technologies and techniques. In this chapter we discuss two such techniques, scanning tunneling microscopy (STM) and atomic force microscopy (AFM). We show how they can be applied to a very wide range of surface studies, often giving information that is not available through any other means. Although STM and AFM are barely 10 years old, these two techniques have permeated much of modern surface science, contributing results to more than 2000 papers already. The nanometer scale details of a surface's electronic and structural properties are what determine its chemical characteristics on all scales, and the unsurpassed resolving power of STM and AFM for both these properties, coupled with ease of use and applicability to a wide range of samples, make them ideally suited for surface studies of all kinds.

STM was invented in 1981 by Gerd Binnig and Heinrich Röhrer working at IBM research laboratories in Zurich [1–3]; they shared the Nobel prize in physics for this work in 1986 [4]. Five years after the invention of STM, Binnig, working with Cal Quate at Stanford University and Christoph Gerber at IBM, announced the development of another, closely related microscope, the AFM [5]. These two original methods have been joined in the last 6 years by a handful of similar techniques, which are now collectively referred to as scanning probe microscopies.

The scanning probe microscopies all share the capability of providing three-dimensional, real-space images of surfaces and surface adsorbates, in some cases with atomic resolution. They are operated by locating a small probe tip over a sample and measuring some local property of the surface at that point. By scanning the tip across the surface in a raster pattern, an image may be constructed representing variations of that property within the raster scan range. The essential difference between the various local probe methods is the distinction between the particular surface properties that each measures. The STM, for example, measures the electrical conductance between the probe tip and the surface; the AFM measures the total force. Both techniques have been shown to operate effectively under a variety of conditions, and to be able to achieve atomic resolution in vacuum, air, liquids, and at opposing temperature extremes.

STM and AFM are now widely used for a large number of different surface science applications and in many different labs throughout the world. In the space of a few pages we cannot hope to review more than a very small part of all the work that is being done. We have, however, attempted to mention at least a few of the areas that are currently of greatest interest, our goal being to afford the reader an appreciation for the diversity of problems the scanning probe microscopies are capable of addressing. We begin with a description of the operational principles and physical theory of STM, followed by those for AFM. We also describe the construction and important experimental considerations of both techniques. We then discuss a few topics of current interest, dividing them into sections on adsorbate studies, electronic property studies, and novel techniques. We conclude with a few remarks on what we believe are promising future directions in STM-AFM research.

## II.  STM OPERATION AND THEORY

### A.  Introduction

STM is based on a quantum mechanical process known as tunneling, which is itself a consequence of the wavelike properties of matter. Tunneling occurs when a system undergoes a transition from an initial state $\psi_i$ with energy $\epsilon_i$ to a final state $\psi_f$ with energy $\epsilon_f \leq \epsilon_i$ along a potential energy surface having at least one point with energy $E > \epsilon_i$ (see Fig. 1). This potential energy surface presents an energetic barrier in the path of the transition from one state to the other, and the points along it with energy greater than $\epsilon_i$ form a classically forbidden region. Quantum mechanics predicts that the wave functions $\psi_i$ and $\psi_f$ extend a short

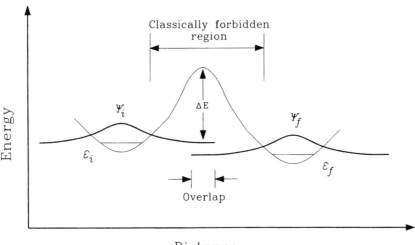

**Figure 1** Tunneling between two states with energies $\epsilon_i$ and $\epsilon_f \leq \epsilon_i$ results from the nonzero overlap of their two wave functions $\psi_i$ and $\psi_f$ within the classical energetically forbidden region. The probability of tunneling depends on the height of the barrier $\Delta E$ and on the spatial separation of the two states. In an STM experiment, $\Delta E$ is related to the work functions of the sample and tip and to the bias voltage applied between them. The spatial separation of the states $\psi_i$ and $\psi_f$ corresponds to the separation of the tip and sample.

distance into this region, exhibiting an exponential decay. Tunneling results from the constructive spatial overlap of the wave functions $\psi_i$ and $\psi_f$ within the classically forbidden region, and the probability of a tunneling event depends on the amount of overlap.

There have been many different theoretical treatments of tunneling. One of the most useful was developed by Simmons [6] who, more than 25 years ago, derived an equation that describes tunneling of electrons in a system consisting of two metallic electrodes separated by a thin insulating barrier. He found that the probability of tunneling could be expressed in terms of the current $I$ of electrons traveling from one electrode to the other. If a small bias voltage $V$ is applied between the electrodes, the tunneling current is

$$I = \int_{\substack{\text{electrode area}}} \frac{\alpha}{s} \sqrt{\phi} \; V \exp(-\beta s \sqrt{\phi}) \, dA \tag{1}$$

where the integral is taken over the area of one electrode, $\phi$ is the average work function of both electrodes, $s$ is the separation between them, and $\beta$ is a constant.

This function is plotted with three different values for the average work function in Fig. 2. It can be seen in this logarithmic plot that there is roughly an order of magnitude change in the tunneling current for each 1 Å change in the electrode separation distance for a 4–5 eV work function value (typical of clean metals in an ultrahigh vacuum environment). Smaller work functions produce a smaller current response to changes in separation distances.

It is this exponential dependence of the tunneling current on electrode separation distance that makes STM feasible and gives it its high vertical and lateral resolution. In STM, a sharpened metallic tip and a conductive sample act as the two electrodes, typically separated from one another by a few angstroms. A (~1 V) bias voltage is applied between them that induces a small but measurable net current flow of $10^{-8}$ to $10^{-12}$ A. Because of the exponential nature of the response signal, the sharpness of the tip is not as important as it may at first seem. It is ultimately the sharpness of the tip that determines the lateral resolution in an STM experiment; even on a relatively blunt tip, however, a few atoms must protrude slightly farther than their neighbors. These atoms dominate the passage of tunneling current, since they are slightly closer to the surface. For

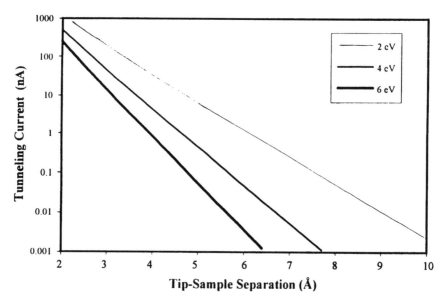

**Figure 2** Plot of tunneling current versus tip–sample separation for three different sample work function values. Note that the tunneling current changes by approximately one order of magnitude for each angstrom change in tip–sample separation for $\phi$ values in the range of 5 eV. The plot is based on an effective tunnel junction area of $\alpha = 10$ Å$^2$ and $\beta = 1.025$ Å$^{-1}$

this reason, atomic resolution is not excessively difficult to achieve when the sample conditions are correct.

## B.  Modes of Operation

An STM is operated by continuously measuring the tunneling current while the tip is scanned across the surface in a raster pattern, as shown schematically in Fig. 3a. The high precision positioning capability required by STM is realized using piezoceramic actuators, which can be made with a response of a few tens of angstroms per volt. A power supply with millivolt stability thus can reproducibly control the motion of the tip to within 0.05 Å. As the tip encounters changes in surface topography or local electronic structure along its trajectory, the tunneling current changes. The nature of the response to these changes is determined by the selected operational mode of the instrument. There are two such modes, constant current and constant height, which we now discuss.

In constant current mode, the $z$ position of the tip is continuously adjusted by fast feedback electronics that maintain a constant predefined tunneling current. In this mode, an STM image is formed by rastering the tip across the surface and recording the changes in $z$ position produced by the feedback electronics, as depicted in Fig. 3b. The result is a topographic image like that in Fig. 4. To accommodate the finite response time of the feedback electronics and to prevent mechanical oscillations from occurring in the microscope, scan speeds are typically limited to about $10^4$ Å/s in this mode.

In constant height mode (Fig. 3c), the feedback electronics are adjusted to give a slow response to changes in tunneling current so that as the tip is scanned across the surface, its $z$ position remains essentially constant. Images are then formed by measuring the current at each point. The result for a flat surface is qualitatively similar to the "topographic image" produced in the constant current mode, although interpretation is less straightforward. This mode is applicable only to sufficiently flat surfaces, on which a slow feedback response can be tolerated. Because continuous corrections to the $z$ position of the tip are not required in constant height mode, the scan speed can be significantly increased. Scan rates of many tens of thousands of angstroms per second are easily achieved, making it possible to complete acquisition of an entire image in less than a second.

## C.  Interpretation of STM Images: The Tersoff–Hamann Approximation

The Simmons equation (Eq. 1) is adequate for describing the tunneling behavior of electrons between relatively large electrodes, or over relatively large areas. But in the case of STM, the localized nature of the probe–sample interaction requires a more detailed description of the electronic structure within the interaction region. In a recent review of tunneling theories related to STM,

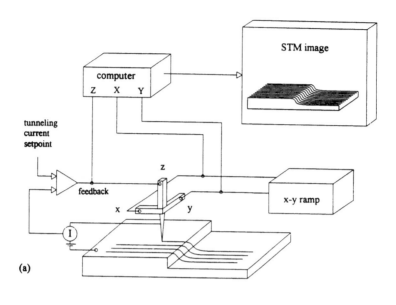

(a)

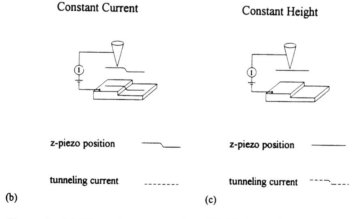

(b)                                                                                 (c)

**Figure 3** (a) Schematic representation of STM electronics and computer interface. The STM tip is moved in a raster pattern across the surface by applying a cyclical voltage ramp to the $x$- and $y$-piezoelectric actuators. The source of contrast in an STM image depends on the operational mode of the instrument. (b) The constant current mode, in which the $z$ position of the tip is continuously adjusted to maintain a constant tunneling current. In this mode, image contrast is derived from the $z$-piezoelectric feedback signal. (c) The second operational mode, constant height mode. In this mode, image contrast is based on the magnitude of the tunneling current collected at constant tip height (no $z$-axis adjustment).

**Figure 4** Constant current STM image of a highly ordered pyrolytic graphite (HOPG) surface. HOPG is widely used as a substrate for STM and AFM studies because it is easily prepared with large (micrometer-sized) atomically flat terraces. The $x$ and $y$ dimensions of this image are 2000 Å. The false color scale represents image contrast information in units of angstroms, having a range of 100 Å.

Feuchtwang and Cutler [7] found that there currently exist three main approaches that attempt to treat the local effects to one degree or another:

One-dimensional or separable, single-particle models that are solved exactly or within the Wentzel–Kramers–Brilluoin (WKB) formalism

Transfer Hamiltonian theories

Theories based on infinite-order perturbation theory for nonequilibrium processes

Among these, the most widely accepted is the approach taken by Tersoff and Hamann [8], who use an approximation based on Bardeen's transfer Hamiltonian formalism [9,10] in which two weakly coupled metal electrodes $\upsilon$ and $\mu$ are separated by an insulator that shares a common Hamiltonian with each. By using this arrangement, first-order perturbation theory yields a simple expression for the tunneling transition matrix, $M_{\upsilon\mu}$, between the electrodes in terms of the current density operator $J$.

Tersoff and Hamann have adopted Bardeen's expression for $M_{\upsilon\mu}$ to derive a simple equation relating the tunneling current to the local density of states

(LDOS) of the surface at the position of the tip. They begin by modeling the tip as a highly localized spherical potential well that has only $s$-like states. From this they show that the tunneling current $I$ is given by

$$I \cong CVR^2 e^{2\kappa R} \, \rho(r_0, E_f) \tag{2}$$

$$\rho(r_0, E_f) = \sum_v |\Psi_v(r_0)|^2 \, \delta(E_v - E) \tag{3}$$

where $\rho(r_0, E_f)$ is the LDOS at the Fermi level at the position of the tip, $R$ is the radius of curvature of the tip, $\kappa$ is the decay length of an electron equal to $h^{-1}\sqrt{(4m_e\pi^2\phi)}$, $\phi$ is the work function, $m_e$ is the mass of an electron, $V$ is the applied voltage, and $C$ is a proportionality constant. The sample states $|\Psi_v (r_0)|^2$ are proportional to $e^{-2\kappa(R+s)}$ where $s$ is the tip–surface separation, shown in Fig. 5. This results in an expression for the tunneling current that is exponentially dependent on tip–sample separation, as expected:

$$I = Be^{-2\kappa s} \tag{4}$$

What makes the Tersoff–Hamann theory appealing is the simple and intuitive explanation it provides for the description of the tunneling current in an STM

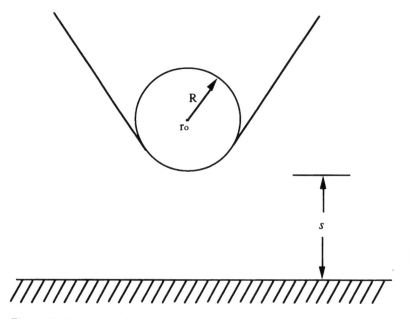

**Figure 5**   In the Tersoff–Hamann theory, the tip is taken to have a radius of curvature $R$ and is separated from the sample by a distance $s$. (Adapted from Ref. 8).

experiment: *the tunneling current is proportional to the surface local density of states about the Fermi level at the position of the tip.* In other words, an STM image is a surface of constant tunneling probability for a given energy and LDOS.

Tersoff and Hamann also gave an expression for the resolution expected from STM. They find that the rms width $W$ of the STM resolution function is approximately [11,12]:

$$W \approx \sqrt{\frac{s}{2\kappa}} \tag{5}$$

For typical experimental values of $s$ and $\kappa$, this corresponds to a resolution of $\sim 2$ Å.

Nakagiri and Kaisuka [13] have performed calculations on the effect on STM resolution of tip radius $R$, distance from surface $s$ and magnitude of the surface work function $\phi$. They considered relatively large (nanometer-sized) surface structures and so were able to obtain approximate results using the Simmons equation. Figure 6 summarizes the results of their calculations. They found that STM resolution depends on $R$, $s$, and $\phi$ as expected, but the dependence on $R$ is most significant. Other workers have performed similar calculations on angstrom-sized structures [14].

Finally, it is worth mentioning one or two difficulties associated with the Tersoff–Hamann theory. One of the essential assumptions of the transfer Hamiltonian approach is that the tip and surface states are decoupled from one another. At small tip–surface separations, however, tip-induced modification of the surface electronic states (and vice versa) is likely to be significant, and perturbation theory may be insufficient to account for the interactions.

The assumption that the tip can be modeled as a featureless spherical potential well has also led to difficulties in some cases. Doyen et al. [15] have discussed the failure of Tersoff–Hamann theory to account for unexpected (vertical) atomic corrugation of Al(111) observed by STM. Studies by Wintterlin [16] and others have shown that STM reveals atomic corrugations with amplitudes of as much as 0.8 Å in the $z$ direction. Tersoff–Hamann theory predicts that the tunneling current, hence the measured variations in surface topography, should be somewhat proportional to the charge density at the position of the tip. But ab initio calculations [17] have shown that Al(111) has a virtually uniform charge density distribution at distances relevant to STM and has no surface states near the Fermi level.

To explain these and other difficulties, a growing number of STM theories have appeared in the literature in recent years, but none has gained the widespread acceptance of Tersoff–Hamann. For a thorough review of Tersoff–Hamann and other theories, see Refs. 18 and 19.

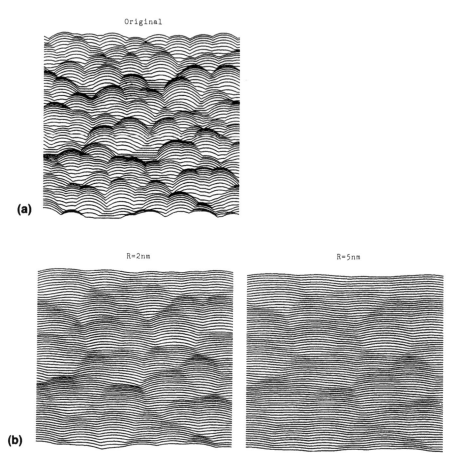

**Figure 6** The spatial resolution of an STM image depends on the radius of curvature of the tip $R$, the distance from the surface $s$, and the magnitude of the surface work function $\phi$. This series of simulated STM images of a hypothetical surface illustrate the effect of these quantities on resolution. (a) The original, ideal representation of a nonuniform surface. (b) The effect of tip radius on resolution. (c) The effect of tip–sample separation. (d) The effect of the sample work function. (Courtesy of N. Nakagiri and H. Kaizuka [13].)

## D. Scanning Tunneling Spectroscopy and Barrier Height Imaging

Returning to Eq. (4), we see that it contains two parameters that depend on the local electronic structure: $B$, which contains information about the sample LDOS (see Eq. 3), and $\kappa$, which is proportional to the square root of the sample work function. These dependencies can be exploited to measure spectral characteris-

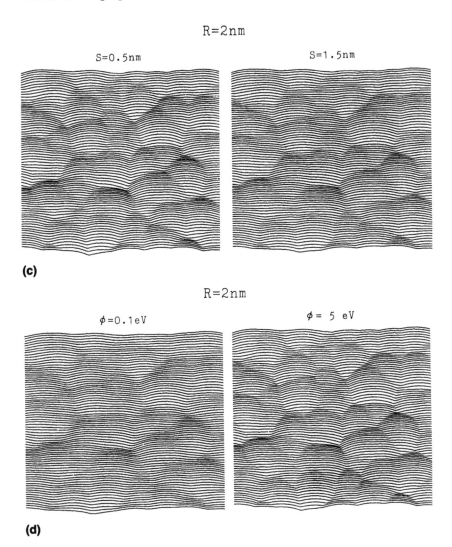

**(c)**

**(d)**

tics of the LDOS and to generate work function profiles, both on extremely small spatial scales.

Experiments designed to measure the LDOS are termed scanning tunneling spectroscopy (STS) experiments. The technique is based on the dependence of the tunneling probability on the LDOS, which is itself a function of energy. The functional form of this dependence can be expressed in terms of measurable quantities as [20]:

$$\rho(r_0, eV) \propto \left[ \frac{dI}{dV} \left( \frac{V}{I} \right) \right]_{V=V_b} \tag{6}$$

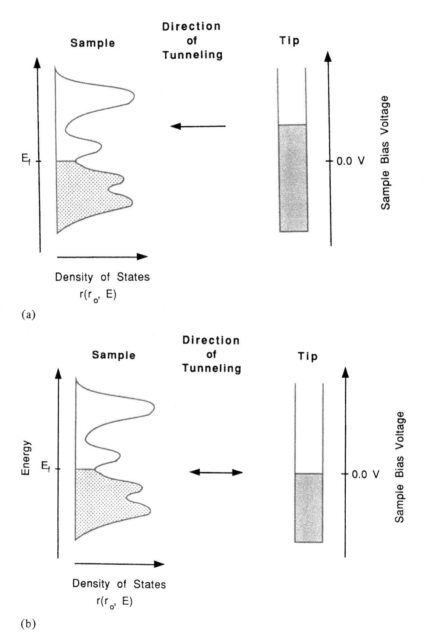

(a)

(b)

**Figure 7**   (a–c) Dependence of tunneling current on sample bias and electronic density of states; arrows indicate direction of electron tunneling for each sample bias polarity. (d) Corresponding scanning tunneling spectrum. The tip density of states is assumed to be constant with energy.

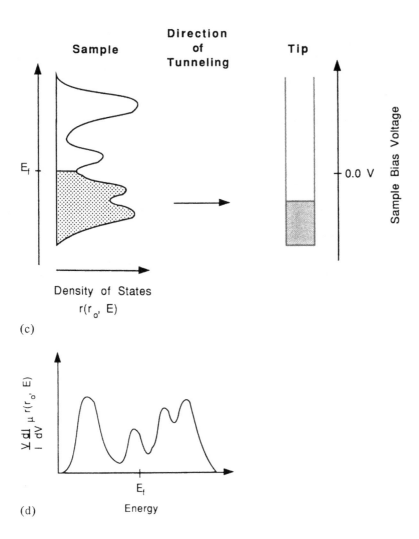

(c)

(d)

where $V_b$ is the bias voltage. Figures 7a–c illustrate this dependence. By measuring the tunneling current over a range of bias voltages (energies), it is possible to generate a spectrum like that shown schematically in Fig. 7d. Such a spectrum represents an "energy map" of the sample LDOS about the Fermi level. It is similar to a valence-level photoemission spectrum and an inverse photoemission spectrum, but on a highly localized scale.

In practice, STS measurements are commonly made following the methods of Feenstra and Fein [21,22] or Hamers, Tromp, and Demuth [23]. The former method is the simplest and is performed by acquiring two images of the same area at reversed bias voltage polarity. The result is two complementary images, one of mostly filled surface states, the other of mostly empty surface states.

The second method, often called current imaging tunneling spectroscopy (CITS), yields more detailed information about the energy profile of the LDOS. CITS experiments are performed by holding the tip stationary at one point over the surface and cycling the bias voltage over a range of values both positive and negative, while the feedback electronics are momentarily disabled. The result is a detailed energy map of the LDOS for one particular point on the surface like that shown in Fig. 7d. With appropriate acquisition electronics and sufficient computer memory, such a spectrum can be acquired at each pixel in an image [23].

Both kinds of STS experiment are performed by maintaining a fixed tip–sample separation and measuring the tunneling current as a function of bias voltage. In another type of experiment, a "work function" measurement, bias voltage is held constant while tunneling current is measured as a function of tip–sample separation $s$. Within the Tersoff–Hamann approximation, a plot of $d(\ln I)/ds$ yields a line whose slope is proportional to $\sqrt{\phi}$, hence the name "work function measurement." Figure 8 demonstrates [24] this relationship for a tungsten tip

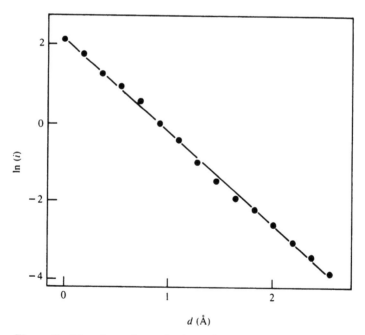

**Figure 8** Plot of experimentally determined ln $I$ versus relative tip–sample separation for a tungsten tip and a gold foil surface in UHV. The slope of the line (5.2 eV) corresponds closely to the work function of gold (5.1 eV). (Adapted from Ref. 24. Copyright 1986 by International Business Machines Corporation; reprinted with permission.)

($\phi$ = 4.6 eV) and a gold surface ($\phi$ = 5.1 eV) measured in ultrahigh vacuum (UHV). The work function calculated from the slope of the line is 5.2 eV, in reasonably good agreement with the expected value.

Actually, "work function measurement" is something of a misnomer, since the work function is by definition a macroscopic property, and STM is sensitive to local conditions. A more appropriate name would be "local barrier height measurement," since the local tunneling barrier height is actually being measured. Nevertheless, work function measurements can give some very useful information about a sample. In one type of experiment, the $z$ position of the tip is sinusoidally modulated about some average height at a frequency outside the feedback loop bandwidth, but within the preamplifier bandwidth, while simultaneously sweeping it across the sample in its usual raster pattern. A lock-in amplifier is used with the method discussed in the preceding paragraph to calculate the local barrier height at each point, and the results are assembled into a barrier height image like that shown in Fig. 9a [25]. This figure, and its companion "conventional" STM image (Fig. 9b), are of the same area on a rough platinum surface in air, illustrating the difference between the kinds of information produced by the two techniques.

The image contrast in the barrier height image (Fig. 9a) is a result of changes in surface topography and chemical composition that perturb the local "work function." Local barrier height measurements also can be used to examine purely adsorbate-related surface properties. It is well known, for example, that the work function of a clean surface $\phi_c$ is modified by the presence of a polar adsorbate according to [26]

$$\phi = \phi_c - \frac{e\mu}{\epsilon_0} \tag{7}$$

where $e$ is the electronic charge, $\mu$ is the dipole moment density of the adsorbate, and $\epsilon_0$ is the permittivity of free space. Because of the exponential dependence of the tunneling current on $\sqrt{\phi}$, small changes in the work function may result in large changes in image contrast. Spong et al. [27] have proposed that this fact may be used to explain the imaging mechanism of certain liquid crystal molecules adsorbed on surfaces.

## III. AFM THEORY AND OPERATION

## A. Introduction

One serious limitation of STM is the requirement that samples be electrically conductive. Although efforts have been made to circumvent this restriction by, for example, using ac instead of dc bias voltage [28], STM is unlikely to provide

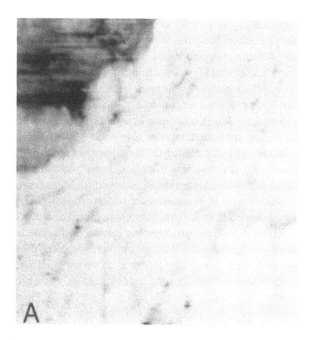

**Figure 9** (A) Barrier height image of a rough platinum surface. (B) Topographic STM image of the same area measuring 1002 Å × 1002 Å. The color scales represent total scales of 1.35 eV (A) and 98 Å (B). (Courtesy of D. Sommerfeld et al. [25].)

the same quality results with insulating samples that it gives with conducting samples. Additionally, although STM is well suited for probing sample electronic and structural properties, it can yield little information about mechanical properties. In this section, we discuss another scanning probe microscope, the AFM, which is capable of doing both. By measuring the attractive or repulsive forces between a probe tip and sample instead of the flow of electric current, AFM can readily image both conductors and insulators. Different experimental configurations also allow for the measurement of different forces including magnetic [29,30], electrostatic [31], frictional [32–34], and colloidal forces [35] with a resolution of less than 5 nN and a spatial resolution of below 5 Å.

AFM is able to achieve superior resolution when compared with other direct-contact surface techniques such as stylus profilometry because of the extremely small probe forces involved. The typical interaction force between an AFM tip and sample is $\sim 10^{-9}$ N, resulting in a very small contact area that in turn enables highly localized measurements. An AFM is operated by scanning a sharpened tip attached to a flexible cantilever across a sample surface and recording deflections of the cantilever caused by changing tip–sample interaction forces.

The cantilever acts as a spring whose restoring tendency works against the tip–sample interaction forces and maintains the tip in an equilibrium position.

## B.   Modeling the AFM Tip and Interaction Potential

There are two important components to any theoretical description of AFM:

The model used to describe the behavior of the tip/cantilever to a changing interaction potential $U$

The description of the interaction potential itself

In most treatments, the cantilever is described in terms of a Hooke's law spring, which responds to an interaction force $F_{cantilever}$ according to

$$F_{cantilever} = -\frac{\partial(U_{cantilever})}{\partial z} = -kz \tag{8}$$

where $U_{cantilever}$ is the potential experienced by a cantilever with force constant $k$ and displacement $z$ from its equilibrium position. The resonance frequency $\omega$ of a cantilever is related to its force constant and mass $m$ by

$$\omega = \sqrt{\frac{k}{m}} \tag{9}$$

It is necessary that $\omega$ be much larger than the scanning rate of the microscope to prevent scanning-induced oscillations in the cantilever. This is accomplished by reducing the mass as much as possible. An actual cantilever with an integral tip is shown in Fig. 10. Integral tip/cantilevers formed from doped $Si_3N_4$ are now commercially available with quite small masses and a variety of spring constants.

The second important component of AFM theory is the model used to describe the interaction potential between the tip and sample. As the sample approaches the tip, the cantilever is deflected from its equilibrium position by an amount necessary to balance the tip–sample interaction force with the restoring force of the cantilever. By assuming a perfectly rigid sample and tip, the overall equilibrium condition may be expressed in terms of opposing forces as

$$-F_{cantilever} = F_{interaction} \quad \text{or} \quad kz = \frac{\partial(U_{interaction})}{\partial s} \tag{10}$$

where $F_{interaction}$ and $U_{interaction}$ are the interaction force and potential between the sample and tip, respectively, and $s$ is the tip–sample separation. The interaction potential may be expressed as a sum of component potentials

$$U_{interaction} = U_{interatomic} + U_{van\ der\ Waals} + \cdots \tag{11}$$

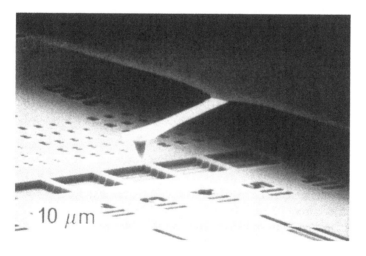

**Figure 10**   Scanning electron micrograph image of an AFM cantilever with integral tip poised above a test pattern in a printed circuit. Courtesy of Olaf Wolter.

where as many additional terms are included as are required to adequately describe the potential for any particular experiment. Thus, for example, if an experiment were configured to measure magnetic forces, a term for the magnetic potential would also be included.

The basic AFM experiment, performed in the absence of magnetic or electrically charged regions, requires only terms for the interatomic and van der Waals potentials [36]. As a result, $U_{\text{interaction}}$ may be approximated using the familiar Lennard–Jones 6–12 potential [37]

$$U_{\text{interaction}} = \frac{\alpha}{s^{12}} - \frac{\beta}{s^6} \tag{12}$$

where $s$ is the separation between the AFM tip and sample, and $\alpha$ and $\beta$ are constants specific to the sample and tip materials. The first term in the Lennard–Jones potential accounts for short-range repulsive interactions and the second term for longer range attractive van der Waals interactions. A plot of the Lennard–Jones potential is shown in Fig. 11, along with the negative of its derivative, which is proportional to the force.

The AFM tip experiences an attractive potential that increases as it approaches the sample until it reaches the distance at which short-range repulsive effects begin to dominate. At this distance, the tip is nearly in contact with the

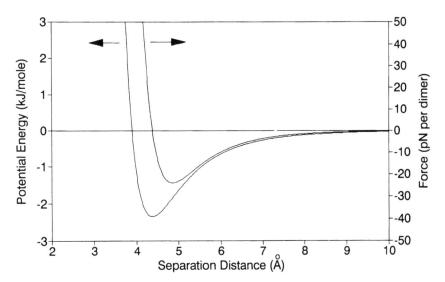

**Figure 11**  The Lennard–Jones 6–12 potential (lower line) and the negative of its derivative (upper line). The Lennard–Jones potential is given in Eq. (12) Values for $\alpha$ and $\beta$ are for Xe-Xe from Barker et al., (1974). *J. Chem. Phys. 61*: 3081.

sample. To describe what happens at such small tip–sample separations, it is necessary to expand our simple model to include the motions of individual tip and sample atoms. Until this point, we have assumed that the tip and sample were perfectly rigid, which greatly simplifies the overall equilibrium condition. When motions of individual atoms are included, however, the equilibrium expression must be expanded to include them all [38]:

$$\frac{\partial U_{\text{interaction}}}{\partial r_i} = 0, \quad r_i = 1, 2, \ldots, N \tag{13}$$

where $r_i$ are the positions of each of the $N$ atoms of the sample and tip.

Theoretical and experimental investigations [39,40] have shown that as the sample comes very close to the AFM tip, the attractive forces between them can cause either the tip or the sample or both to deform, resulting in the sudden formation of an adhesive contact between them, or a so-called jump to contact. In one such study, performed by Landman et al. [39], the behavior of a nickel tip and a gold surface was investigated using molecular dynamics simulations and AFM. Their theoretical simulations indicated that an instability should occur when the tip approached to within about 4.2 Å of the gold surface, resulting in a sudden jump to contact. The jump to contact behavior in this case was caused chiefly by a sudden deformation of the softer gold sample being pulled toward the tip, leading to adhesive contact formation between them. When the tip is withdrawn from the sample after contact, inelastic deformation of the sample occurs in the form of a thin neck that reaches atomic thickness and eventually breaks, leaving a single layer of gold atoms on the end of the nickel tip.

The behavior described above can be tested experimentally by measuring the total force experienced by an AFM tip as a function of separation from the sample. Figure 12, taken from the work of Landman et al., shows two "force curves" measured on the same system these investigators modeled theoretically. The curves represent the force between the tip and sample at each point along a closed loop. The loop begins with the tip far from the sample, approaches until contact (Fig. 12a) or until the tip has indented the sample (Fig. 12b), and ends with the tip withdrawn to its original position. The arrows indicate the temporal sequence of the experiment.

Note that as a function of separation, the force behaves in approximately the way expected from a simple Lennard–Jones potential, growing more attractive as the sample approaches the tip until the distance has become small enough for short-range repulsive forces to become important. The apparent hysteresis between the approach and retract segments of the force curves is caused by adhesion between the sample and tip. As the sample is withdrawn from the tip, it must overcome an additional attractive force originating from adhesive contact.

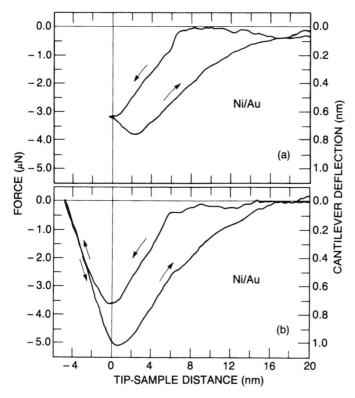

**Figure 12**   AFM force curves for a gold surface and nickel tip in a dry nitrogen environment. Arrows indicate the direction of the approach and retract cycle for (a) contact followed by separation and (b) indentation followed by separation. (Courtesy of U. Landman et al. [39], copyright 1990 by the AAAS.)

The data shown in Fig. 12 were obtained in dry nitrogen with a tip and sample relatively free from contaminants such as water vapor. When such precautions are not taken, however, water vapor from the air may condense on the sample, introducing an additional perturbation of the force curve caused by capillary action. Another effect, not shown in the force curves of Fig. 12, occurs when the magnitude of the gradient of the attractive force $\partial^2(U_{\text{interaction}})/\partial s^2$ becomes greater than the cantilever force constant $k$. When this happens, an instability develops in the cantilever, resulting in a jump to contact very similar to the jump to contact discussed above in connection with tip–sample deformation. Cantilever instability can be avoided by choosing a sufficiently large value for $k$. However, a value too large will reduce the deflection of the cantilever in response to an applied force, resulting in a loss of sensitivity, and ultimately damage to the sample or tip.

## C.  Modes of Operation

There are two basic operational modes for AFM: contact (also called repulsive) and noncontact (also called attractive). As their names imply, these two modes differ in the distance maintained between the tip and the sample during image collection. In noncontact mode, the tip is located tens to hundreds of angstroms from the sample and experiences a weak attractive potential. In contact mode, it also usually experiences a net attractive potential, although the last group of tip atoms may be within the repulsive regime. In addition to these two basic modes, AFM may be operated in constant force or constant height mode. Just as for STM, the choice is determined by the parameters of the feedback loop.

When discussing AFM operational modes, it is important to distinguish between the potential experienced by the last tip atom and the potential experienced by the bulk of the tip. Although the last tip atom will begin to experience a strongly repulsive force before it penetrates the surface of the sample, the bulk of the tip will remain in the attractive regime and the overall potential may be attractive. The terms "attractive" and "repulsive," when used to describe the mode in which an AFM experiment is performed, generally refer to the potential felt by the last tip atom, not the entire tip.

The definition of the onset of contact is not clear at these small distances. One way to define tip–surface contact experimentally is by modulating the $z$ position of the sample during a new translation toward the tip. The weak coupling between the tip and sample will cause an induced motion of the tip even at large separations. When a sudden increase in the amplitude of modulation of the cantilever occurs, however, the tip is said to be in contact with the surface [41]. The point of contact is near the bottom of the potential well of Fig. 11.

The nearness of tip and sample in contact mode permits significantly higher resolution imaging than is possible in noncontact mode (~5Å vs. ~50 Å), and is highly sample dependent. This is true in part because the force gradient is steeper in this region and in part because in noncontact mode, the longer range van der Waals forces interact with a larger area of the tip. The increased resolution provided by contact mode, however, is accompanied by three principal disadvantages:

1.  The interaction of the tip and sample becomes much more complex as nearby sample atoms react to the closer presence of the tip, requiring a self-consistent approach to provide a satisfactory description [42].
2.  The forces exerted by the tip on the sample can distort it, producing misrepresentative results. If the sample is a weakly bound adsorbate, the tip may sweep it out of the way, making imaging impossible.

3.  The tip may be damaged or destroyed by an inadvertent collision with the sample.

Note that some of these disadvantages are shared by STM, which can exert a significant influence on the sample as well [43].

Noncontact mode is best suited for samples that are particularly delicate, exhibit their important features through large-scale structures, or have rough topography. In this mode, the tip is typically located far from the sample surface, where the force gradient is less steep and noise arising from thermal fluctuations can make measurements of cantilever deflection difficult. To alleviate this problem, the cantilever is often driven to vibrate normal to the surface while it is simultaneously scanned in a raster pattern. Changes in the attractive force between the tip and sample effectively alter the cantilever spring constant, producing a change in the frequency of vibration that can be measured and converted into the corresponding force [44].

## IV.  EXPERIMENTAL CONSIDERATIONS FOR STM AND AFM

### A.  Introduction

The first STM employed a superconducting magnetic levitation system for vibration isolation and could record only single line scans. The current generation of instruments can be held in the palm of one hand and can be operated through sophisticated, user-friendly computer interfaces. The evolution of instrumentation and data acquisition and processing software in the years since 1981 has made STM and AFM accessible to hundreds of researchers throughout the world. When considered against the power and versatility of these methods, the relatively low cost and experimental simplicity of STM and AFM are remarkable. With only a modest knowledge of electronics and access to a simple mechanical shop, it is possible to build one of these instruments from scratch. Indeed, it is still sometimes *desirable* to do so for some very specialized applications, although there are presently many commercial vendors of STMs and AFMs covering all levels of sophistication, performance, and cost.

It is useful to consider first the relative merits of building versus buying an instrument. Building an instrument can take a considerable amount of time if one is starting from an untested design with little experience, while buying an instrument can enable one to collect data quickly in some form. However, commercial instruments by their very nature tend to be generally useful for a number of jobs, but perhaps not particularly for a specific type of analysis. There is also something of a "black box" character to commercial instruments, with proprietary software and electronic features that result in a general lack of adaptability. This can lead to problems with data interpretation if it is uncertain exactly what an instrument is measuring. Finally, when instruments break, as they invariably do, the ability to fix what is broken in an expedient fashion, based on a thorough knowledge of the equipment, is an important consideration.

In this section we will discuss some of the experimental and technical aspects of building, maintaining, and operating STMs and AFMs in a number of different environments including air, liquids, and UHV. The discussion is illustrated with design examples taken from this and other laboratories. The reader is also referred to Chapter 5, by Orr, for a more thorough treatment of the practical and experimental considerations of STM and AFM.

## B. Noise Isolation

One of the primary considerations when building an STM or an AFM is the means by which the experiment is decoupled from problematic noise sources present in the laboratory. STMs and AFMs measure signals collected over subnanometer-sized regions, and these signals can be orders of magnitude weaker than those measured in other spatially averaging surface science techniques. As a result, STM and AFM are sensitive to extremely minor sources of noise. Several types of noise must be avoided, each in its own way. These include mechanical, acoustic, and electronic noise, which we now address.

In STM and AFM, it is necessary to maintain a tip–sample separation with a stability of a few hundredths of an angstrom. This presents a problem, since the floor of an average building moves up and down by several micrometers at a frequency of less than 20 Hz. The ideal solution is to construct a special low acceleration building [45], but the cost of this solution is generally prohibitive. A more practical solution is to employ a system of vibration isolation whose function is to decouple the microscope from external mechanical noise sources. The general design places the microscope on a massive plate suspended at the bottom of a set of springs or elastic supports. By making the resonance frequency of this suspension system much lower than the frequency of the building vibrations, one can effectively decouple the two. Some researchers employ highly sophisticated vibration isolation systems that include multiple nested suspension systems along with a series of stacked metal plates separated by Viton and eddy current damping. Others simply suspend cinder blocks from the ceiling using bungee cords. A useful material for this suspending application is the elastomer Viton because of its UHV compatibility and excellent damping characteristics.

Acoustic noise can also interfere with STM and AFM operation, since the nature of these instruments requires the feedback loop to respond to oscillations in the audible frequency range. The most popular method of decoupling this type of noise is to enclose the instrument in an acoustically shielded box or room. Again, this can be done at extremes of cost on each end of the scale, although inexpensive acoustic foam alone is very effective at reducing noise levels.

Electronic noise in STM is most problematic in connection with the measurement of extremely small tunneling currents, and in "grounding loops" produced in the voltages applied to piezoelectric scanning elements. A grounding loop occurs if electrical grounds for different components of the microscope are at different potentials. The STM tunneling current preamplifier must be very sensitive to small signals (down to the picoampere range or even lower in some applications), and therefore it is sensitive to noise pickup as well. One very effective method of decoupling unwanted pickup noise (capacitive, inductive, and resistive) is to make the input lead from the tip to the preamplifier as short as possible by locating the preamplifier near the tip. Careful shielding of this area is also helpful. Another method, which uses the concept of a differential input signal, is employed in this lab and others [46,47]. Two separate input wires, one coming directly from the STM tip (the signal + noise lead) and the other making a nearly identical path, but not in electrical contact with the tip (the noise lead), are electronically subtracted in the first stage of the preamplifier. Together these methods can reduce electronic noise to 0.1 pA or less over a 5–10 kHz bandwidth.

## C. Scanning Heads

The mechanical quality of the STM or AFM scanning head ultimately determines the performance level of the microscope, although the materials and its construction account for only a very small fraction of the overall cost of a complete microscope system. Two general design criteria are mechanical stiffness and high resonance frequency. Increasing the stiffness or rigidity of the microscope results in a corresponding reduction of vibration-related noise. By making the microscope's resonance frequency much higher than any possible mechanical noise frequency, it is less likely that mechanical noise will affect the scanning head, regardless of the method of vibration isolation used. The resonance frequency depends inversely on the mass, so that construction materials with a high stiffness-to-mass ratio are preferred.

Thermal drift is also an important mechanical consideration. Very small temperature fluctuations of the materials in the scanning head can cause large problematic dimension changes in the tip–sample gap and in the lateral position of the tip on the sample. Materials with low thermal expansion coefficients are therefore used wherever possible. Furthermore, it is possible to design a scanning head so that, in principle, its geometry automatically compensates for thermal expansion and contraction [48–50]. One such design, employed in the laboratories of the authors, is shown in Fig. 13a. A coaxial arrangement of piezoelectric scanning tubes accomplishes the temperature compensation, with the outer tube controlling the motion of the sample and the inner tube controlling the motion of the tip. If the inner tube expands in response to a temperature rise

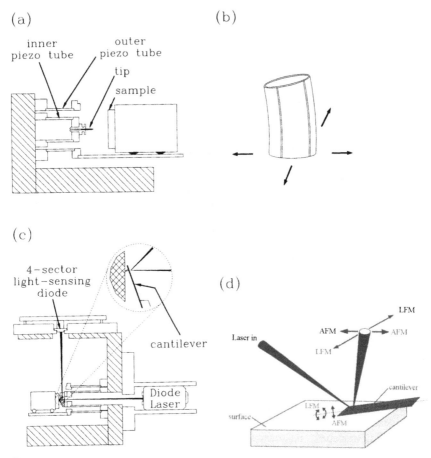

**Figure 13** Microscope designs used in the laboratories of the authors. (a) Schematic diagram of a scanning tunneling microscope. (b) Exaggerated illustration of how motion is achieved using a piezoelectric tube. (c) Schematic diagram of an atomic force microscope. (d) Lateral force microscopy is performed by measuring cantilever torsional motion.

(thereby decreasing the tip–sample gap), the outer tube expands by the same amount (thereby increasing the tip–sample gap), maintaining a constant tip–sample separation. The same cylindrical symmetry also compensates for thermal expansions and contractions along the lateral dimensions.

Perhaps the most technically demanding aspect of the STM or AFM experiment is the initial approach of the sample to the tip. To begin an experiment, the tip–sample gap must be reduced from millimeters to angstroms in a reasonable amount of time and without crashing the tip into the sample. Over the years, researchers have devised numerous ways of accomplishing this. The earliest designs incorporated fine micrometers in an arrangement using the lever reduction principle. A single micrometer cannot accomplish the necessary motion smoothly enough, since even the finest graduations produce erratic motion on an angstrom scale. When combined with a second micrometer in a lever reduction scheme, however, a smooth approach is possible. This concept has been adapted by some of the commercial instrument makers for use with a stepper motor in an automated process. Other designs have employed various forms of "walkers," in which alternating cycles of gripping/releasing and expansion/contraction of piezoelectric elements are used to provide linear motion in much the same way as a commercial "inchworm" device.

The desire to operate STMs and AFMs in remote and harsh environments such as UHV has prompted the development of other methods of accomplishing the tip–sample approach, especially those that do not employ micrometers (although micrometer-based designs have been used with great success by some researchers). In addition to environment compatibility problems, micrometers introduce difficulties including feedthrough alignment and engagement/ disengagement complications, as well as a reduction of the instrument's mechanical resonance frequency. The tip–sample approach of the microscope shown in Fig. 13 is accomplished in a completely different way. The technique implemented in this design relies on the interplay between static friction and inertia. The sample sits on a quartz platform that is attached directly to the outer piezoelectric tube. By applying a saw-toothed signal to this outer tube, the sample can be selectively "jerked" toward or away from the tip in small increments. The sample approach process is terminated when an arbitrary current (STM) or cantilever deflection (AFM) level is exceeded.

Following the coarse approach described above, the STM or AFM experiment can begin. The heart of the microscope is the piezoelectric scanning element, which positions the tip with subangstrom resolution in the $x$, $y$, and $z$ dimensions. Most researchers employ piezoelectric scanning elements with a tubular geometry [51], although the first STMs and AFMs employed a tripod arrangement of piezo bars. Since these bars are mechanically less rigid on a gram-for-gram basis, most researchers have switched to the tubular geometry. Figure 13b illustrates how this motion is achieved. A typical response for a piezoelectric element 1–2 cm in size is 25 Å/V. A power supply that is capable of producing a voltage stable to 1 mV can therefore produce a piezoelectric element position stable at the 0.025 Å level.

## D. Electronics

The basic electronic circuitry required for STM and AFM operation is very simple, although it can be made more or less sophisticated depending on the requirements of the experiment. Although simple, the electronic circuitry is the single most expensive module in the set up costs of an STM or AFM. The electronics are divided into four modules:

The signal sensing module
The feedback module acting along the $z$ axis
The scanning module acting along the $x$ and $y$ axes
The offset module, which determines the location of the image on the sample surface

The signal sensing module is a tunneling current preamplifier in STM and a force sensing module in AFM. In the standard AFM experiment, force is measured by monitoring cantilever deflection, as discussed above. This can be accomplished in a variety of ways [52–54], although the most common approach involves measuring the motion of a laser beam reflected off the backside of the cantilever [54]. In this approach, the position of a reflected laser beam is measured by a light sensing diode divided into two sectors. When the cantilever is at its equilibrium (relaxed) position, the laser spot falls approximately between the two sectors. A deflection of the cantilever causes the position of the spot to change by a proportional amount. Very small cantilever motions are detectable as a result of the long ($\sim 10$ cm) path length between the cantilever and light sensing diode, which geometrically converts small cantilever motions into relatively large deflections in the position of the laser spot (refer to Fig. 13c).

The photocurrent generated by the light sensing diode is converted to an amplified and conditioned voltage that is the input of the feedback loop. This conditioned voltage is a difference signal between the two sectors $A$ and $B$ of the photodiode. A signal proportional to the light intensity difference $I_A - I_B$ can be used as the input of the feedback loop. As the cantilever undergoes deflections resulting from various forces acting on it in the direction normal to the surface, this signal will change and can be used to adjust the tip–surface distance. An additional refinement of the $I_A - I_B$ signal can be made by a real-time normalization factor proportional to the laser intensity, $I_A + I_B$. The final signal $(I_A - I_B) / (I_A + I_B)$ is the input of the feedback loop.

Additional contrast mechanisms are possible in the AFM, just as they are in the STM. One of these is derived from the frictional force that acts *in* the plane of the surface, as opposed to normal to it, and is the basis for the lateral force microscopy (LFM) technique. Electronically, the LFM signal sensing module is the same as that of the AFM, but it must be capable of sensing torsional, rather than vertical cantilever deflections. If the AFM sensing unit is made with a four- instead of two-quadrant photodiode, the additional sectors can be used to detect these deflections, as shown in Fig. 13d. The appropriate arrangement divides a circular photo diode into four equal sectors with sector $A$ opposite $B$, and $C$ opposite $D$. The LFM signal is then generated by $(I_C - I_D) / (I_C + I_D)$. Since AFM and LFM signals are independent to a first approximation, they can be collected and displayed simultaneously (see below). The AFM signal is used for feedback and the LFM signal is collected as an open-loop signal. Simultaneous collection allows a pixel-by-pixel comparison of "topography" with "friction."

We now discuss the remaining three electronic modules. Aside from the signal detection module just described, all other electronic components of STM and AFM are identical.

In STM, the experimental signal of interest is the tunneling current, and in AFM it is the cantilever deflection signal. The feedback module compares these signals to a preset target signal and outputs an error signal related to their difference. The error signal is then converted to a correction signal and applied to the $z$-piezoelectric element. The magnitude of the correction signal contains a mathematical component of the error signal. The relative contributions of the proportional, integral, and differential components can be varied, and usually these are tuned to reach an optimal and stable condition. These settings may be a sensitive function of the condition of the tip and sample. The feedback module must also be capable of momentary interruption to perform spectroscopic measurements. This is achieved most simply by a track-and-hold circuit on the output of the feedback loop. A computer is used to open the loop long enough to collect a spectroscopic signal, after which the track-and-hold circuit reestablishes the output of the loop at its value prior to interruption.

The $x$ and $y$ motions of the tip are controlled by the scanning module, which is responsible for lateral positioning. It must be capable of producing a raster pattern that usually consists of a quickly varying $x$ ramp and a slowly varying $y$ ramp. The scanning module must be capable of varying the range of these scan ramps, and their velocity. More sophisticated scanning modules have the capability to rotate the absolute orientation of the $x$-scan direction on the surface. The simplest example of such a rotation exchanges the fast scan direction $(x)$ with the slow scan direction $(y)$. This capability is useful for determining the effect of the scanning action of the tip on the surface structures, so that tip-induced artifacts can be distinguished from native surface features.

The last important component of STM/AFM electronics is the offset module. It consists of a stable source of high voltage that can be applied independently of the scanning voltages to a piezoelectric offset element. Together, the $x$- and $y$-offset voltages determine the position of the center of the scan area or image window prior to the acquisition of an image. The ability to move the image window around by several micrometers requires that several hundred volts be applied to the offset elements. These voltages are applied to the outer piezoelectric tube in Fig. 13a for STM and the inner piezoelectric tube in Fig. 13c for AFM. By avoiding large changes in the scanning element's dimension (through application of these voltages to a separate offset element), problems of hysteresis and depolarization are significantly reduced. Researchers are continuing to develop more advanced scanning methods, such as the "hopping mode"[55], which may alleviate tip-induced problems.

## E. Image Acquisition and Processing Software

Almost all STM and AFM experiments are controlled by a computer, although this was not true when the first instruments were made. A fast personal computer can easily control, acquire, store, manipulate, and present data from STM and AFM. Indeed, successful presentation, analysis, and interpretation of image data requires a flexible software interface. Several commercial instrument manufacturers will provide software along with their instruments, and these programs are usually adequate. Only a few will also supply the source code so that users can customize the software to their own needs.

Most visual image displays employ a false color scale so that the contrast information in an image (topography, barrier height, force, frictional force, etc.) can be displayed in a top-view format. The display should be capable of generating at least 256 on-screen colors for sufficiently fine resolution of the signal along the $z$ axis, since most instruments collect data with 12 or more bits of precision.

The lateral resolution of an image is ultimately limited by the pixel acquisition and display density in the $x$ and $y$ dimensions for large-scale images and by tip size for smaller scales. There is, of course, a tradeoff between resolution and image memory requirements and collection time. Most instruments collect 256, 512, or 1024 pixels along $x$ and $y$. If an insufficient number of pixels are collected, periodic features on the sample may become "aliased" with the sampling frequency, generating an artificial long-range periodicity. To confirm the period of a repeating structural feature, therefore, it is important to vary the collection frequency.

Most software includes image processing utilities along with basic acquisition and presentation capabilities. There are many different image processing techniques, and they should always be used with caution when applied to AFM and STM images. One of the simplest and least perturbative manipulations of raw image data is to subtract a least-squares fit plane from the image. By "tilting" the image in this way, one leaves more of the dynamic range of the color scale to display the details of the image features instead of the slope of the image, which may be related to sample mounting.

Various other image processing techniques have been borrowed from fields such as electron microscopy and aerial cartography. Two-dimensional Savitsky–Golay smoothing and median filtering are useful for removing noise spikes due to tip crashes and other common STM/AFM noise sources. A considerably more severe type of filtering is frequency domain filtering, or Fourier filtering, in which an entire range of spatial frequencies is removed from a transformed image in the reciprocal space domain. The filtered image is then back-transformed into the real-space domain. Great care must be exercised when using these filters, since they can artificially enhance any spatial frequency over the others if improperly used. Raw images should always be presented with any filtered images for comparison.

## V. CHEMICAL PROPERTIES OF ADSORBATES

## A. Growth Modes

The growth of mono- and multilayers of one metal on another in the controlled environment of UHV or in an electrochemical cell is a process well suited for study by STM and AFM. Because of their unique physical, chemical, and electronic properties, thin metal films, from submonolayer to hundreds of multilayers, are fertile areas of research. The ability to understand atomic level details of their formation and stability will be an important step toward designing advanced molecular scale electronic devices and novel catalytic systems. In addition, it will provide new fundamental knowledge important in its own right.

An example illustrating work being done on growth modes and nucleation processes with scanned-probe microscopy is shown in Fig. 14, which comes from the work of Jurgen Behm and his collaborators [56]. This study followed the nucleation of gold on a single-crystal Ru(0001) surface in UHV. The high vertical (z-axis) resolution of STM enables it to observe the presence of individual atoms and islands of gold. Behm et al. found that Au islands extend and grow preferentially along the Ru(0001) crystallographic axes, forming dendritic appendages, at coverages less than 0.8 monolayer (ML). Higher coverages show a markedly different growth pattern, lacking the dendritic features so evident in the first monolayer. Annealing to temperatures between 500 and 1000 K also causes the islands to coalesce into more compact shapes.

For in situ electrochemical AFM experiments on similar metal–metal systems, a very different kind of growth is generally observed, in which ordered and uniformly spaced atoms are deposited. It is possible to reduce and deposit aqueous metal ions at potentials more positive than their standard reduction potentials in a process known as underpotential deposition (UPD). UPD is possible if the first monolayer is bound more strongly to the substrate than successive bulk layers are bound to each other. A. Gewirth and his collaborators have studied a number of (UPD) systems in several different electrolytes. Figure 15 shows an atomic resolution in situ AFM image from their investigations of copper deposited onto a Au(111) single crystalline surface [57]. An ordered $(\sqrt{3} \times \sqrt{3})$ $R30°$ structure is visible in one area, and its registry with the underlying

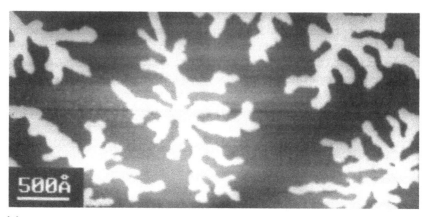

**(a)**

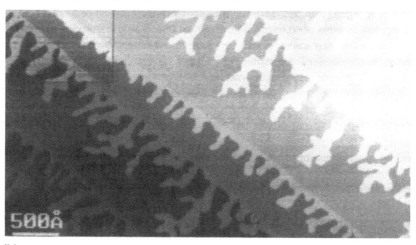

**(b)**

**Figure 14**   STM images of dendritic Au film growth on Ru(0001) in UHV. Coverage is 0.3 ML. (a) Two-dimensional islands with distinct dendritic character. (b) Film growth in a stepped region of the Ru surface with Au islands extending away from step edges; individual atom height steps are visible. The image size is indicated by the scale bar. (Courtesy of G. Pötschke et al. [56].)

Au(111) surface is known from extrapolation from the uncovered Au(111) area at the bottom of the image. Because this experiment and others like it are directly coupled to electrochemical methods such as voltammetry and coulometry, accurate coverage measurements and other direct correlations are possible [58]. An interesting aspect of this work is the implication of specific ion adsorption,

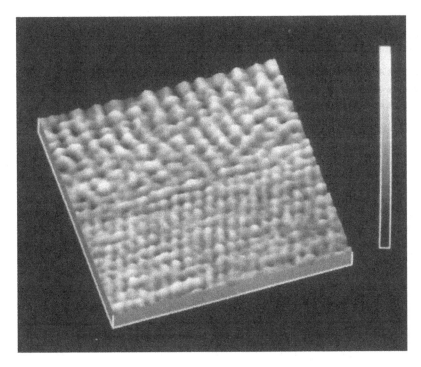

**Figure 15** AFM image of a partial monolayer of copper on gold in 0.1 M sulfuric acid. Image size is 60 Å × 60 Å. Atomic resolution of Cu and Au can be seen in the upper and lower halves of the image, respectively. (Courtesy of S. Manne et al. [57].)

which leads to different UPD overlayer structures in similar systems. Direct observation of these ions has not yet been made.

## B. Surface Reactions and Surface Dynamics

The first application of STM to in situ studies of surface reactions was made by Avouris et al. [59] in 1987. They employed an STM specially adapted for use in ultrahigh vacuum in combination with a number of other spatially averaged spectroscopic techniques (x-ray photoelectron, spectroscopy and ion scattering) to investigate the reactions of NO and $NH_3$ with Si(100)2 × 1. The importance of their work extended beyond the obvious scientific merits, however, demonstrating not only that STM is a powerful tool for studying surface reactions but also that it could be used very effectively in combination with other more established surface analytical techniques. Knowledge of the detailed electronic and structural properties of a surface is a prerequisite for understanding and predicting chemical behavior. STM is ideally suited for measuring these properties, and as a result a number of studies similar to that of Avouris et al. have now been published. We discuss a few of them in this section, which we group into three areas:

Studies of surface reactive sites, or the nature of surface reactions
Studies of dynamic properties of surface adsorbates
Studies of STM-induced chemical processes

The first group, composed almost exclusively of surface reaction studies carried out on silicon, has included investigations of $NH_3$ [60], $H_2O$ [61], and $Cl_2$ [62] on Si(111)7 × 7, as well as other reactions on Si(100)2 × 1 [59] and Si(001) [63]. The work of Wolkow and Avouris [60] serves as a good example.

The Si(111)7 × 7 surface is shown in Fig. 16a. It is particularly well suited for surface reactivity studies because it has been the subject of numerous studies with non-STM methods, and its large-scale properties are well understood. In addition, it contains dangling bonds of different types on several different sites within the unit cell, and it is of interest to know how they participate in its sur-

**Figure 16**   The reaction of $NH_3$ with Si(111)7 × 7 is site specific. (A) STM image of the unoccupied states of the clean Si surface. (B) The same surface after partial reaction with $NH_3$. The darkened areas are reacted sites whose electronic states have been sufficiently perturbed to render them invisible under the conditions of bias polarity and tip–sample separation at which the image was acquired ($V_{bias}$ = 0.8 V). (C) The same area under different tunneling conditions ($V_{bias}$ = 3.0 V), which enable the reacted sites to be imaged. (Courtesy of R. Wolkow and Ph. Avouris [60].)

face chemistry. Wolkow and Avouris approached these questions by considering the reaction of Si(111)7 × 7 with $NH_3$. To determine how these dangling bonds affect reactivity and which sites are most important, they compared the results of STM and STS measurements made on clean surfaces and on silicon surfaces that had been exposed to $NH_3$ for various lengths of time. Their comparison revealed that interactions and charge transfer between various sites significantly influence reactivity and that the chemical characteristics of a particular site depend strongly on the site's location within the unit cell.

Figure 16b shows the Si(111)7 × 7 surface after reaction with ~1 L of $NH_3$. Note that approximately half the atoms appear to be "missing." The "missing" Si atoms have not, in fact, been removed from the surface. Rather, the darkened areas represent locations at which the reaction with $NH_3$ has altered the local electronic structure. Under the conditions of bias voltage and tunneling current used to collect this image, the perturbed Si atoms are invisible because they have reacted with $NH_3$. Figure 16c demonstrates that the reacted atoms become visible again when the bias voltage is raised to +3 V. Greater exposures lead to the reaction of a different set of Si atoms within the unit cell. The work of Wolkow and Avouris demonstrates the ability of STM to reveal, atom by atom, the progress of a surface reaction.

Another group of studies uses STM to investigate time-dependent surface phenomena, such as self-diffusion rates [64] and other surface dynamics [65]. As described earlier, STM can be operated with quite fast scan rates on sufficiently flat surfaces. Several researchers, among them Buchholz and Rabe [66, 67], have exploited this capability to capture images of dynamic processes in real time as revealed by STM using a video camera. They applied fast imaging techniques to the study of dynamic behavior in systems of dialkyl-substituted benzene molecules deposited onto graphite. These molecules form large arrays in which distinct grain boundaries separate domains with different crystalline alignments. Their fast scan approach allowed Buchholz and Rabe to observe a range of interesting dynamic phenomena, including the creation, diffusion, and annealing of grain boundaries and defects involving individual molecules.

By capturing STM images with a video camera, it is possible to analyze dynamic processes in real time and thereby gain new insight into the specific motions and sequence of events. The last group of studies we discuss goes one step further, however, by using STM to probe surface reactivity and dynamics and influence or directly cause chemical changes of surface species. Workers in this area use either the local electric field between the tip and sample (which can be as large as 1 V/Å) or low energy tunneling electrons to exert a highly localized influence on surface species. These investigations are indicative of a growing realization among the STM/AFM community of the potential for the local probe microscopies to perform controlled, nanometer scale manipulations of surfaces and surface adsorbates. In a later section, we will discuss the direction of these efforts toward physical manipulations, but here we focus on those with chemical consequences. There have been several reported cases of STM-induced chemical modifications, including demonstration of the ability to control the reaction of $H_2O$ with Si(111)7 × 7 [61] and the field-induced electropolymerization of glutaraldehyde [68].

The present authors have been studying the chemical and physical properties of molecules that exhibit liquid crystalline behavior in the bulk phase [69]. One such molecule that has been studied by a number of groups (see below also) is 4-octyl-4'-cyanobiphenyl (or 8CB), shown in Fig. 17. Under the appropriate conditions on a flat surface like highly ordered pyrolytic graphite (HOPG), these molecules self-assemble into ordered arrays of two-dimensional crystalline phases. It is thought that STM images the molecular monolayer in direct contact with the surface, even when thick liquid films are present, often revealing molecular level resolution. The interpretation of these images is a matter of debate, but the generally accepted procedure is to assign the bright areas to biphenyl functionality (the "heads") and the dark areas to the alkyl "tails."

In studies of these molecular systems in the authors' labs, abrupt changes in STM bias voltage magnitude and polarity during imaging have induced rearrangements of the molecular rows, often extending over hundreds of angstroms. Figure 18a shows the same area before and after an abrupt reversal in bias voltage polarity. The single prominent diagonal feature is a defect in the HOPG sub-

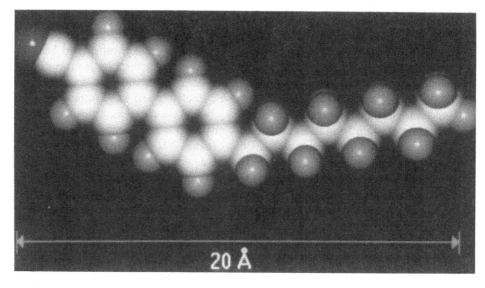

**Figure 17** The molecule 4-octyl-4'-cyanobiphenyl (8CB) is a room temperature smectic-A liquid crystal that spontaneously forms highly ordered two-dimensional patterns upon contact with a suitable surface. These patterns are readily imaged by STM.

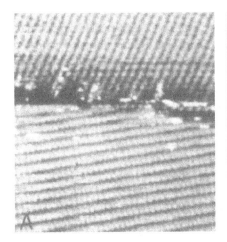 

**Figure 18** The effect of a sudden, brief reversal in bias voltage polarity on the alignment of rows of the liquid crystal 8CB deposited onto graphite, illustrated by the same area before (A) and after (B) the polarity change. The STM tip was located just above the top edge of the image when the bias reversal was applied. The direction of the molecular lines above the defect that horizontally bisects the image has apparently changed by 70°, although the lines before the defect were unaffected. Both images are 1000 Å × 1000 Å, and they were acquired by −0.7 V in constant current mode.

strate, probably a step several atomic layers high. The striped features are due to the molecular rows, where, as stated above, the bright features represent the biphenyl functionality and the dark features the alkyl tails. Notice that before and after the reversal of bias voltage polarity, the features below the HOPG step are oriented in the same direction, whereas the orientation of the rows above the HOPG step changes after the pulse. The tip was located above the upper half of the image when the pulse was applied, and it is significant that these changes did not propagate beyond the step edge.

In a similar observation, Foster and Frommer [70] have reported that brief pulses in bias voltage can pin down or remove portions of what may be individual dimethyl phthalate molecules on a graphite surface. This remarkable claim immediately suggests a whole different set of STM experiments designed to perform nanometer scale chemistry using a form of "molecular pruning."

## C.  Biological Molecules

Two years after inventing the STM in 1982, Binnig and Röhrer demonstrated the applicability of this instrument to studies of biological molecules by presenting the first STM images of DNA [71]. It was somewhat surprising that STM could be used to image biological macromolecules at all, since these particles are generally thought to be electrically insulating in the bulk. Nevertheless, such molecules were visible, and the work of Binnig and Röhrer initiated a line of investigation that has proved to be one of the most prolific and challenging areas of STM-related research.

It was initially hoped that the high resolution capabilities of STM could be used to answer detailed structural questions about biological molecules that other, older methods had left unaddressed. Conventional microscopic methods such as transmission electron microscopy (TEM) and scanning electron microscopy (SEM), while giving high structural resolution [72] in many cases, suffer from the need to coat the molecules with a conducting film, which can potentially lead to distortions. STM is able to resolve uncoated molecules, and the first images of biological specimens presented by Binnig and Röhrer were of bare molecules. Some of the early studies [73] continued to use coating as a way to understand the details of the imaging mechanism and spreading properties for these insulating molecules, although one would prefer to use uncoated samples, if possible.

Beebe et al. [74] were among the first groups to report high resolution STM images recording the structural details of uncoated DNA out of solution. Figure 19a shows a portion of an uncoated double-stranded DNA molecule deposited from a saline solution onto HOPG. The accompanying schematic (Fig. 19b) indicates the positions of the major and minor grooves in the DNA duplex. The interpretation of this and other images has been questioned in subsequent research (see below).

Between 1987 and 1989 Lindsay et al. [75–78] went one step further by showing that strands of recombinant A-DNA could be imaged in situ, producing high quality STM images without coating. Their work marked the culmination of a string of successes that produced widespread interest and encouraged high expectations for contributions of STM to biological structural studies. However, these initial hopes have been tempered by difficulties in distinguishing between biomolecules and natural surface features [79–81] (at least on graphite), and by problems in securely attaching specimens to the substrate. The realization of these difficulties has produced a period of reassessment among workers in the field. During the last 2 years, attention has focused on more fundamental problems such as finding suitable substrate materials [46], confirming the presence of biomolecules with other techniques, and gaining a better understanding of the details of the STM imaging mechanism for biological materials.

The work of Amrein et al. provides a good example of the kind of research that is necessary to answer some of these fundamental questions. These investigators have conducted comparative studies of the suitability of STM, SEM, and TEM for analysis of biological specimens [82, 83]. Figure 20 shows an STM image of a crystalline array of the outer capsid surfaces of bacteriophage T4 type III polyheads, prepared by freeze-drying and coating with a thin Pt-Ir-C film [84]. In addition to revealing the hexagonal symmetry of the capsomere mor-

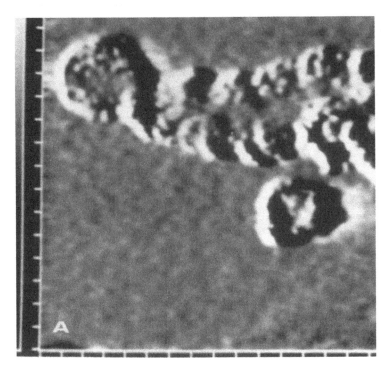

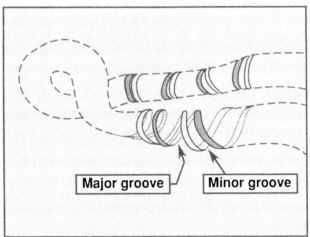

**(B)**

**Figure 19**   (A) Processed STM image of a portion of calf thymus DNA on graphite. (B) Schematic illustration of the DNA fragment in (A). If the original interpretation of this image is correct, the major and minor grooves of the DNA helix are visible. In a subsequently scanned image, the DNA-like feature was swept away, revealing a smooth background. It remains possible that the feature is due to a transiently stable HOPG flake. (Courtesy of T. P. Beebe, Jr., et al. [74]. Copyright AAAS.)

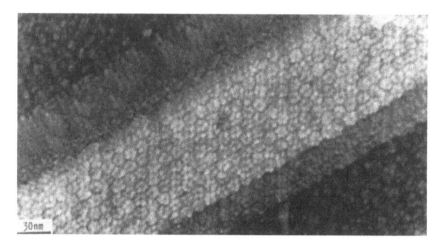

**Figure 20** Processed STM image of an array of the outer capsid surfaces of bacterio-phage T4 type III. The polyhead array is the bright region traversing the center of the image from lower left to upper right and composed of small hexagonal units. On both sides of this bright band is an artificial border produced by an interaction of the polyhead geometry with the local geometry of the tip, as explained in Fig. 21. The image size is indicated by the scale bar. (Courtesy of M. Amrein et al. [73].)

phology, this image shows an example of the so-called multiple-tip effect, which is occasionally observed with samples having abrupt changes in vertical relief. Note the individual capsomeres along the top border of the polyhead array. In this region, the geometry of the tip is mapped onto that of the surface, resulting in a distortion of the STM image. The caption of Fig. 21 explains the multiple-tip effect in detail.

Efforts to find the best conditions for applying STM to biological studies are ongoing, and progress continues to be made. AFM has also proven useful for biological studies, in part because it can image insulating samples without the need for a conductive coating. The first in situ atomic resolution AFM images in 1989 represented an important step toward wide applicability of AFM to bio-logical studies [85]. In the past few years there have been in situ studies on a large number of biomolecules ranging from DNA [86, 87] to immunoglobin [88]. AFM has also been applied to the study of dried biological materials with equal success [89,90].

## D.  Self-Ordering Molecules

Another group of adsorbates that has received increased attention in recent years consists of molecules that have the ability to self-assemble into ordered patterns upon contact with a suitable surface. The first such system reported by STM [91,

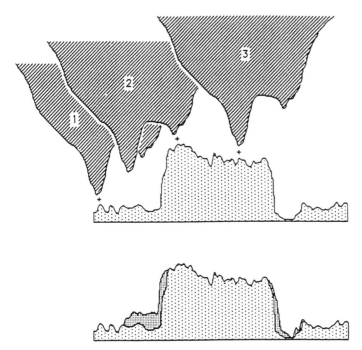

**Figure 21** The artificial border above and below the polyhead in Fig. 20 may be explained by considering how the geometry of the tip interacts with the local topology of the surface. The manufacture of STM tips entails as much art as science, and occasionally a tip is produced that has multiple protrusions of nearly equal extension. When such a tip is brought into tunneling range, the protrusion closest to the sample will carry the majority of the current because of the exponential dependence of tunneling probability on distance. If the surface is not uniform, however, unexpected results may occur and the resulting image will reflect what is known as the "multiple-tip effect," of which Fig. 20 is an example. The upper part of Fig. 21 illustrates how multiple-tip protrusions can be used to explain the artificial border around the polyhead in Fig. 20. As the tip moves across the sample from left to right, the protrusion responsible for carrying the dominant portion of the tunneling current changes from the longer whisker (1) to the shorter one (2), and back to the longer (3), as indicated by the " + ." The resulting image is presented in the bottom half of the figure. This effect can be significant not only for biological samples, but also for any surface possessing large topographic relief. (Courtesy of M. Amrein et al. [73].)

92] was $CO + C_6H_6$ on Rh(111). In the past few years, many more self-ordering molecules have been observed, and this group has grown to include liquid crystals, long chain $n$-alkanes and their alcohol and fatty acid derivatives, and lipids. These samples are often easily prepared by either the Langmuir–Blodgett technique, direct application, or direct application followed by, for example, exposure to elevated temperatures. Figure 22 shows as an example the ordered patterns formed on graphite by the alkane $n$-$C_{32}H_{66}$, which has been studied by McGonigal et al. [93].

Ordered or periodic systems such as these offer two advantages over isolated organic adsorbates:

1. They form unique patterns that can be easily recognized and usually cannot be mistaken for features due to the substrate.
2. Relatively strong molecule–molecule and/or molecule–substrate forces help hold the molecules fixed in place so that they are more readily imaged with STM and AFM. They thus provide a relatively reproducible system on which to conduct experiments with molecular resolution.

These properties have suggested experiments designed to study grain boundaries [66] and the effect of different substrates on packing structures [94]. The present authors have investigated the effects of elevated temperatures during sample preparation of the liquid crystal 8CB deposited onto HOPG. We have observed new two-dimensional structures that are in some ways analogs of the well-known three-dimensional mesophases. Figure 23 shows three distinct structures for 8CB along with proposed models to explain them.

The bases of DNA are another group of molecules that form self-ordering arrays and that are particularly relevant to biological STM. A number of investigators have reported the ability to image monolayer films of the free-

**Figure 22** STM image of $n$-$C_{32}H_{66}$ deposited onto graphite. These molecules form large, ordered arrays on the surface, in part because of the favorable spacing of the graphite lattice with the distance between methylene groups. Each bright line is one molecule. Image size is 100 Å × 100 Å. (Courtesy of G. C. McGonigal et al. [93].)

standing DNA bases (i.e., not as part of the DNA molecule). The base guanine (Fig. 24) is an example. Although it is not obvious from the contrast pattern in the image that the molecule being imaged is guanine, all the bases assemble in unique patterns. Interest in such experiments has stemmed from the desire to apply STM and AFM to DNA sequencing. Although the end goal of these experiments is probably far off in the future (if attainable at all), several groups, including this one, will continue to systematically address the problems of STM/ AFM biomolecule imaging.

Self-assembled monolayers are in many ways ideally suited for studying the microscopic nature of intermolecular forces. STM and AFM can probe the local effect on packing structures of surface defects, sample impurities, and external perturbations such as temperature and electric fields. Systematic studies offer the potential of explaining the detailed interplay of molecule–molecule and molecule–substrate forces on a scale heretofore unknown. Combining these methods with spatially averaging surface science techniques may be a very powerful approach, and it is being used by more and more researchers.

## VI. ELECTRONIC PROPERTIES OF SURFACES
### A. Semiconductors

Studies of semiconductor surfaces have provided some of the best examples illustrating the usefulness of STM for spatially and energetically resolving surface electronic states. By employing the spectroscopic methods introduced earlier, a number of workers have presented elegant results measuring band gap energies and giving surface LDOS profiles for a number of different samples.

The upper part of Fig. 25 shows two constant current STM images of Si(111)2 × 1 taken at negative and positive voltages for the left and right frames, respectively. The cross sections displayed in the lower part of Fig. 25 indicate that the maxima in one frame correspond to the minima in the other. This interesting effect can be understood by considering the arrangement of the filled and empty Si surface electronic states. The Si(111)2 × 1 surface has singly occupied dangling bonds at each Si atom, as indicated in Fig. 26a. When the atoms on a simple semiconductor surface such as this one are arranged in energetically or structurally distinct ways, the surface states may transform in a way known as "buckling."

Buckling occurs when the half-filled bonds with lower energy accept electron density from the bonds with higher energy. For Si(111)2 × 1, the result is shown in Fig. 26b. The atoms labeled "B" have accepted electron density from the "A" atoms, splitting the surface states into two different groups: empty and filled states. When the STM bias voltage is set so that the tip is positive with respect to the sample, electrons predominantly tunnel to the tip from the filled surface states, B. Conversely, when the tip is negative with respect to the

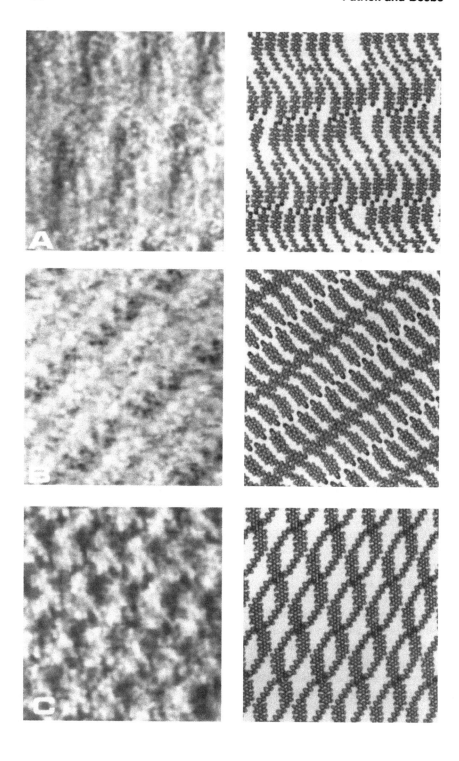

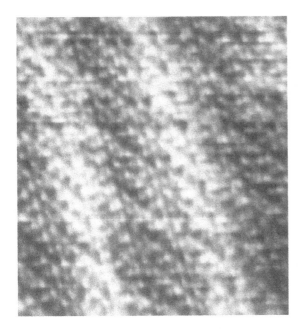

**Figure 24** Periodic structure composed of guanine molecules adsorbed onto HOPG. The image measures 100 Å × 100 Å. (Image obtained in the authors' laboratories by C. R. Clemmer.)

sample, electrons predominantly tunnel from the tip to the empty surface states, A, reversing the image contrast. The phase difference in the cross sections of Fig. 25b is a result of this phenomenon.

The results shown in Fig. 25 were obtained using an STS method similar to that of Feenstra et al., described earlier. If the method of Hamers et al. is used instead, detailed information about the energy profile of the LDOS and the band gap may also be collected. Figure 27 shows a Si(111)(7 × 7) surface with the unit cell enclosed by a solid border. Avouris and Wolkow [95] obtained the accompanying spectra by acquiring current–voltage curves at three different locations within the unit cell, as indicated by the arrows. The peaks on the positive bias side represent spectral features of the empty surface states, and those on the

**Figure 23** Another group of molecules that form ordered patterns on some surfaces consists of liquid crystal molecules. (A)–(C) Three distinct packing structures of the same liquid crystal, 8CB, (left) along with proposed models to explain them (right, not on same scale). All images measure approximately 100 Å × 100 Å and were collected in the constant height mode. (This work is from the authors' laboratories.)

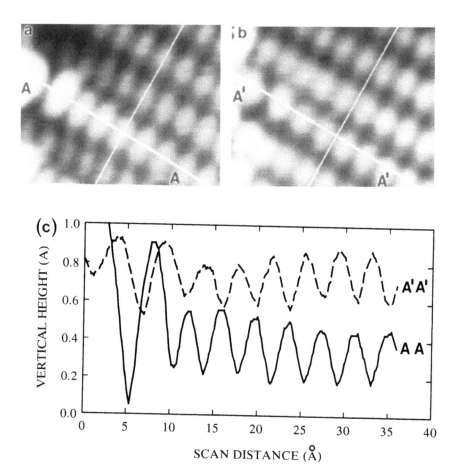

**Figure 25**   (a,b) Two constant current STM images of Si(111)(2 × 1) acquired at bias voltages of −0.7 and +0.7 V for the left and right images, respectively. By operating at opposite bias polarity, the STM can selectively image either empty or filled surface states. (c) Cross sections of the two images, drawn along the lines $AA$ and $A'A'$. Note that the maxima in one cross section correspond to the minima in the other, revealing the alternating pattern of filled and empty surface states. (From J. A. Stroscio, R. M. Feenstra, and A. P. Fein, Electronic structure of the Si(111)2 × 1 surface by scanning-tunnel microscopy, *Phys. Rev. Lett.* 57: 20, 2579 (1986), courtesy of the authors.)

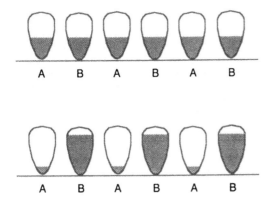

**Figure 26** On certain semiconductor surfaces, a process known as "buckling" can occur in which electron density is transferred from one set of half-filled orbitals to another set. The upper part of this figure shows a semiconductor surface with a complete set of half-filled orbitals. In the lower part, electron density has been transferred from the set of orbitals labeled "A" to the set labeled "B," creating one group of filled orbitals and one groups of empty orbitals.

negative side, features of the filled states. The right half of Fig. 27 shows how these surface chemical bond states are perturbed upon reaction with $NH_3$, as discussed above.

## B. Superconductors

The Nobel prize winning discovery in 1987 of a new class of high $T_c$ superconducting compounds by Bednorz and Müller [96] inspired renewed interest among inorganic chemists and solid state scientists to try to better understand the principles of superconductivity. It is noteworthy that the application of STM to these efforts was actually initiated 30 years before its invention by the work on electron tunneling and superconductivity of another Nobel laureate, Ivar Giaver [97]. Giaver measured the dependence of tunneling current on voltage between a superconducting strip of Pb and a second strip of Al separated by a thin oxide film. Lead becomes superconductive at 7.2 K, but aluminum only below 1.2 K. By performing his measurements at the temperature of liquid helium (4.2 K), Giaver originated the tunneling spectroscopy experiment now performed with STM in which the tunneling current is measured between a normally conductive tip and a superconductive sample.

The primary advantages of employing STM-STS in superconductor band gap and density-of-states measurements are (1) gaining information about the structural characteristics of the superconductor surface in the region of the measurement, and (2) exploring the spatial dependences of the surface electronic states and superconducting properties.

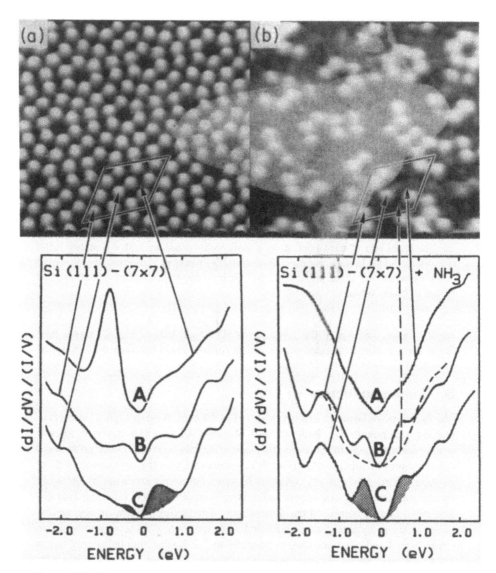

**Figure 27**   The upper part of this figure presents STM images of the unoccupied states of the Si(111)(7 × 7) surface with the unit cell indicated by a solid line. Spectra acquired at three different points within the unit cell are given in the lower portion. Following the reaction of this surface with $NH_3$ (right), the appearance of the image changes, as does the spectrum of the atomic sites that have reacted. Refer also to Fig. 16. (Courtesy of Ph. Avouris and R. Wolkow [95].)

Both these advantages were demonstrated by the work of de Lozanne et al. [98], who used a low temperature STM to perform STM-STS measurements on the superconductor $Nb_3Sn$. In addition to measuring the superconducting gap, they were able to detect extremely small scale ($\sim 13$ nm) spatial variations in superconductivity. Their work represented the first observation of such local variations.

## C. Light Emission and Absorption in the STM Tunnel Junction

In this section, we discuss an adaptation of STM that uses photons in conjunction with tunneling electrons to provide additional information about the sample in the vicinity of the tip. We focus on two basic experiments, one involving detection of photons emitted from the tunnel junction, and the other involving the effect of illumination of the tip–surface junction region on tunneling current. These two techniques are distinguished from photon scanning tunneling microscopy (PSTM) and near-field optical microscopy (NFOM), discussed elsewhere in this book, by the fact that their localized response is due to the localized nature of STM itself, not to any particular optical arrangement of the apparatus.

The first basic experiment examines the photons emitted from the tunnel junction in an STM experiment [99–101]. The photons detected in this way are produced by a luminescent process in which high energy tunneling electrons make radiative transitions into unoccupied sample surface states. Because of the inelastic nature of the transition, spectral analysis of the emitted photons can provide information about surface states and energy dissipation processes in tunneling.

Three approaches have been adopted to analyze STM photon emission. The first monitors the monochromatic light emitted as a function of bias voltage at a particular location on the sample. A second method consists of collecting an entire optical spectrum at a particular bias voltage and sample position. The third approach yields a simultaneously collected topographic image along with a map of the monochromatic light intensity emitted at each point in an STM image. An example of this approach is shown in Fig. 28. The left and right portions of the figure are topographic and photon map images, respectively, of an annealed silver film on Si(111). These images are taken from the work of Berndt et al. [102], which also contains examples of the two other methods of analyzing photon emission.

The second basic experiment uses an external light source, typically a laser, to illuminate a semiconductor sample and produce a photovoltage that can vary by 5 mV or more on an atomic scale [103]. The induced photovoltage augments the STM bias voltage, producing a change in the tunneling current according to Eq. (2). Variations in photovoltage may thus be recorded as variations

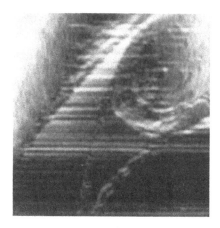

**Figure 28** STM topographic (left) and photon map (right) images of an annealed Ag film on Si(111). Image sizes are 600 Å × 600 Å. The contrast information in the right-hand image represents photon emission intensity, which correlates well with the topography of the surface. (Courtesy of R. Berndt et al. [102].)

in tunneling current. A topograph collected in this way is called a photocurrent image.

Another effect arising from semiconductor illumination is a change in the density-of-states band structure. This change manifests itself as an alteration of the band shape and a shift in energy, as discussed in detail elsewhere [104, 105]. Figure 29, from the work of Kuk et al., shows the effect of increasing illumination intensity on the density of states for different sites on Si(111) 7 × 7.

## D. Electrochemistry

The first demonstration that the STM could achieve atomic resolution while operating in solution [106] was important in several respects, especially for the community interested in electrochemical systems. Many processes, such as corrosion and certain deposition methods, can be studied much more effectively in situ. In addition, in situ work had been performed in the biological area (see above) to control the adsorption of macromolecules by changing the surface potential. Researchers are also using the in situ STM to measure electron transfer rates at interfaces, in work that may lead to the development of many new applications [107, 108]. AFM has also contributed promising electrochemical results on several surfaces including gold [109], platinum [110], and graphite [111], to name a few.

Investigation of the structure of transition and noble metal surfaces both in UHV and in solution has been an important area of interest for many years. The atomic level details of reconstructed surfaces have been very difficult to ascertain, even using a combination of experimental and theoretical techniques, most of which are limited to vacuum environments [112]. Surface reconstructions at the electrochemical interface, long surmised because of indirect electrochemical measurements [113], can now be studied directly by in situ STM and AFM. Because it is also possible to independently potentiostat the working electrode (the surface) and isolate the current flowing in that circuit from the tunneling current, STM images, and more simply, AFM images, can be collected as a function of electrochemical potential. Studies of this kind have yielded a surprisingly rich and complex picture of the atomic structure of single-crystal electrode surfaces. Figure 30 shows a series of STM images illustrating the deposition and subsequent removal of copper from a polycrystalline Pt surface [114]. Starting from a clean surface (Fig. 30a), copper was electrodeposited at −0.05 V [vs. a saturated calomel electrode (SCE)]. Then ~3 ML (Fig. 30b) and ~8 ML (Fig. 30c) of Cu were deposited, resulting in a noticeable change in surface roughness. Applying 0.065 V (vs. SCE) restored the surface to its initial condition (Fig. 30d), demonstrating the reversibility of the process. Other researchers are investigating similar structural transformations on gold surfaces in air [115] and in UHV [116].

## VII. NOVEL TECHNIQUES

## A. Nanolithography

For most of the examples discussed to this point, it has been implicitly assumed that the probe tip (STM or AFM) is a more or less passive observer and that if appropriate considerations are taken, its influence on the system under investigation may be neglected. This is of course not always the case, and the same characteristics that make STM and AFM so useful for microscopic surface observations also make them useful for nanometer scale surface modifications. The ability to locate a small probe tip with extreme accuracy and, having located it, to exert some controllable influence at that point has led to strong interest in the development of STM and AFM as tools for nanometer scale modification. The technological and scientific implications for nanometer scale lithography, or nanolithography, are abundant. Improved integrated circuit performance, the capability to create and study ultra-small-scale structures with interesting quantum mechanical properties, and the ability to influence chemical reactivity on an atomic scale, are a few of the areas that may benefit from STM and AFM nanolithographic technologies.

In this section, we discuss the applicability of STM and AFM as nanolithographic tools. For the most part, their lithographic action is achieved in three different ways: by a mechanical force, by thermal or large electrical energies,

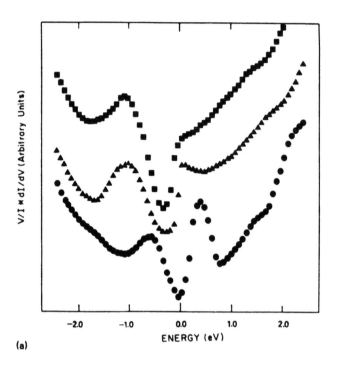

(a)

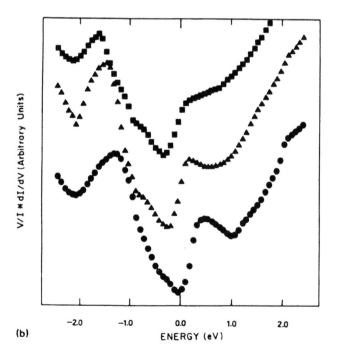

(b)

and by low energy electrons. For convenience, however, we will group the various lithographic studies together in the following way:

Direct modifications of the substrate
Modifications involving adsorbed molecules or atoms
Controlled deposition of material from a surrounding liquid or gas

Direct modification of clean surfaces or films has been reported for several different materials, among them graphite [117], semiconductors [118], and $TiO_2$ films on Pt substrates [119]. The simplest method for producing lithographic features (and also the most common, if not always intentionally) is to push the tip into the surface, creating a small depression at the point of contact. Schneir and Hansma [120] have applied this method to gold surfaces coated with different nonpolar liquids. Their technique is quite reliable, but it has the disadvantage of working properly only for surface materials that are softer than the tip. Gold satisfies this requirement for most choices of tip materials, however the large self-diffusion rates for gold surfaces result in lithographic features that have unacceptably short lifetimes for practical application.

The obvious solution to this problem is to choose a different substrate. Graphite is one candidate that recommends itself through its other superior features, including ease of preparation and near-atomic flatness. In 1989 Albrecht et al. discovered that brief pulses in the STM bias voltage could reproducibly create small ($\sim$40 Å) craters in the topmost layer of graphite single crystals. The etching mechanism in this case appears to be very different from that used by Schneir and Hansma to modify gold surfaces. Studies in vacuum [121] have revealed that etching does not take place in the absence of water (or possibly just in the absence of hydrogen-containing compounds), suggesting that a chemical reaction may be responsible. Siekhaus [122] has proposed that a reaction between $H_2O$ and graphite C atoms to produce CH and $H_2$, followed by desorption or diffusion of the products away from the reaction zone, may be responsible for the etching action.

Another interesting approach has been developed by Casillas et al. for etching thin films [119]. These workers found that by linearly cycling the bias voltage while holding the STM tip stationary, it is possible to selectively etch $TiO_2$ deposited onto a Pt substrate. Figure 31 shows an example of a 5 nm disk created in this way. The $TiO_2$ film in the area of the disk has been removed, exposing the Pt substrate underneath. Such small structures used as nanoelectrodes may possibly lead to a new generation of sensors. The etching method of Casillas et al.

---

**Figure 29** Scanning tunneling spectra of a Si(111) 7 × 7 surface over (a) adatom sites and (b) rest atom sites at three different laser illumination intensities. Solid circles represent $I = 0$; triangles, $I = 100$; squares, $I = 175$. Notice that the density-of-states energy maxima shift with illumination intensity. (Courtesy of Y. Kuk et al. [103].)

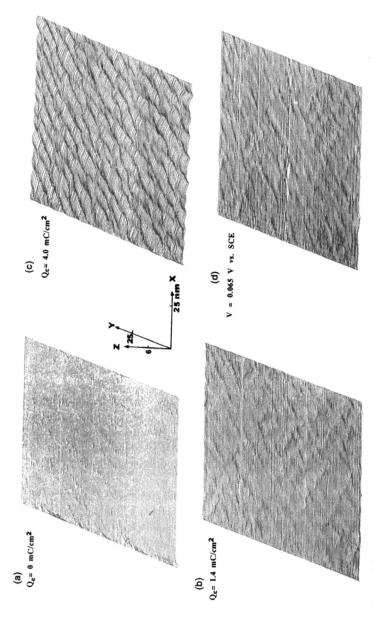

**Figure 30** STM images of Cu deposition and subsequent removal on a polycrystalline Pt substrate in 0.1 M HClO$_4$ containing 1 mM Cu(ClO$_4$). (a) Clean Pt surface. (b) After deposition of 1.4 mC/cm$^2$ of Cu ($\approx$3 ML). (c) After deposition of 4.0 mC/cm$^2$ of Cu ($\approx$8 ML). (d) After removing Cu by reversing the electrochemical potential. The image size is indicated by the scale bars. (Courtesy of F. F. Fan and A. J. Bard [114] reprinted by permission of The Electrochemical Society, Inc.)

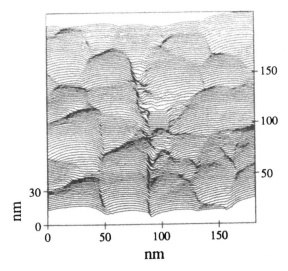

**Figure 31** STM image of a 5 nm Pt disk (see flat circular region in center) etched into a TiO$_2$ film deposited onto a Pt/mica substrate. (Courtesy of N. Casillas et al. [119]. Reprinted by permission of the publisher, The Electrochemical Society, Inc.)

includes as an additional advantage the ability to control the size of the disk created by adjusting parameters related to the cycling of the bias voltage. The second group of nanolithographic techniques, which we now discuss, has extended this capability to the ultimate limit, demonstrating the ability to modify small groups of molecules, and even atoms.

In a *tour de force* of nanomanipulation, Eigler and Schweizer [45] dramatically illustrated the capacity of STM to position individual xenon atoms with extremely high precision. By changing the tip–sample bias voltage, they were able to pick up and selectively deposit xenon atoms to form the pattern "IBM" in nanometer-tall letters on a Ni(110) surface at low temperatures.

Another example is provided in the work of Albrecht et al. [123], who prepared and studied monolayers of the polymer poly(octadecylacrylate) on graphite using the Langmuir–Blodgett technique. These molecules link together to form thin bundles of chains, or fibrils, which align in parallel groups on the surface. By applying short voltage pulses, the investigators observed local modification of the fibrils in a way that suggested they had been severed. The polymer polymethyl methacrylate is known to exhibit similar behavior upon exposure to electron beams, indicating that the observations of Albrecht et al. may be a scaled-down version the same well-known mechanism, using low energy electrons from the STM tip to induce localized changes in the polymeric structure.

The last group of methods we discuss also uses the low energy tunneling electrons of STM, but in a slightly different way. Instead of acting on a species already present on the surface, low energy electrons are used to deposit material from a covering fluid or gas. Deposition rates can in principle be fine-tuned by varying the current, which determines the flux of electrons across the tunnel junction, providing an avenue for controlling both the location and the amount of material deposited.

The work done by Foster et al. [70] on STM-activated deposition and removal of dimethyl phthalate (DMP) and di-2-ethylhexylphthalate (EHP) molecules on graphite serves as a good example. In their case, the source of material was a covering fluid, and they used small voltage pulses to achieve deposition. These workers demonstrated through high resolution images the capability of creating, and then removing, small features from the surface, which appear to be individual molecules. Others have achieved similar results using gaseous sources of materials and have written patterns using tungsten and carbon [124–126] with line widths less than 10 nm.

Nanolithography with STM and AFM is a rapidly developing field, and the few examples discussed here represent a very small portion of all the work being done. For a more comprehensive review, see Shedd and Russell [127].

## B. Creation and Imaging of Charged Regions and Magnetic Domains

Several techniques that have the ability to probe surface charges and magnetic domains have recently become the newest members of the growing family of local probe microscopy methods. They are modifications of the standard STM and AFM experiments that allow direct observation, and in some cases creation, of nanometer-sized charged regions and magnetic domains. We will begin with a discussion of the charge-related methods.

Nejoh [128] has measured what appears to be the incremental charging of a single molecule (a liquid crystal) using STM. A simplified explanation of his approach follows. Suppose that an STM tip is located directly over a molecule on a conducting surface, and that the electrical resistances between the molecule and the tip, and between the molecule and the surface, are such that the molecule behaves like a capacitor toward the tunneling current. Then as electrons are transferred from the tip to the molecule, the molecule becomes charged, and the energy required to transfer each successive electron becomes larger. This behavior should manifest itself in a plot of current against voltage ($I$–$V$) as a series of plateaus, or steps, the height of each being representative of the minimum energy required to transfer that electron. This phenomenon, known as the Coulomb staircase [129], was observed by Nejoh to occur when the tip was located above an adsorbed molecule.

Any STM-based method for the creation or imaging of charged surface regions is inherently limited to use on semi-insulating samples since (1) a perfect
insulator would give zero tunneling current, and (2) a perfect conductor (i.e., a
metal) would dissipate charge too rapidly. Because AFM does not use tunneling
current and is not hampered by these limitations, it is more amenable to charge
measurement experiments, and correspondingly more positive results have
been reported.

There have been several different versions of the AFM-based charge measurement technique, but they are all for the most part based on the same approach:
charge is deposited by applying a charge to the AFM tip and then touching it to
the surface, and charge is detected through a change in the force gradient produced by Coulombic attraction between the surface charge and its image charge
in the cantilever. Terris et al. [130] have used an AFM-based charge measurement approach to deposit and image electric charges on a polycarbonate surface.
Their approach was sensitive to both the magnitude and the polarity of the
charge, enabling them to distinguish between positively and negatively charged
regions. An example of their work appears in Fig. 32, which shows several
charged areas. These regions had been created prior to imaging by applying a
charge to the cantilever and touching it to the surface.

As in the case of the various STM- and AFM-based techniques for probing
electric charges, there are several different methods for detecting sample magnetic domains. The group of methods based on STM generally adopts one of two
different approaches: image contrast is derived either from a magnetic force between the tip and the sample [131] or from spin analysis of the tunneling electrons [132,133] (referred to as spin-polarized STM, or SPSTM). One of the first
attempts at creation of magnetic structures was made by McCord and Awschalom [134], who used STM to deposit iron pentacarbonyl into the input coil of a
small planar dc superconducting quantum interference device (SQUID). The
$Fe(CO)_5$ source was a surrounding gas, yielding magnetic dots with diameters
between 10 and 30 nm.

AFM-based approaches (called magnetic force microscopy, or MFM) measure the magnetic force between a suitable sample and a magnetic tip. Magnetic
forces are very long-ranged, allowing MFM experiments to be performed in a
genuine noncontact mode. Figure 33 shows the first reported result of MFM used
to image a static magnetic field [135]. The black dots are magnetic regions written thermomagnetically, using a laser in combination with a magnetic bias field.

## VIII. CONCLUSION

In the 1950s, the physicist Richard P. Feynman was one of the first to seriously
speculate on the ultimate results of miniaturization. He suggested that someday
it might be possible to manipulate individual molecules and even atoms to build

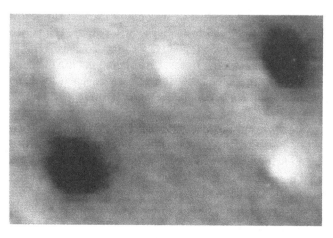

2μm

**Figure 32** AFM image of charged regions on a polycarbonate surface. The regions were created by selectively depositing charge using the AFM tip. The three white regions are positive and the two black regions are negative. The image size is indicated by the scale bar. (Courtesy of B. D. Terris et al. [130].)

nanometer-sized machines with all sorts of interesting properties [136]. We believe that one of the most predictive discoveries of STM and AFM was the realization that these techniques could be used not only to observe samples, but to purposefully manipulate them as well. Today articles are beginning to appear in the literature with titles such as "Scanning Tunneling Engineering" [137] and "Nanoelectronics" [138]. Certainly this will continue to be an active and productive area of STM/AFM-related research, and although the development of a true Feynman machine has not yet been announced, it may very well be within the next 10 years. In the mean time, STM and AFM will continue to contribute results to many different areas of pure and applied surface science and technology.

Two of the most visible areas of research probably will involve biological specimens and chemical reactions. D. Rugar and P. Hansma have speculated that it may someday be possible to observe the processes by which viral entities, such as the AIDS virus, attack cells [44]. High resolution images of DNA have already been reported. If STM and AFM biological techniques continue to develop at their current rate of progress, it may soon be possible to use them to

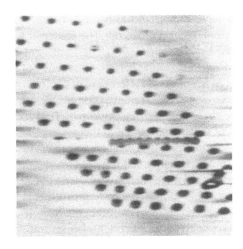

⊢———◀ 10μm

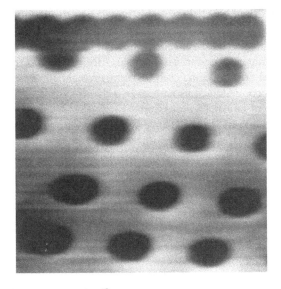

⊢————◀ 5μm

**Figure 33**  AFM image of magnetic domains on a TbFe thin film written using a laser in combination with a magnetic bias field. The image size is indicated by the scale bars. (Courtesy of Y. Martin et al. [135].)

sequence DNA. STM and AFM may promote the discovery of new chemical reactions brought about atom by atom, using a probe tip. The precursor to these experiments has already been done by Eigler et al. [45].

A large number of other systems are also under investigation, and only a few of them have been discussed here. Many of these systems are of purely scientific interest, but a growing number have technological implications as well. Increasing reliability and ease of use have already allowed a transition of STM and AFM from use only in the laboratories of those who can build them, to the laboratories of those who can purchase commercial instruments, to more diversified commercial applications such as product quality control.

Finally, we remark on the trend toward increasing diversification in STM and AFM instrumentation. It is now possible to independently measure a large variety of different surface properties at the atomic level, ranging from frictional forces to local variations in superconductivity. There are also several ways to purposefully manipulate samples with atomic level accuracy. As researchers experiment with new techniques, the list of STM/AFM capabilities will continue to grow, and the usefulness of these techniques will grow along with it.

In August 1991, researchers from around the world gathered in Interlaken, Switzerland to celebrate 10 years of STM. During his remarks to the opening session, Nobel laureate Heinrich Röhrer recounted how, on the same day in 1981 he and his colleagues at IBM announced the invention of STM, others were coincidentally introducing the company's newest product, the IBM personal computer. Ironically, STM and AFM are now being used for commercial applications in the production of advanced computer chips and other technology related to the PC and its next generation of advanced processors.

Today both the STM and the original IBM PC have evolved into machines far more powerful and far more useful than even their inventors could have imagined. As STM moves into its second decade and AFM approaches the end of its first, we can safely predict that they will become important tools in the physics, chemistry, biology, and engineering communities, and that they will continue to contribute to the solution of many difficult technological, scientific, and clinical problems.

## ACKNOWLEDGMENTS

The authors thank Carol Rabke-Clemmer, Andrew Leavitt, Lisa Wenzler, Bahareh Patrick, and Clayton Williams for their insightful comments and suggestions, and Taejoon Han for assistance in figure preparation.

## REFERENCES

1.  Binnig, G., Röhrer, H., Gerber, C., and Weibel, E. (1982). Tunneling through a controllable vacuum gap, *Appl. Phys. Lett. 40*(2): 178.

2. Binnig, G., Röhrer, H., Gerber, C., and Weibel, E. (1982). Surface studies by scanning tunneling microscopy, *Phys. Rev. Lett. 49*: 57.
3. Binnig, G., and Röhrer, H. (1986). Scanning tunneling microscopy, *IBM J. Res. Dev.*, *304*: 355.
4. Binnig, G., and Röhrer, H. (1987). Scanning tunneling microscopy—From birth to adolescence (Nobel Lecture), *Angew. Chem. Int. Ed. Engl. 26*: 606.
5. Binnig, G., Quate, C. F., and Gerber, C. (1986). Atomic force microscope, *Phys. Rev. Lett. 569*: 930.
6. Simmons, J. G. (1963). Generalized formula for the electric tunnel effect between similar electrodes separated by a thin isolating film, *J. Appl. Phys.*, *34*: L1793.
7. Feuchtwang, T. E., and Cutler, P. H. (1986). Tunneling and scanning tunnel microscopy, *Phys. Scripta*, *35*: 132.
8. Tersoff, J., and Hamann, D. R. (1983). Theory and application for the scanning tunneling microscope, *Phys. Rev. Lett. 50*(25): 1998.
9. Bardeen, J. (1961). Tunnelling from a many-particle point of view, *Phys. Rev. Lett. 6*: 57.
10. Barniol, N., Farrés, E., Martin, F., Suñe, J., Placencia, I., and Aymerich, X. (1990). Simple STM theory, *Vacuum*, *41*: 379.
11. Tersoff, J., and Hamann, D. R (1985). Theory of the scanning tunneling microscope, *Phys. Rev. B. 31*(2): 805.
12. Hansma, P. K., and Tersoff J. (1987). Scanning Tunneling Microscopy, *J. Appl. Phys.*, *612*: R1.
13. Nakagiri, N., and Kaizuka, H. (1990). Simulations of STM images and work function for rough surfaces, *Jpn. J. Appl. Phys. 294*: 744.
14. Garcia, N., Ocal, C., and Flores, F. (1983). Model theory for scanning tunneling microscopy: Application to Au(110)(1 × 2), *Phys. Rev. Lett. 50*: 202.
15. Doyen, G., Koetter, E., Vigneron, J. P., and Scheffler, M. (1990). Theory of scanning tunneling microscopy, *Appl. Phys. A51*: 281.
16. Wintterlin, J., Wiechers, J., Brune, H., Gritcsh, T., Höfer, H., and Behm, R. J. (1989). Atomic-resolution imaging of close-packed metal surfaces by scanning tunneling microscopy, *Phys. Rev. Lett. 62*: 59.
17. Mednick, K., and Kleinman, L. (1980). Self-consistent Al(111) film calculations, *Phys. Rev. B*, 22: 5768.
18. Coley, T. R., Goddard, III, W. A., and Baldeschwieler, J. D. (1991). Theoretical interpretation of scanning tunneling microscopy images: Application to the molybdenum disulfide family of transition metal dichalcogenides, *J. Vac. Sci. Technol. B*, *9*(2): 470.
19. Huang, Z. H., Feuchtwang, T. E., Cutler, P H., and Kazes, E. (1990). The Wentzel–Kramers–Brillouin method in multidimensional tunneling: Application to scanning tunneling microscopy, *J. Vac. Sci. Technol. A*, *8*(*1*): 177.
20. Feenstra, R. M., Stroscio, J. A., and Fein, A. P. (1987). Tunneling spectroscopy of the Si(111)2 × 1 surface, *Surf. Sci. 181*: 95.
21. Feenstra, R. M., and Fein, A. P. (1985). Surface morphology of GaAs(110) by scanning tunneling microscopy, *Phys. Rev. B*, *32*: 1394.
22. Feenstra, R. M., Thompson, W. A., and Fein, A. P. (1986). Real-space observation of π-bonded chains and surface disorder on Si(111)2 × 1, *Phys. Rev. Lett. 56*(6): 608.

23. Hamers, R. J., Tromp, R. M., and Demuth, J. E. (1986). Surface electronic structure of Si(111)-(7 × 7) resolved in real space, *Phys. Rev. Lett.* *56*: 1972.
24. West, P., Kramer, J., Baxter, D. V., Cave, R. J., and Baldeschwieler, J. D. (1986). Chemical applications of scanning tunneling microscopy, *IBM J. Res. Dev. 30* (5): 484.
25. Sommerfeld, D. A., Cambron, T. R., and Beebe, T. P., Jr. (1990). Topographic and diffusion measurements of gold and platinum surfaces by STM, *J. Phys. Chem.* *94*: 8926.
26. Weber, R. E., and Peria, W. T. (1969). Work function and structural studies of alkali-covered semiconductors, *Surf. Sci. 14*: 13.
27. Spong, J. K., Mizes, H. A., LaComb, L. J., Jr., Dovek, M. M. Frommer. J. E., and Foster, J. S. (1989). Imaging of organic molecules using tunneling microscopy, *Nature*, *338*: 137.
28. Kochanski, G. P. (1989). Nonlinear alternating-current tunneling microscopy, *Phys. Rev. Lett. 62*(19): 2285.
29. Martin, Y., and Wickramasinghe, H. K. (1987). Magnetic imaging by "force microscopy" with 1000 Å resolution, *Appl. Phys. Lett. 50*(20): 1455.
30. Rugar, D., Mamin, H. J., Guethner, P., Lambert, S. E., Stern, J. E., McFadyen, I., and Yogi, T. (1990). Magnetic force microscopy: General principles and application to longitudinal recording media, *J. Appl. Phys. 68*(3): 1169.
31. Martin, Y., Abraham, D. W., and Wickramasinghe, H. K. (1988). High-resolution capacitance measurement and potentiometry by force microscopy, *Appl. Phys. Lett. 52*(13): 1103.
32. Mate, C. M., McClelland, G. M., Erlandsson, R., and Chiang, S. (1987). Atomic-scale friction of a tungsten tip on a graphite surface, *Phys. Rev. Lett. 59*(17): 1942.
33. Erlandsson, R., Hadziioannou, G., Mate, C. M. McClelland, G. M., and Chiang S. (1988). Atomic scale friction between the muscovite mica cleavage plane and a tungsten tip, *J. Chem. Phys. 89*(8): 5190.
34. Cohen, S. R., Neubauer, G., and McClelland, G. M. (1990). Nanomechanics of Au-Ir contact using a bidirectional atomic force microscope, *J. Vac. Sci. Technol. A*, *8*(4): 3449.
35. Ducker, W. A., Senden, T. J., and Pashley, R. M. (1991). Direct measurement of colloidal forces using an atomic force microscope, *Nature*, *353*: 239.
36. Although this is correct for the case of a clean sample surface in UHV, an experiment performed in air, or in the presence of an adsorbate, may require additional terms to fully describe the interaction.
37. Eyring, H., Henderson, D., and Jost, W., Eds. (1971). *Physical Chemistry*, Academic Press, New York.
38. Gould, S. A. C., Burke, K., and Hansma, P. K. (1989). Simple theory for the atomic-force microscope with a comparison of theoretical and experimental images on graphite, *Phys. Rev. B*, *40*(8): 5363.
39. Landman, U., Luedtke, W. D., Burnham, N. A., and Colton, R. J. (1990). Atomistic mechanisms and dynamics of adhesion, nanoindentation, and fracture, *Science*, *248*: 454.
40. Landman, U., and Luedtke, W. D. (1991). Nanomechanics and dynamics of tip–substrate interactions, *J. Vac. Sci. Technol. B*, *9*(2): 414.

41. Burnham, N. A., and Colton, R. J. (1992). Force microscopy, in *Scanning Tunneling Microscopy: Theory and Application*, (Bonnell, D., Ed.), VCH Publishers, New York, 1992.
42. Sarid, D. (1991). *Scanning Force Microscopy*, Oxford University Press, New York.
43. Yamada, H., Fujii, T., and Nakayama, K. (1988). Experimental study of forces between a tunnel tip and the graphite surface, *J. Vac. Sci. Technol. A*, 6(2): 293.
44. Rugar, D., and Hansma, P. (1990). Atomic force microscopy, *Phy. Today*, October, p. 23.
45. Eigler, D. M., and Schweizer, E. K. (1990). Positioning atoms with the scanning tunneling microscope, *Nature*, 344: 524.
46. Clemmer, C. R., and Beebe, T. P., Jr. (1992). A review of graphite and gold surface studies for use as substrates in biological scanning tunneling microscopy studies, *Scanning Microsc*, 6(2): 319.
47. We acknowledge the contribution to this design by Joe Katz and the Salmeron STM lab at Lawrence Berkeley Laboratories.
48. Lyding, J. W., Skala, S., Hubacek, J. S., Brockenbrough, R., and Gammie, G. (1988). Variable-temperature scanning tunneling microscope, *Rev. Sci. Instrum.* 59(9): 1897.
49. Lyding, J. W., Skala, S., Hubacek, J. S., Brockenbrough, R., and Gammie, G. (1988). Design and operation of a variable temperature scanning tunneling microscope, *J. Microsc.* 152(2): 371.
50. Zeglinski, D. M., Ogletree, D. F., Beebe, T. P., Hwang, R. Q., Somorjai, G. A., and Salmeron, M. B. (1990). An ultrahigh vacuum scanning tunneling microscope for surface science studies, *Rev. Sci. Instrum.* 61(12): 3769.
51. Binnig, G., and Smith, D. P. E. (1986). Single-tube three-dimensional scanner for scanning tunneling microscopy, *Rev. Sci. Instrum.* 57(8): 1688.
52. Erlandsson, R., McClelland, G. M., Mate, C. M., and Chiang, S. (1988). Atomic force microscopy using optical interferometry, *J. Vac. Sci. Technol. A*, 6(2): 266.
53. Sarid, D., Iams, D., Weissenberger, V., and Bell L. S. (1988). Compact scanning-force microscope using a laser diode, *Opt. Lett. 13*: 1057.
54. Meyer, G., and Amer. N. M. (1988). Novel optical approach to atomic force microscopy, *Appl. Phys. Lett. 53*: 1045.
55. Blackford, B. L., Dahn, D. C., and Jericho, M. H. (1987). High-stability bimorph scanning tunneling microscope, *Rev. Sci. Instrum. 58*(8): 1343.
56. Pötschke, G., Schröder, J., Günther, C., Hwang, R. Q., and Behm, R. J. (1991). An STM investigation of the nucleation and growth of thin Cu and Au films on Ru(0001), *Surf. Sci. 251/252*: 592.
57. Manne, S., Hansma, P. K., Massie, J., Elings, V. B., and Gewirth, A. A. (1991). Atomic-resolution electrochemistry with the atomic force microscope: Copper deposition on gold, *Science*, 251: 183.
58. Chen, C. H., Vesecky, S. M., and Gewirth, A., J. (1992). In situ AFM of underpotential deposition of Ag on Au(111), *J. Am. Chem. Soc. 114*: 451.
59. Avouris, P., Bozso F., and Hamers, R. J. (1987). The reaction of Si(100)2 × 1 with NO and $NH_3$: The role of surface dangling bonds, *J. Vac. Sci. Technol. B*, 5: 1387.

60. Wolkow, R., and Avouris, P. (1988). Atom-resolved chemistry using scanning tunneling microscopy, *Phys. Rev. Lett. 60*(11): 1049.

61. Avouris, P., and Lyo, I. (1991). Probing and inducing surface chemistry with the STM: The reactions of Si(111)-7 × 7 with $H_2O$ and $O_2$, *Surf. Sci. 242*: 1.

62. Boland, J.J., and Villarrubia, J. S. (1990). Identification of the products from the reaction of chlorine with the silicon(111)-(7 × 7) surface, *Science, 248*: 838.

63. Hamers, R. J., Avouris, P., and Bozso, F. (1987). Imaging of chemical-bond formation with the scanning tunneling microscope: $NH_3$ dissociation on Si(001), *Phys. Rev. Lett. 59*(18): 2071.

64. Lin, T., and Chung, Y. (1989). Measurement of the activation energy for surface diffusion in gold by scanning tunneling microscopy, *Surf. Sci. 207*: 539.

65. Whitman, L. J., Stroscio, J. A., Dragoset, R. A., and Celotta, R. J. (1990). Scanning-tunneling-microscopy study of InSb(110), *Phys. Rev. B, 42*(11): 7288.

66. Rabe, J., and Buchholz, S. (1991). Direct observation of molecular structure and dynamics at the interface between a solid wall and an organic solution by scanning tunneling microscopy, *Phys. Rev. Lett. 66*(16): 2096.

67. Buchholz, S., and Rabe, J. P. (1991). Confirmation, packing, defects, and molecular dynamics in monolayers of dialkyl-substituted benzenes, *J. Vac. Sci. Technol. B, 9*(2): 1126.

68. Heckl, W. M., and Smith, D. P. E. (1991). Electropolymerization of glutaraldehyde observed by scanning tunneling microscopy and its implications for scanning tunneling microscopy imaging of organic material, *J. Vac. Sci. Technol. B, 9*(2): 1159.

69. Patrick, D. L. and Beebe, T.P., Jr. Manuscript in preparation.

70. Foster, J. S., Frommer., J. E., and Arnett, P. C. (1988). Molecular manipulation using a tunnelling microscope, *Nature, 331*: 324.

71. Binnig, G., and Röhrer, H. (1984). In *Trends in Physics* (Janta, J., and Pantoflicek, J., Eds.), European Physical Society, The Hague, pp. 38–46.

72. Amrein, M., Strasiak, A., Gross, H., Stoll, E., and Travaglini, G. (1988). Scanning tunneling microscopy of recA-DNA complexes coated with a conducting film, *Science, 240*: 514.

73. Amrein, M., Durr, R., Winkler, H., and Travaglini, G. (1988). STM of freeze-dried and Pt-Ir-C coated bacteriophage T4 polyheads, *Science, 240*: 514.

74. Beebe, T. P., Jr., Wilson, T. E., Ogletree, D. F., Katz, J. E., Balhorn, R., Salmeron, M. B., and Siekhaus, W. J. (1989). Direct observation of native DNA structures with the scanning tunneling microscope, *Science, 243*: 370.

75. Lindsay, S. M., and Barris, B. (1988). Imaging deoxyribose nucleic acid molecules on a metal surface under water by scanning tunneling microscopy, *J. Vac. Sci. Technol. A, 6*(2): 544.

76. Lindsay, S. M., Thundat, T., Nagahara, L., Knipping, U., and Rill, R. L. (1989). Images of the DNA double helix in water, *Science, 244*: 1063.

77. Lindsay, S. M., Nagahara, L. A., Thundat, T., Knipping, U., Rill, R. L., Drake, B., Prater, C. B., Weisenhorn, A. L., Gould, S. A. C., and Hansma, P. K. (1989) STM and AFM images of nucleosome DNA under water, *J. Biomol. Struct. Dyn. 7*(2): 279.

78. Lindsay, S. M., Nagahara, L A., Thundat, T., and Oden, P. (1989). Sequence, packing and nanometer Scale structure in STM images of nucleic acids under water, *J. Biomol. Struct. Dyn. 7*(2): 289.

79. Clemmer, C. R., and Beebe, T. P., Jr. (1991). Graphite: A mimic for DNA and other biomolecules in scanning tunneling microscope studies, *Science*, *251*: 640.
80. Salmeron, M., Beebe, T., Odriozola, J., Wilson, T., Ogletree, D. F., and Siekhaus, W. (1990). Imaging of biomolecules with the scanning tunneling microscope: Problems and prospects, *J. Vac. Sci. Technol. A*, *8*(1): 635.
81. Chang, H., and Bard, A. J. (1991). Observation and characterization by scanning tunneling microscopy of structures generated by cleaving highly ordered pyrolytic graphite, *Langmuir*, *7*: 1143.
82. Wepf, R., Amrein, M., Bürkl, U., and Gross, H. (1991). Platinum/iridium/carbon: A high-resolution shadowing material for TEM, STM and SEM of biological macromolecular structures, *J. Microsc. 163*(1): 51.
83. Amrein, M., Wang, Z., and Guckenberger, R. (1991). Comparative study of a regular protein layer by scanning tunneling microscopy and transmission electron microscopy, *J. Vac. Sci. Technol. B*, *9*(2): 1276.
84. Amrein, M. Dürr, R., Winkler, H., Travaglini, G., Wepf, R., and Gross, H. (1989). STM of freeze-dried and Pt-Ir-coated bacteriophage T4 polyheads, *J. Ultrastruct. Mol. Struct. Res. 102*: 170.
85. Drake, B., Prater, C. B., Weisenhorn, A. L., Gould, S. A. C., Albrecht, T. R., Quate, C. F., Cannell, D. S., Hansma, H. G., and Hansma, P. K. (1989). Imaging crystals, polymers, and processes in water with the atomic force microscope, *Science*, *243*: 1586.
86. Hansma, H. G., Vesenka, J., Siegerist, C., Kelderman, G., Morrett, H., Sinsheimer, R. L., Elings, V., Bustamante, C., and Hansma, P. K. (1992). Reproducible imaging and dissection of plasmid DNA under liquid with the atomic force microscope, *Science*, *256*: 1180.
87. Weisenhorn, A. L., Egger, M., Ohnesorge, F., Gould, S. A. C., Heyn, S. -P., Hansma, H. G., Sinsheimer, R. L., Gaub, H. E., and Hansma, P. K. (1991). Molecular-resolution images of Langmuir–Blodgett films and DNA by atomic force microscopy, *Langmuir*, *7*(1): 8.
88. Lin, J. N., Drake, B., Lea, A. S., Hansma, P. K., and Andrade, J. D. (1990). Direct observation of immunoglobulin adsorption dynamics using the atomic force microscope, *Langmuir*, *6*(2): 509.
89. Gould, S. A. C., Drake, B., Prater, C. B., Weisenhorn, A. L., Manne, S., Hansma, H. G., Massie, J., Longmire, M., Elings, V., Northern, B. D., Mukergee, B., Peterson, C. M., Stoeckenius, W., Albrecht, T. R., and Quate, C. F. (1990). From atoms to integrated circuit chips, blood cells, and bacteria with the atomic force microscope, *J. Vac. Sci. Technol. A*, *8*(1): 369.
90. Bustamante, C., Vesenka, J., Tang, C. L., Rees, W., Guthold, M., and Keller, R. (1992). Circular DNA molecules imaged in air by scanning force microscopy, *Biochemistry*, *31*(1): 22.
91. Ohtani, H., Wilson, R. J., Chiang, S., and Mate, C. M. (1988). Scanning tunneling microscopy observations of benzene molecules on the Rh(111)-(3 × 3) ($C_6H_6$ + 2CO) surface, *Phys. Rev. Lett. 60*: 2398.
92. Chiang, S., Wilson, R. J., Mate, C. M., and Ohtani, H. (1988). Real space imaging of co-adsorbed CO and benzene molecules on Rh(111), *J. Micros. 152* (2): 567.

93. McGonigal, G. C., Bernhardt, R. H., Yeo, Y. H., and Thompson, D. J. (1991). Observation of highly ordered, two-dimensional *n*-alkane and *n*-alkanol structures on graphite, *J. Vac. Sci. Technol. B, 9*: 1107.

94. For a comparison of anchoring structures of 8CB on graphite and $MoS_2$, see the commentary by Hara, M., Iwakabe, Y., Tochigi, K., Sasabe, H., Garito, A. F., and Yamada, A. (1990). *Nature, 344*:228.

95. Avouris, P., and Wolkow, R. (1989). In *Chemical Perspectives of Microelectronic Materials* (Gross, M. E., Yates, J. T., Jr., and Jasinski, J., Eds.), *MRS Symp. Proc.*, Vol. 131.

96. Bednorz, J. G., and Müller, K. A. (1986). Possible high $T_c$ superconductivity in the Ba-La-Ca-O system, *Z. Phys. B64*: 189.

97. Giaver, I. (1974). Electron tunneling and superconductivity (Nobel Lecture), *Rev. Mod. Phys. 46*(2): 245.

98. de Lozanne, A. L., Elrod, S. A., and Quate, C. F. (1985). Spatial variations in the superconductivity of $Nb_3Sn$ measured by low-temperature tunneling microscopy, *Phys. Rev. Lett. 54*(22): 2433.

99. Gimzewski, J. K., Riehl, B., Coombs, J. H., and Schlittler, R. R. (1988). Photon emission with the scanning tunneling microscope, *Z. Phys. B, 72*: 497.

100. Abraham, D. L., Veider, A., Schönenberger, C., Meier, H. P., Arent, D. J., and Alvarado, S. F. (1990). Nanometer resolution in luminescence microscopy of III–V heterostructures, *Appl. Phys. Lett. 56*: 1564.

101. Coombs, J. H., Gimzewski, J. K., Reihl, B., Sass, J. K., and Schlittler, R. R. (1988). Photon emission experiments with the scanning tunneling microscope, *J. Microc. 152*: 325.

102. Berndt, R., Schlittler, R. R., and Gimzewski, J. K. (1991). Photon emission scanning tuneling microscope, *J. Vac. Sci. Technol. B, 9*(2): 573.

103. Kuk, Y., Becker, R. S., Silverman, P. J., and Kochanski, G. P. (1991). Photovoltage on silicon surfaces measured by scanning tunneling microscopy, *J. Vac. Sci. Technol. B, 9*(2): 545.

104. Hamers, R. J., and Markert, K. (1990). Atomically resolved carrier recombination at Si(111)-(7 × 7) surfaces, *Phys. Rev. Lett. 64*: 1051.

105. Kuk, Y., Becker, R. S., Silverman, P. J., and Kochanski, G. P. (1990). Optical interactions in the junction of a scanning tunneling microscope, *Phys. Rev. Lett. 65*: 456.

106. Sonnenfeld, R., and Hansma, P. K. (1986). Atomic resolution microscopy in water, *Science, 232*: 211.

107. White, Henry, personal communication.

108. Schmickler, W., and Widrig, C. (1992). *J. Electrochem. Soc.* (accepted for publication).

109. Otsuka, I., and Iwasaki, T. (1988). STM imaging of noble metal electrodes in solution under potentiostatic control, *J. Microsc. 152*: 289.

110. Fan, F., and Bard, A. J. (1988). Scanning tunneling microscopic studies of platinum electrode surfaces, *Anal. Chem. 60*: 751.

111. Gewirth, A. A., and Bard, A. J. (1988). In situ scanning tunneling microscopy of the anodic oxidation of highly oriented pyrolytic graphite, *J. Phys. Chem. 92*: 5563.

112. Van Hove, M. A., Weinberg, W. H., and Chan, C. -M. (1986). *Low-Energy Electron Diffraction*, Springer-Verlag, New York.

113. Adamson, A. A. (1990). Physical Chemical of Surfaces, Wiley, New York.

114. Fan, F., and Bard, A. J. (1989). In situ scanning and tunneling microscopy of polycrystalline platinum electrodes under potential control, *J. Electrochem. Soc. 139*(11): 3216.

115. Haiss, W., Lackey, D., and Sass, J. K. (1991). Atomic resolution scanning tunneling microscopy images of Au(111) surfaces in air and polar organic solvents, *J. Chem. Phys. 95*(3): 2193.

116. Barth, J. V., Brune, H., Ertl, G., and Behm, R. J. (1990). Scanning tunneling microscopy observations on the reconstructed Au(111) surface: Atomic structure, long-range superstructure, rotational domains, and surface defects, *Phys. Rev. B, 42*(15): 9307.

117. Albrecht, T. R., Dovek, M. M., Kirk, M. D., Lang, C. A., Quate, C. F., and Smith D. P. E. (1989). Nanometer-scale hole formation on graphite using a scanning tunneling microscope, *Appl. Phys. Lett. 55*(17): 1727.

118. Becker, R. S., Golovchenko, J. A., and Swartzentruber, B. S. (1987). Atomic-scale surface modifications using a tunnelling microscope, *Nature, 325*: 419.

119. Casillas, N., Snyder, S. R., and White, H. S. (1991). STM fabrication of platinum disks of nanometer dimensions, *J. Electrochem. Soc. 138*(2): 641.

120. Schneir, J., and Hansma, P. K. (1987). Scanning tunneling microscopy and lithography of solid surfaces covered with nonpolar liquids, *Langmuir, 3*: 1025.

121. Quate, C. F. (1990). In Scanning Tunneling Microscopy and Related Methods (Behm, R. J., et al., Eds.), Kluwer Academic Publishers, Dordrecht, pp. 281–297.

122. Siekhaus, W. J. Referenced by Berndt, R., Baratoff, A., and Gimzewski, J. K. (1990). In Scanning Tunneling Microscopy and Related Methods (Behm, R. J., et al., Eds.), Kluwer Academic Publishers, Dordrecht, pp. 269.

123. Albrecht, T. R., Dovek, M. M., Lang, C. A., Grutter, P., Quate, C. F., Kuan, S. W. J., Frank, C. W., and Pease, R. F. W. (1988) Imaging and modification of polymers by scanning tunneling and atomic force microscopy, *J. Appl. Phys. 64*(3): 1178.

124. Baba, M., and Matsui, S. (1990). Nanostructure fabrication by scanning tunneling microscope, *Jap. J. Appl. Phys. 29*(12): 2854.

125. Ehrichs, E. E., Yoon, S., and de Lozanne, A. L. (1988). Direct writing of 10 nm features with the scanning tunneling microscope, *Appl. Phys. Lett. 53*: 2287.

126. McCord, M. A., Kern, D. P., and Chang, T. H. P. (1988). Direct deposition of 10 nm metallic features with the scanning tunneling microscope, *J. Vac. Sci. Technol. B, 6*: 1877.

127. Shedd, G. M., and Russell, P. E. (1990). The scanning tunneling microscope as a tool for nanofabrication, *Nanotechnology, 1*: 67.

128. Nejoh, H. (1991). Incremental charging of a molecule at room temperature using the scanning tunnelling microscope, *Nature, 353*: 640.

129. Tsukada, M., Shima, N., Nobayashi, K., Inada, K., and Mizokami, T. (1990). Numerical simulations of the single electron tunneling processes in the scanning tunneling spectroscopy through metal fine particles, *Prog. Theor. Phys. Suppl. 101*: 221.

130. Terris, B. D., Stern, J. E., Rugar, D., and Mamin, H. J. (1989). Contact electrification using force microscopy, *Phys. Rev. Lett. 63*(24): 2669.

131. Allenspach, R., Salemink, H., Bischof, A., and Weibel, E. (1987). Tunneling experiments involving magnetic tip and magnetic sample, *Z. Phys. B, 67*: 125.

132. Allenspach, R., and Bishof, A. (1989). Spin-polarized secondary electrons from a scanning tunneling microscope in field emission mode, *Appl. Phys. Lett. 54*: 587.

133. Wiesendanger, R., Güntherodt, H. J., Güntherot, G., Gambino, R. J., and Ruf, R. (1990). Observation of vacuum tunneling of spin-polarized electrons with the scanning tunneling microscope, *Phys. Rev. Lett. 65*: 247.

134. McCord, M. A., and Awschalom, D. D. (1990). Direct deposition of magnetic dots using a scanning tunneling microscope, *Appl. Phys. Lett. 57*(20): 2153.

135. Martin, Y., Rugar, D., and Wickramasinghe, H. K. (1988). High-resolution magnetic imaging of domains of TbFe by force microscopy, *Appl. Phys. Lett. 53*(3): 244.

136. Feynman, R. P. (1961). There's plenty of room at the bottom, in *Miniaturization* (Gilbert, H. D., Ed.), Reinhold, New York.

137. Schneiker, C., Hameroff, S., Voelker M., He, J., Dereniak, E., and McCuskey, R. (1988). Scanning tunnelling engineering, *J. Microsc. 152*(2): 585.

138. Bate, R. T. (1989). Nanoelectronics, *Solid State Technol. . November, p. 101.*

# 7

# Near-Field Optical Microscopy, Spectroscopy, and Chemical Sensors

**Raoul Kopelman and Weihong Tan**   *University of Michigan, Ann Arbor, Michigan*

## I. INTRODUCTION

Optical microscopy and spectroscopy have long been key techniques in medicine, biology, chemistry, and materials science. Among their advantages are:

1. *Universality.*   All materials and samples attenuate light and have spectroscopic states.
2. *Noninvasiveness.*   Most often the sample is not altered in a microscopic and/or spectroscopic investigation. Moreover, biological samples usually can be studied in their native environment. Most chemical reactions are not perturbed by light of long enough wavelength.
3. *Real-time observation.*   Biological phenomena, chemical reactions, crystallization, and so on can be observed under the microscope as they happen *in situ* (even with one's eyes); spectroscopic measurements can be performed on line in an industrial process or other setting.
4. *Energy and chemical state resolution.*   The obvious advantages of spectroscopy and photochemistry, at ambient temperature, can be trivially added to the optical methods mentioned above. By contrast, this is not easily accomplished with other techniques such as electron microscopy or x-ray crystallography.
5. *Safety.*   Optical and spectroscopic analyses usually are very safe, and precautions are mostly limited to wearing optically protective eyeglasses.
6. *Low price.*   Optical microscopy is much cheaper than, say, electron microscopy; optical spectroscopy is usually a bargain compared with, say, NMR instruments. Obviously, there are exceptions.

7. *Speed, zoom, and human factors.* Optical techniques are usually fast and can be extended even into the femtosecond time domain. They can be used from astronomical to microscopic distances. Preliminary or concomitant observations can be made using our most developed sense—sight, and in living color—even without the brokerage services of an analog or digital interface.

One can obviously add to the list of positives the use of polarization, electro-optical effects, and the recent phenomenal advances with optical fiber probing techniques. However, on the negative side there are two major factors:

1. *Spatial resolution.* The well-known diffraction limit, which is on the order of half a wavelength. This opened the way to competing technologies such as electron microscopies.
2. *Attenuation.* Penetration of samples is limited, opening the way to x-ray, magnetic resonance, acoustical, and other imaging techniques.

This chapter describes some newly emerging techniques of near-field microscopy that do overcome the historical obstacle number 1: the diffraction limit. Eventually, with the aid of optical fibers, these approaches may even lead to techniques that can penetrate deeply into the human body and perform measurements with molecular precision. Furthermore, optical probes can be combined with scanning tunneling and atomic force microscopy (STM and AFM) probes (see Chapter 6).

## II. PRINCIPLES OF NEAR-FIELD OPTICAL TECHNIQUES

Conventional (''far-field'') optical techniques are based on focusing elements such as a lens (see Chapter 1). This leads to the ''diffraction limit'' of about $\lambda/2$ (half the wavelength) [1]. However, light can be *apertured* down to much smaller sizes, with no obvious theoretical limit. The simplest example is the passage of light through a small hole. The rules governing the ''near-field'' regime of evanescent waves cause whatever light passes through to be confined to the dimension of the aperture in the immediate vicinity outside this aperture [2–6]. This principle has been discovered and rediscovered several times (in 1928 [2], in 1956 [3], in 1972 [4], and in the 1980s [5,6]). Actually, the photon ''scanning probe technique'' has preceded all others, such as STM. (see Chapters 5 and 6). However, only in the 1980s was the principle followed by optical experiments (the 1972 work demonstrating it with ''microwave'' radiation [4]). More recently came the idea of active subwavelength-sized light sources [7], which can be based on a fluorescing material or even on a light guide (e.g., optical fiber) transmitting light from a far-field light source (e.g., laser). As pointed out in Chapter 5, scanning probes give an image due to raster scanning over a sample. Today's optical methods have indeed borrowed some of the electromechanic and computer control scanning techniques from STM, AFM, and so on. Figure 1 shows schematically a near-field apparatus, active or passive, containing source, sample, and detector.

The light source, which is the "heart" of the technique, must be small, intense, durable, and spatially controlled. As pointed out above, its size determines the resolution, provided it can be scanned close to the sample, as close as its size (technically 20–80% of it). In addition, to perform successful microscopy or nanospectroscopy, the issue of contrast is of the highest importance. Figures 2 and 3 show examples of the most common passive and active light sources for near-field scanning optical microscopy (NSOM).

In the past, the requirements of smallness and intensity were in direct conflict. The passive light sources are typically apertures letting light through, and when the size of the aperture gets to be significantly below that of a wavelength, most of the light will be diffracted back, rather than transmitted [7–10]. For instance, for a hole in a metal plate, with apertures of 500 Å or less, the intensity goes down superexponentially with aperture. This occurs because to be opaque

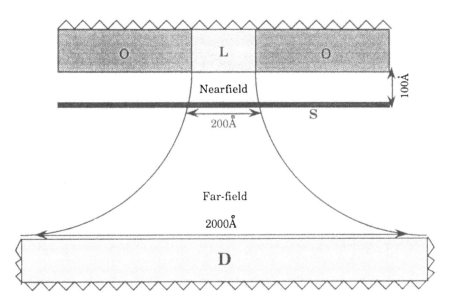

**Figure 1**   Near-field optics scheme: D, far-field detector system (e.g., lens); L, active or passive light source; O, opaque material; S, sample (support not shown).

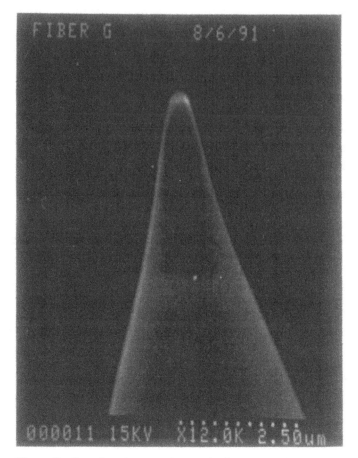

**Figure 2**   Scanning electron micrograph of an optical fiber tip.

enough to define a hole, the metal surrounding the hole must be at least about 500 Å thick [9]. With a given minimal thickness, the emerging light intensity $I$ is described by [10]:

$$I \sim r^6 \exp\left(\frac{-L}{r}\right) \tag{1}$$

where $L$ is the length (thickness) of the aperture. Reducing $r$ by one order of magnitude reduces $I$ by about 20 orders of magnitude! The foregoing statements assume that the wavelength $\lambda$ is much larger than the aperture

$$\lambda >> r \tag{2}$$

We note that Eq. (1) is valid only when Eq. (2) is valid. $L$ must be significantly larger than the penetration length $l$ of the radiation into the metal. The shortest known penetration length (for visible light) is a couple of hundred angstroms (in aluminum). The smallest practical value for $L$ is 500 Å (using aluminum).

Recently, several methods have been suggested for overcoming the above-mentioned severe losses of light in subwavelength cavities.

1.  Transforming the photons into physically smaller energy quanta, such as excitons [7], followed by the retransformation of these energy quanta into photons (evanescent photons). Despite significant inherent losses in such a process (see below), the dependence on aperture size is significantly less steep, that is, linear with cross section [8]

$$I \sim r^2 \tag{3}$$

Therefore, effectively, there is no loss of brightness ($I / \pi \, r^2$) with reduced size.

2.  Overcoming Eq. (1) by overcoming Eq. (2). In practice this means an effective reduction of $\lambda$. It is well known that inside dielectric materials $\lambda$ is reduced from the vacuum (or air) value by a factor equal to the refractive index $n$. One can thus choose appropriate transparent materials with high $n$. The resultant reduction in $\lambda$ effectively pushes down the diffraction limit—a well known trick (e.g., the use of lenses with oil rather than air). In general, $n$ has a value of about 1.5 or 2, at best. However, by using materials at the edge of their absorption wavelengths (or, equivalently, shifting the wavelength to this absorption edge) one may effectively increase $n$ by an order of magnitude.

3.  Using the waveguide (or "coax") principle. The use of appropriately designed optical interfaces creates an "optical fiber" in the subwavelength regime of evanescent waves [8,11].

In a more general approach, the generation of photons by an active light source is decoupled from the problem of transmitting the photons through a cavity. In principle, the light source may generate light from electrical, thermal, chemical, nuclear, or exciton processes or from light of a different wavelength [7,11]. This opens the way to a large number of potential schemes. It may even decouple the monotonic relation between size and brightness (see above).

As described above, in past schemas intensity was severely limited by the requirement to reduce the size of the light source. While this situation has been somewhat alleviated with the use of active sources (compared to "empty" holes), other problems have arisen. One such concern is the durability of the active material. For instance, the anthracene crystals used in the first exciton light source [7] deteriorated significantly within minutes (resulting in long-time amplification factors of about 3, compared to several orders of magnitude, initially). Photochemistry, thermochemistry, and so on are the obvious "culprits": it is well known that in the presence of oxygen and ultraviolet light, anthracene

**(a)**

**Figure 3** Nanometer light sources. (a) Optical micrograph of micropipette light source: perylene crystal tip in an aluminum-coated micropipette. Magnification about 600X. The 50 nm supertip is unresolved but overexposed. (b) Conventional optical micrograph (about 600 X) of an optical fiber tip light source: aluminum-coated optical fiber tip excited with 442 nm He–Cd laser line. Note that the tip is unresolved and overexposed.

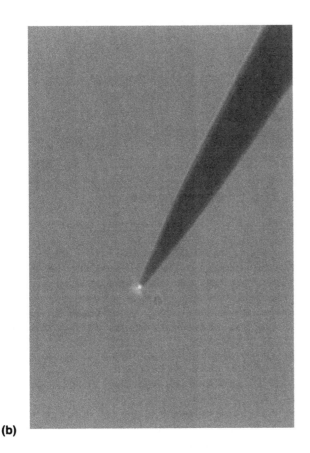

**(b)**

is oxidized to anthraquinone. Obviously, one can either avoid the oxygen, reduce the temperature, or replace the material with a more stable one. The problem is thus transformed to one of molecular engineering, a common occurrence with modern optical as well as other materials.

Spatial control (e.g., scannability) is obviously related to the mechanical durability of the light source (tip). However, it mainly involves geometrical, mechanical, piezoelectric, optoelectronic, and other design considerations (Fig. 4), which are all intimately related to the nature of the sample. For instance, a perfectly smooth sample could be scanned by a metal plate with a hole in it. Obviously, this is not true for a rough sample. Staying within near-field range (a fraction of $\lambda$) without crashing into the sample is no trivial problem. This has led in the past to the use of metal-coated micropipettes with empty cavities [12,13] or with crystals at their tips [7,8,13]. Similarly, sharpened optical fibers [14,15],

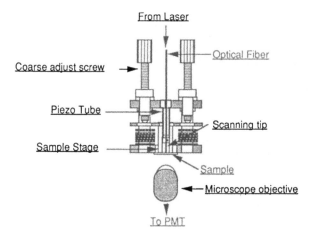

**Figure 4**   Pictorial representation of the NSOM at the University of Michigan. The sample stage is supported by three springs and tilted by adjusting three fine threaded screws spaced 120° apart radially. The sample plane sweeps out an arc determined by the relative screw adjustments, this allows both coarse $x$-$y$ translation as well as $z$ translation. The scan is actuated by a piezoelectric tube with fine Z control; it scans the near-field probe in a raster scan. The signal from the photomultiplier tube (PMT) is plotted as a function of position to obtain the NSOM image. (Courtesy of S. Smith.)

often metal-coated, have been used more recently, both as light collectors [14] and as light sources [15].

The concept of active light source enables a totally new mode of NSOM, based not on the blocking or absorption of photons but rather on quenching directly the energy quanta that otherwise would have produced photon. For instance, a thin, localized gold film (or cluster) can quench an excitation (or exciton) that would have been the precursor of photons. This example is further discussed below. Furthermore, a single atom or molecule on the sample could quench (i.e., by energy transfer) the excitations located at the tip of the light source. For simplicity, we assume that the active part of the light source is a single atom, molecule, or crystalline site, serving as the "tip of the tip." This quenching energy transfer from the excitation source's active part (donor) to the sample's active part (acceptor) may or may not qualify technically as an NSOM technique. However, it is our best hope, currently, for single atom or molecule resolution and sensitivity, as explained below. Semantically, we could call this technique quantum optics microscopy. Alternatively, it has been called scanning exciton microscopy (SExM).

Scanning exciton microscopy is conceptually quite similar to STM. The excitons "tunnel" from the tip to the sample. However, there is no driving voltage or field. Rather, it is the energy transfer matrix element which controls the transfer efficiency. Its unusual matrix elements allow for the highest sensitivity to distance, higher than that of STM and comparable to that of AFM. In addition, the most striking result of this direct energy transfer is its ultrahigh sensitivity to isolated or single molecular chromophores. The quantum optics energy transfer is highly efficient within the range of the "Förster radius" (defined as the distance at which there is a 50% chance for transfer). Thus, a single excitation could be "absorbed" by the sample acceptor. In contrast, based on the Beer-Lambert law, about a billion photons are needed [13] to excite a single acceptor in the absence of other acceptors. Furthermore, since the distance range is limited to about 8 nm for direct energy transfer, scanning exciton microscopy is as much a near-field technique as STM or AFM: that is, SExM is very sensitive in the single-digit nanometer range and much less sensitive beyond 10 nm. However, in combination with conventional NSOM, the range can be extended to about 100–200 nm (and, adding far-field optical microscopy, to micrometers and beyond). We thus have a technique able to "zoom in" from macroscopic to nanoscopic distances. Obviously such "zooming in" enhances the speed of operation. It also allows for a much more universal range of samples, from metal spheres and clusters to soft, in vivo biological units. Just as classical optical microscopists use staining to enhance contrast, and electron microscopists use various markers, with NSOM one can use fluorophores, metal clusters, and so on to enhance contrast, sensitivity, and resolution.

In the next section, we distinguish between traditional NSOM and scanning exciton microscopy. We note that traditional NSOM can be carried out with active light sources based on exciton transport and quenching. However, in this case the exciton transport and quenching occur only inside the source, while the sample is exposed to real evanescent photons. This has been occasionally called [8,16] "molecular exciton microscopy" (MEM). In contrast, scanning tunneling exciton microscopy (STExM) has also been called, occasionally, "molecular tunneling exciton microscopy" (MTEM). The reader has thus been warned. The different regimes of optical and exciton microscopy are shown in Table 1.

## III.  TRANSMISSION AND FLUORESCENCE NSOM

The two most important modes of traditional optical microscopy are transmission and fluorescence ("luminescence" may be a better term, including both fluorescence and phosphorescence). For chemical and biological applications, these will probably continue to be the most important modes, especially if one includes nanospectroscopy (see below).

### A.  The Micropipette Tip

The original idea of NSOM depended on a light transmission through a subwavelength aperture (see above), that is, a nanofabricated hole in a thin metal sheet

**Table 1**  Optical Microscopies

| Optical technique | Interaction | Resolution limit (Å) | Implementation |
|---|---|---|---|
| Far-field optics | Conventional, diffraction-limited | 2500–5000 | Lens |
| Near-field optics | Evanescent wave, intensity-limited | 100–200 | Fiber-optic tip; micropipette tip |
| Molecular exciton microscopy (MEM) | Excitation transport | 3.5–10* | Molecular donor supertip |
| | Spin-orbit coupling | 2.5* | Molecular sensor supertip |

* Limited by the interaction (e.g., triplet excitation exchange) cut-off.

or film. Obviously, a flat sheet or film (with a nanohole) can be used as a scanning probe only with completely flat samples. This led to the idea [5,9,12], among others, of using a glass micropipette such as invented by Nobel laureates Neher and Sackman [17] for intracell electrical measurements and as used for microelectrochemical sensors [18]. The micropipette is metallized on the outside with an aluminum or gold coating thick enough to block the light. Pulling such micropipettes is mostly done today with a commercial puller (e.g., Sutter P-87 and P-2000). Recently these devices have been improved on with computerized multistep control of the pulling procedure [19] and with infrared lasers replacing the electric heater strips. Thus, one can now reproducibly pull robust and efficient micropipettes with orifices as small as 10 nm (inner diameter), optically streamlined (e.g., short "shank"), and clean enough on the outside to facilitate the metal coating. The latter procedure (vacuum deposition of metals) is well known but far from trivial.

The micropipette approach has been extended in several ways. For example, the tip has been filled with a photoactive material [7]. Alternatively, the "pretip" region is specially bend, to double as a "force probe" [20]. These extensions have kept the micropipette tip in competition with the optical fiber tip described next.

## B.  The Optical Fiber Tip

Compared to a hollow tip, such as an empty pipette, this optical fiber tip is a "semiactive" photon tip. Generally it is orders of magnitude brighter, easily coupled to an optical fiber (it is part of it), and at least as sturdy mechanically.

On the other hand, it is not transparent to short-wave radiation (deep UV or soft x-rays). Also its preparation is a bit trickier (see Fig. 5). Not only does the higher melting point silica require the temperature provided by an infrared laser, but the metallization process is more demanding, both on the pulling process and on the deposition process. For instance, to avoid metallizing (and thus blocking) the photon tip, the tip surface must be essentially orthogonal to the pulled fiber wall. Furthermore, the deposition angle must be carefully adapted to the tip geometry. Single-mode optical fibers have been pulled successfully by Betzig et al. [15] and also in our lab (and probably in several others by now). Figure 5 shows a setup.

Using a classical optics description, we note that the higher index of refraction ($n$) of the fiber material reduces the photon wavelength inside it to $\lambda = \lambda_0 / n$, compared to the wavelength in vacuum or air ($\lambda_0$). This reduces significantly the diffraction of the light at the orifice. In principle, as $\lambda$ approaches the optical absorption of the dielectric, $n$ increases and eventually becomes a complex quantity [21]. Alternatively, one can use a quantum approach and consider the exciton-polariton resonance or quasi-resonance [22,23]. Thus the optical fiber tip exhibits a crossover with wavelength from a passive to an active photon tip (see below).

## C.    Resolution, Contrast, and Other Parameters

The best resolution to date has been claimed by Betzig et al. [15] to be about 12 nm (with 514 nm light). Presumably this was achieved with a 20 nm diameter aperture. A signal of 50 nW has been claimed for an 80 nm aperture [15].

There are several contrast methods: absorption, refractive index, reflection, and fluorescence (luminescence), and not all of them are well understood. One can also count [15] polarization and spectroscopy as separate modes of contrast. Furthermore, there are a large number of quantum effects, such as energy transfer, energy down-conversion, and energy quenching (see below). The simplest optical contrast mechanisms in the near-field regime (e.g., refractive index) are not yet well understood [24], and thus they are under intensive study. Actually,

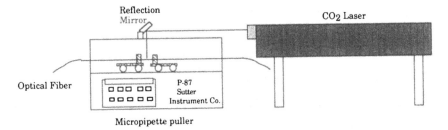

**Figure 5**   Apparatus for pulling fiber tips. (Courtesy of S. Smith.)

the microscopic quantum effects are better understood than the mesoscopic (near-field) optical interactions.

The mechanical stability of optical tips is excellent (days to months). The photochemical stability for optical fiber tips is also excellent, and under very intense illumination it is the heat that damages the aluminum coating at the tip. The price of a tip is very low ($10, according to a calculation by Betzig et al.). The spectral range extends from about 300 nm to 2000 nm (depending on optical fiber quality). However, with micropipettes it extends from the x-ray regime to microwaves.

## D.  Sample Requirements

The most important consideration, sample roughness, is limited by the probe shape in the most obvious way (as for all scanning probes). The sample thickness is an important factor for all transmission (forward-scattering) modes of operation, but not for the reflectance (backscattering) and some "collection" modes. The near-field approach couples the optical resolution with the distance from the probe: the higher the desired resolution, the thinner the required sample. Examples of simple sample scans are shown in Fig. 6. On the other hand, the contrast mode (absorption, refractive index) may limit the thinness of the sample. Even the fluorescence mode may be limited by the thinness of the sample (i.e., the absorption cross section). However, this drawback can be overcome by intensity, by auxiliary fluorophores, or by quantum mechanisms (energy transfer). Thus the various luminescence modes appear to be the most promising modes of forward-scattering near-field microscopy.

## E.  Luminescent Tags

In view of the foregoing discussion, the selection of luminescent tags is as important for NSOM as it is for traditional fluorescence microscopy. This is an actively studied aspect of biology [25] and related fields. The tags are usually tailor-made either for a polychromatic lamp (e.g., xenon) or for monochromatic sources (e.g., lasers), with the obvious use of spectral filters. In addition to lamps and lasers, near-field microscopy sources include luminescent probes, such as crystals, dye-balls, or luminescent polymers, with broadband characteristics falling somewhere between those of lamps and lasers. These nanosources are located right next to the sample, and thus no filter can be inserted between them (only between the sample and the detector). Furthermore, the spectral absorption and photostability of the probes are the major aspects of consideration, in addition, of course, to their chemical and/or biological selectivity.

## IV.  COLLECTION MODE NSOM

Various reflection and collection mode NSOM techniques have been devised [15,20]. One of the most elegant designs (Fig. 7) is a combined collection–feed back method [14–16,23]. The sample is placed on top of a prism, which is illuminated in a total internal reflection mode. A thin quartz tube ("collection probe") with a tip at its end scans the surface of the prism. In the absence of sample, the collection tip will collect light only if it touches the prism or is slightly above it (at about $\lambda/20$). It is well known [21] that light penetrates slightly from the prism into the air above it. The collection decreases exponentially with the gap size. [This is the classical analog of quantum mechanical tunneling through a barrier.] With the addition of a sample to the prism, the gap decreases and more light is passed into the collector, depending on the local optical density of the sample. If the sample is homogeneous, the local optical density is determined by the local thickness.

This method has excellent sensitivity in the vertical ($z$) direction—a resolution of about 2 Å. However, in the horizontal ($x$, $y$) directions the resolution is only about 0.1 μm. The horizontal resolution depends on the size of the tip. These tips have been produced by chemical etching, and smaller tips could be produced by other methods (see above). We note that this is a zero-background method, with all its advantages. On the other hand, the sample must be extremely thin and quite transparent. It is also easy to confuse sample thickness variation with optical density variation.

Alternatively, the sample may fluoresce or contain luminescent tags. This changes the contrast method and may improve the sensitivity. This method has also been used for spatially resolved fluorescence and Raman spectroscopy (see below), with a spatial resolution of about 0.1 μm. Historically, the biggest advantage of this technique has been its inclusion of feedback, via the intensity of the light leaking into the tip. Other feedback methods are described below (Section VIII).

## V. SOURCE–SAMPLE INTERACTION MODE

In traditional optical, x-ray, and electron microscopy, the source first produces photons or electrons and then the photons or electrons interact with the sample. Obviously, this is not the case for scanning probe microscopies. Similarly, when a near-field active light source is close (much less than $\lambda$ from the sample), source–sample interactions arise. We list several examples.

### A. Excitation Transfer Interactions

A molecular nanocrystal light source (say, anthracene) will transfer its energy to a nearby chromophore molecule (such as rhodamine) of the sample (such as a Langmuir–Blodgett film). Thus the light source becomes an energy donor and the sample molecule an energy acceptor. The result is a fluorescence emission typical of the sample. Superficially this may appear to be no different from the ordinary radiative process in which photons are first emitted from the source and then absorbed by the sample, causing it to fluoresce. However, in reality one has

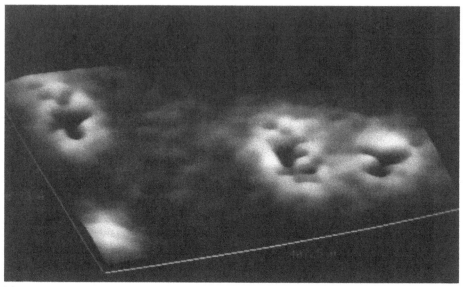

**(a)**

**Figure 6**  NSOM scanning images. (a) Channel-pore polycarbonate membrane (Nucle-pore, 300 nm diameter). (b) Calibration images of a grating replica (chromium, 20 nm thickness) with a spacing of 0.463 μm. The upper images are our NSOM fast scans (4 min total). The lower image is our transmission electron microscope comparison scan. (Courtesy of E. Monson and S. Smith.)

a nonradiative Förster–Dexter energy transfer process [13]. The latter may be much more efficient than the radiative process, and it will exhibit a very different quantitative behavior. For instance, the dependence of fluorescence on the distance from the probe is much steeper (fourth to sixth power), compared to a very weak geometry dependence for the radiative process. Also, the light polarization dependence may differ for the two cases.

## B.  Excitation Quenching and Transformation Interactions

A simple example consists of a molecular light source (e.g., anthracene) and a metallic sample (e.g., silver). At subwavelength ($< \lambda$) distances, the photon flux of the source is modulated and quenched by the sample. This quenching effect [13,20] can be considered to be a special case of the energy transfer mechanism, one without radiative reemission.

A special case of quenching is the Kasha effect [26]. This effect is based on an interatomic spin-orbit coupling, where one atom (the heavy one, atomic

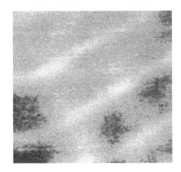
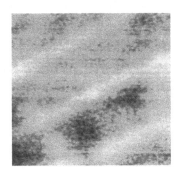

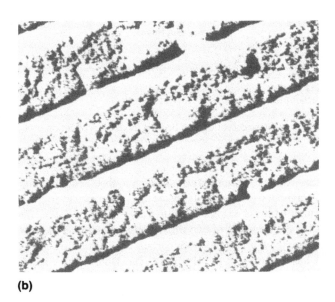

**(b)**

mass > 40) has a high degree of spin-orbit coupling ("relativistic effect") and the other atom (the light one) has very little of it on its own. The effect is "internal" when both atoms belong to the same molecule and "external" when they are not. This effect is empirically observed spectroscopically as a change in intensity or lifetime of an absorption or emission. For instance, a molecule such as anthracene has a fluorescence quantum efficiency of near unity because the rules governing spin selection prevent the first excited singlet state from transferring its energy to the lower lying first excited triplet state. However, upon "induced" spin-orbit coupling, such a singlet–triplet transfer becomes possible internally (with the aid of vibrational quanta that absorb the extra energy) or radiatively. Empirically one observes a "quenching" of the fluorescence (both

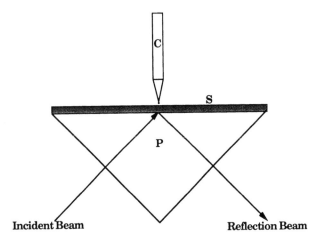

C: Collection Probe;
P: Prism;
S: Sample.

**Figure 7**  Schematic drawing of evanescent light collection mode: C, collection probe; P, prism; S, sample.

intensity and lifetime are reduced). These effects had been observed only when both molecules were neighbors (or collided) inside the same phase (e.g., in liquid solution). The interaction is extremely short range (e.g., ≤5 Å). This effect can be employed at the interface of two distinct phases, that is, the scanning supertip and the sample, say DNA, where certain base atoms are chemically substituted with heavy atoms (e.g., Hg or I). The fluorescing supertip molecules (active center) will be quenched by the substituted DNA base. As a result, the source may emit light at a longer wavelength (i.e, phosphorescence rather than fluorescence). This effect has been suggested for in situ DNA sequencing (see below).

## C.  Single Molecule Sensitivity

As discussed above, nearly a billion photons of the right frequency must hit a highly absorbing molecule to cause a single excitation. Assuming that this molecule also has excellent quantum efficiency and that the detector system requires about 100–1000 emitted photons/s, one needs about $10^{11}$–$10^{12}$ photons/s to emanate from the light source. This is now just possible with an 0.1 μm optical fiber (see above). However, a tough problem still remains—to filter out the billion times stronger excitation light. This calls for very demanding spectral, time, and other optical filtration.

An alternative method of single-molecule excitation [13] has been utilized by nature's photosynthesis and demonstrated in synthetic systems [27]. An antenna made of hundreds, thousands, or millions of dye molecules absorbs the light and transmits the excitation to the desired "active center." The transmission is done via excitation transfer (exciton transport), which usually is just a multi-step Förster or Dexter energy transfer—from one antenna molecule to the next, and eventually from the nearest antenna molecule to the active center (acceptor) molecule.

An antenna of a million molecules needs only about 1000 photons to be excited once. This excitation may be transmitted to the acceptor molecule with 90% efficiency under favorable conditions [28]. Thus, the excitation sensitivity has been improved by five orders of magnitude. Furthermore, the problem of background discrimination has been reduced by a factor of a million. Even under less favorable antenna transfer conditions, the improvement is striking.

In practice, our working scheme is given by Fig. 8. The antenna is a molecular crystallite or aggregate (such as anthracene). One of its molecules ("active center") is closest to the single sample molecule, such as tetracene or perylene. This distance is 10–20 Å and is much smaller than the Förster radius [13]. The antenna (about $10^6$ molecules) transfers excitations to the active center, which in turn transfers energy to the single sample molecule. Overall, only $10^4$–$10^6$ photons are now required for a single excitation of the sample molecule (i.e., $10^3$–$10^5$ times less than with direct excitation).

Switching terminology, the antenna tip is an exciton tip. First the excitons are trapped by the active center. Then the exciton "tunnels" from the tip's active center to the sample molecule, causing the latter to emit a photon. Only about

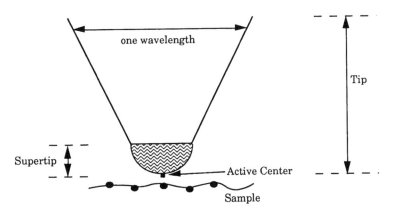

**Figure 8**  Schematic drawing of tip, supertip, and active center: a fiber-optic tapered tip, a crystallite antenna supertip, and a single-molecule active center. The sample shows acceptor molecules or moieties.

one exciton is needed for this single-molecule excitation! However, $10^3$–$10^5$ pho tons were needed to produce that exciton. Still, the exciton approach, exemplifying a direct source–sample interaction, presents a significant improvement in single–molecule sensitivity.

## D.  Single-Molecule Spatial Resolution

In STM, a single atom at the tip is the active center. This is not achieved by atomic precision fabrication but by "natural selection." The tip is produced with ordinary scissors, and the electrons find a defect site of atomic size (see Chapters 5 and 6). If the same principle of defect site applies to the molecular exciton tip, then molecular spatial resolution has been achieved. Such resolution is the aim of present work in the field. Specialized molecular engineering approaches with supermolecules are also envisaged.

Another way of both sensing a single molecule and locating it with high precision is based on the Kasha effect [26]. Here the sample molecule contains a heavy atom that transforms the excitation of the tip's active center molecule. By singlet-to-triplet conversion, the source–sample interaction has effectively transformed the color of the source emission! We believe that this source–sample Kasha effect may yield one of the most sensitive and highest resolution optical near-field methods. We have thus suggested [26,29] its use for in situ gene sequencing. The molecular exciton microscope scans a long DNA strand, after heavy-atom substitution in only one type of base (such as A), and detects these bases individually while scanning along the immobilized strand.

## VI.  NANOSPECTROSCOPY

Microspectroscopy has traditionally complemented optical microscopy. It has developed from visual color observations to high spectral and temporal resolution. NSOM is similarly accompanied by NSOS (near-field scanning optical spectroscopy). Various claims have been made, from the observation of two-color fluorescence spots [15] to the observation of Raman and fluorescence spectra. The latter spectra differ significantly between sample locations that are only 0.25 or 0.1 μm away from each other [30,31]. Figure 9 gives an example of spatially resolved spectra from a molecularly doped polymer film. The experimental setup is given in Fig. 10.

## VII.  SUBWAVELENGTH OPTICAL SENSORS

Microspectroscopy has often been utilized for chemical analysis and biological intracell analysis. For instance, a reagent is introduced into the cell (e.g., with a micropipette under the microscope) and the color or spectrum of the cell

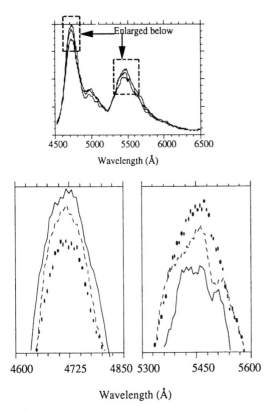

Wavelength (Å)

**Figure 9**  Spatially resolved fluorescence spectra. Top: Fluorescence from a perylene–polymethyl methacrylate film as a result of excitation with 442 nm light via a submicrometer fiber tip (aperture around 200 Å). The fiber tip was positioned piezoelectrically and the spectrum recorded. These spectra correspond to three positions in a line above the sample with 1000 Å separating the successive positions. As the tip is positioned closer to the blue region of the film, the fluorescence intensity in this region steadily increases, while it simultaneously decreases in the yellow. Bottom: Peak maxima in the blue and yellow regions enlarged to show the effect. (Courtesy of D. Birnbaum.)

changes and provides information about the pH or the calcium content of the cell. More recently, fiber-optic chemical sensors (FOCS) have been introduced for such measurements [32]. However, their spatial resolution has been limited by the physical size of the optical fiber, typically 100 μm. A spin-off from recent NSOM technology has been the development of submicrometer, subwavelength FOCS [33,34] (Fig. 11). The chemical preparation of these sensors by photopolymerization is based on near-field optical excitation, which limits the size of the produced probe [33]. In addition, the sensing occurs in the near-field regime

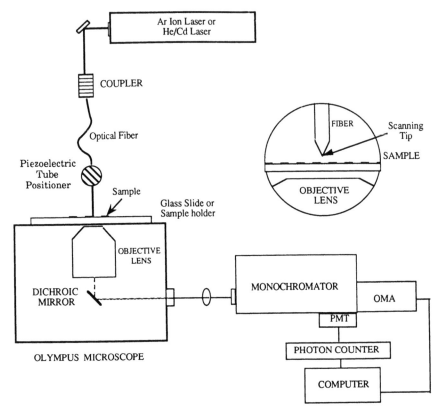

**Figure 10** Schematic drawing of apparatus for near-field scanning optical spectroscopy (NSOS).

of the optical excitation, thus highly increasing the sensitivity per photon and per sensor molecule. This near-field operation has decreased the volume needed for nondestructive analysis to well below a femtoliter [34]. We give here as an example a pH-measuring FOCS.

Submicrometer pH sensors have been tested using Nuclepore porous polycarbonate membranes, such as shown in Fig. 6a. The hole sizes available in these membranes range from 0.02 to 10 μm. For a 10 μm hole membrane, the hole depths are about 6 μm, deep enough to hold pH buffer solutions inside. The membranes are first immersed in a pH buffer solution, then taken out and put on the microscope viewing stage, where the sensor is aligned with a specific hole using the microscope. By positioning the Z translational stage, the sensor could be inserted into one of the holes. The sensors have been tested with different buffers over the pH range of 4 to 9. They operate well even when

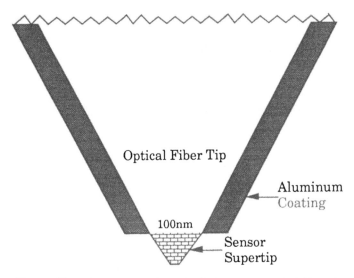

**Figure 11** Schematic drawing of submicrometer optical fiber chemical and biological sensor.

cycled several times from pH 4 to 9 and then back to 4. Figure 12 shows the response measurements over this range (pH 4–9). The fluorescence intensity ranges from about 15,000 counts per second (cps) at pH 4 to about 50,000 cps at pH 9 for a submicrometer sensor. This sensor has very high fluorescence intensity, even though it is extremely small. A major reason is the high collection efficiency in our microscope set-up. Another contributing factor is the excitation efficiency of the submicrometer optical fiber sensor in the nearfield range.

The first biological application of submicrometer FOCS [34] has been demonstrated for 10- and 12-day-old rat embryos [35], placed on the stage of an inverted fluorescence microscope, as shown in Fig. 13. The FOCS consist of an aluminized fiber tip with a copolymer supertip containing the pH-sensitive dye FLAC [33]. One or two laser (Ar$^+$) lines are used for excitation. The emission spectrum is taken with an optical multichannel analyzer (OMA). The analysis is based on ratios of fluorescence intensities at different wavelengths of the same spectrum, or on ratios of fluorescence intensities at different wavelengths of two different spectra obtained by two different excitations (ratios of ratios), providing for internal calibration [34,36]. An internal calibration curve based on fluorescence intensity ratios is shown in Fig. 14. By using internal calibration, the embryonic pH values were 7.55 for 10-day rat embryos and 7.27 for 12-day rat embryos, respectively. These values are in good agreement with the reported results for "homogenized" embryo samples, consisting of more than 1000 crushed embryos. In contrast, only a single embryo was needed for the pH

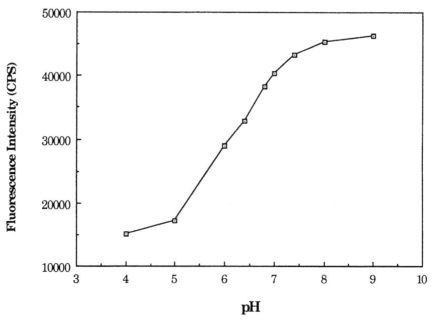

**Figure 12** pH response of a submicrometer optical fiber pH sensor. The submicrometer pH sensor was tested using Nuclepore porous polycarbonate membranes as sample containers.

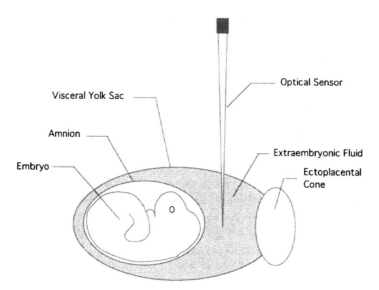

**Figure 13** Schematic representation of the rat conceptus with its associated tissues, showing placement of the ultramicrofiberoptic sensor inside the extraembryonic fluid compartment. (Courtesy of C. Harris and B. Thorsrud.)

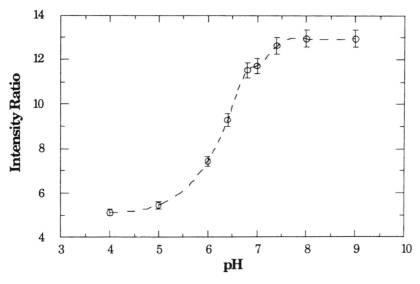

**Figure 14** Fluorescence intensity ratio versus pH. A 20 μm optical fiber pH sensor was excited by 488 nm $Ar^+$ laser at room temperature. The ratio was calculated as $I_{540nm}/I_{610nm}$, where $I_{540nm}$ and $I_{610nm}$ are the fluorescence intensities at 540 and 610 nm from the same fluorescence spectrum, respectively.

measurements via miniaturized FOCS. In addition, chemical dynamic alterations in pH of intact rat conceptuses have been measured in response to several variations in their environmental conditions [19,37]. The results are reasonable and self-consistent. This is the first time such an experiment has been done on a single and live embryo. The ability of the sensors to measure pH changes, in real time, in the intact rat conceptus, demonstrates their potential applications for dynamic analysis in small multicellular organisms and single cells.

Compared to the conventional devices [32], a thousandfold miniaturization of immobilized optical fiber sensors, a millionfold or more sample size reduction and at least a hundredfold shorter response time have been achieved by combining nanofabricated optical fiber tips with near-field photopolymerization. Also, the submicrometer sensors have improved the detection limits by a factor of a billion [36]. These miniaturized optical sensors have potential applications in spatially and temporally resolved chemical analysis and kinetics, inside single biological cells and multicellular organisms.

## VIII. FEEDBACK AND MULTIFUNCTION PROBES

All modern scanning probe microscopies contain feedback as an essential part. Feedback keeps the probe close enough to the sample but prevents "crashing" into it. Traditional NSOM lacked feedback except for the topographical scanning and collection method described in Section IV. Recently, a number of feedback methods have been applied. Lewis and Lieberman [20] have used STM feedback between the metallic coating of the micropipette and the sample. Obviously, this requires the sample to be an electrical conductor. In addition, the optical and electrical probing are centered away from each other, roughly by the probe radius. Quade [38] has microfabricated a combined AFM/NSOM probe. However, its resolution is low, 5 μm. More recently, Lewis and his coworkers developed bent micropipettes that also serve as combined AFM/NSOM probes and have submicrometer resolution [private communication].

Very recently, a lateral ("dithering") force feedback has been developed [39,40] in which the micropipette or optical fiber tip is dithered piezoelectrically. This lateral swinging modulates the light reaching the detector (see below). The modulation frequency is related to the resonating frequencies of the tip. As the tip comes within the (attractive) van der Waals force field of the sample, this frequency changes. A typical dithering amplitude is 2–5 nm. This method appears to be most promising. A schematic presentation is given in Fig. 15.

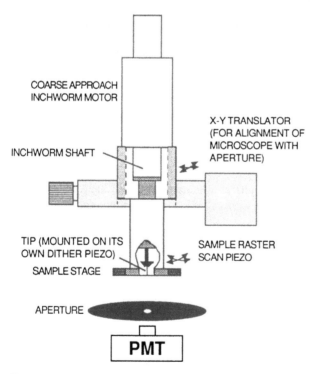

**Figure 15** Lateral force–NSOM feedback scheme. (Courtesy of G. Merrit and S. Smith.)

## IX.  APPLICATIONS, PRESENT AND FUTURE

The following advances toward nanometer-resolved microscopy, spectroscopy, and chemical sensor probes promise to push chemical analysis much closer to one of its ultimate goals—the noninvasive detection of a single molecule, radical, or ion, and the determination of its precise coordinates and the characterization of its structural conformation, as well as its internal dynamics and energetics, as a function of time and environmental perturbations.

### A.  Chemical Sensors

The first applications of micrometer and submicrometer pH sensors (Section VII) have been made in toxicology—the effects of biological growth and chemical environment on the pH of rat embryos [34,36,37]. Following the in vivo pH of a single embryo in time is much preferable to using a thousand embryos at every time point, "homogenizing" them, and taking pH readings in vitro. We expect much work along this direction. The improvement of the spatial resolution by a factor of about a thousand, the sensitivity by a factor of about a billion, and the response time by two or more orders of magnitude promises a busy future. Work on other sensors such as oxygen and calcium is in progress.

### B.  Optical Microscopy

The most obvious applications of optical microscopy are in the direction of chemical nanoanalysis as well as in the biological, environmental, and medical fields. Some first steps have been made in the direction of biological applications [15]. More work is in progress in a number of labs, including our own. The possibility of monitoring processes in real time is most promising. Examples are the intracell motility and the very slow dynamics of DNA molecules (second and subsecond time scale). Schemes for higher spatial resolution and chemical identification have been mentioned above in the context of in situ gene sequencing.

### C.  Nanospectroscopy, Quantum Optics, and Nanolithography

Applications to spatially resolved Raman and particularly fluorescence microscopy have also been mentioned (Section VI). This again opens the door to a wide spectrum of applications in biology as well as in molecular materials and surface science [41,42]. Further applications are envisioned for solid state physics and quantum optics and electronics [42,43,44]. Nanometer light and exciton sources have also been used in preliminary probe-to-sample distance regulated, Förster energy transfer studies; probe-to-sample interfacial Kasha effect (external heavy atom effect) studies and supertip development for near-field scanning optical microscopy and molecular exciton microscopy [19,43,44]. Technological applications of nanometer light sources and photochemical probes to nanolithography [20,25] and optical memories [15] have been mentioned as well.

## ACKNOWLEDGMENTS

We first thank Michael Morris for stimulating us to write this chapter at the present exploding stage of this field. We thank our colleagues at Michigan, who are partners in many of the described enterprises: Duane Birnbaum, Eric Evans, Craig Harris, John Langmore, Vladimir Makarov, Greg Merrit, Eric Monson, Brad Orr, Steve Parus, Zhong-you Shi, Michael Shortreed, Steve Smith, and Bjorn Thorsrud. We also thank our collaborators in Jerusalem—Aaron Lewis and Klony Lieberman. We further acknowledge useful discussions with our colleagues Dan Axelrod and Mark Meyerhoff at Michigan; Eric Betzig and Jay Troutman at ATT Bell; Bruce Warmack at Oak Ridge Labs, and David Walt at Tufts. Finally, we acknowledge the support and encouragement from Jerry Goldstein and Matesh Varma from the Office of Health and Environmental Research, U.S. Department of Energy. Financial support came from DOE grants DE-FG02-90ER60984 and 61085. The excitonic light source research and development was supported by National Science Foundation grant DMR-9111622; we thank David Nelson, Office of Solid-State Chemistry and Polymers, for his interest.

## REFERENCES

1. Abbe, E. (1873). *Arch. Mikroskop. Anat. 9*:413.
2. Synge, E. H. (1928). A suggested method for extending microscopic resolution into the ultra-microscopic region, *Phil. Mag. 6*:356.
3. O'Keefe, J. A. (1956). Resolving power of visible light, *J. Opt. Soc. Am. 46*:359.
4. Ash, E. A. (1972). Super-resolution aperture scanning microscope, *Nature, 237*:510.
5. Lewis, A., Isaacson, M., Muray, A., and Harootunian, A. (1984). Development of a 500 Å spatial resolution light microscope, *Ultramicroscopy, 13*:227.
6. Pohl, D. W., Denk, W., and Lanz, M. (1984). Optical stethoscopy: Image recording with resolution $\lambda/20$, *Appl. Phys. Lett. 44*:651.
7. Lieberman, K., Harush, S., Lewis, A., and Kopelman, R. (1990). A light source smaller than the optical wavelength, *Science, 247*:59.
8. Kopelman, R., Lewis, A., and Lieberman, K. (1989). Exciton microscopy and scanning optical nanoscopy, in *X-Ray Microimaging for the Life Sciences* (Attwood, D., and Barton, B., Eds.), Lawrence Berkeley Laboratory, Berkeley, CA, p. 166.
9. Lewis, A., Betzig, E., Harootunian, A., Isaacson, M., and Kratschmer, E. (1988). Near-field imaging of fluorescence, in *Spectroscopic Membrane Probes*, Vol. II (Loew, L. M., Ed.), CRC Press, Boca Raton, FL, p. 81.
10. McDonald A. (1972). Electric and magnetic coupling through small apertures in shield walls of any thickness, *IEEE Trans. Microwave Theory Technol. MTT-20*:698.

11. Kopelman, R., and Lewis, A. (1992). A nanometer dimension optical device with microimaging and nanoillumination capabilities, U.S. Patent 5, 148, 307.

12. Betzig, E., Isaacson, M., and Lewis, A. (1987). Collection mode near-field scanning optical microscopy, *Appl. Phys. Lett.* *51*:2088.

13. Kopelman, R., Lewis, A., Lieberman, K., and Tan, W. (1991). Evanescent luminescence and nanometer-size light source, *J. Lumin.* *48/49*:871; Kopelman, R., Lewis, A., and Lieberman, K. (1990). Subwavelength molecular optics: The world's smallest light source? *Mol. Cryst. Liq. Cryst.* *183*:333.

14. Reddick, R. C., Warmack, R. J., Chilcott, D. W., Sharp, S. L., and Ferrell, T. L. (1990). Photon scanning tunneling microscopy, *Rev. Sci. Instrum.* *61*:3669.

15. (a)Betzig, E., Troutman, J. K., Harris, T. D., Weiner, J. S., and Kostelak, R. L. (1991) Breaking the diffraction barrier: Optical microscopy on a nanometric scale, *Science*, *251*:1468. (b) Betzig, E., and Troutman, J. K. (1992). Near-field optics: Microscopy, spectroscopy, and surface modification beyond the diffraction limit, *Science*, *257*:189.

16. Kopelman, R., Lewis, A., and Lieberman, K. (1989). Molecular exciton microscopy, *Biophys. J.* *55*:450a.

17. Sakmann, B., and Neher, E., Eds. (1983). *Single-Channel Recording*, Plenum Press, New York.

18. Thomas, R. C., (1978). *Ion-Sensitive Intracellular Microelectrodes*, Academic Press, New York.

19. Tan, W., (1993). Nanofabrication and applications of subwavelength optical probes: Chemical and biological sensors, light sources and exciton probes, Ph. D. Thesis, The University of Michigan, Ann Arbor, Michigan.

20. (a) Lewis, A. and Lieberman, K. (1991). The optical near field and analytical chemistry, *Anal. Chem.*, *63*:625A. (b) Lewis, A., and Lieberman, K. (1991). Near-field optical imaging with a nonevanescently excited high-brightness light source of sub-wavelength dimensions, *Nature*, *354*:214.

21. Born, M., and Wolf, E. (1959). *Principles of Optics*, Pergamon Press, London.

22. Agranovich, V. M., and Galanin, M. D. (1982). *Electronic Excitation Energy Transfer on Condensed Matter*, North Holland, Amsterdam.

23. Aavikso, J. Freiberg, A., Lipmaa, J., and Reinot, T. (1987). Time resolved studies of exciton polaritons, *J. Lumin.* *37*:313.

24. Troutman, J. K., Betzig, E., Weiner, J. S., DiGiovani, D. J., Harris, T. D., Hellman, F., and Gyorgy, E. M. (1992). Image contrast in near-field optics, *J. Appl. Phys.* *71*:465.

25. Herman, B., and Jacobson, K. (1989). *Optical Microscopy for Biology*, Wiley-Liss, New York.

26. Kopelman, R. (1991). Exciton microscopy and reaction kinetics in restricted spaces, in *Physical and Chemical Mechanisms in Molecular Radiation Biology* (Glass, W. A., and Varma, M., Eds.), Plenum Press, New York: *Basic Life Science*, series, Vol. 58, p. 475.

27. Fauman, E. B., and Kopelman, R. (1989). Excitons in molecular aggregates and a hypothesis on the sodium channel gating, *Mol. Cell. Biophys.*, *6*:47

28. Francis, A. H., and Kopelman, R. (1986). Excitation dynamics in molecular solids, *Topics in Applied Physics*, Vol. 49, *Laser Spectroscopy of Solids*, 2nd ed. Yen, W. M. and Selzer, P. M., Eds.), Springer-Verlag, Berlin, p. 241.

29. Kopelman, R., Langmore, J., Orr, B., and Makarov, V. (1992). Scanning molecular exciton microscopy: A new approach to gene sequencing, *Human Genome Program Reports (1991–1992)*, U.S. Department of Energy, Washington, D.C., pp. 130–131.

30. Warmack, R. J., and Ferrell, T. L. (1992). Private communication.

31. Kopelman, R., Smith, S., Tan, W., Zenobi, R., Lieberman, K., and Lewis, A. (1992). Spectral analysis of surfaces at subwavelength resolution, *SPIE, Int. Soc. Opt. Eng. 1637*:33.

32. Seitz, W. R. (1988). Chemical sensors based on immobilized indicators and fiber optics, *CRC Crit. Rev. Anal. Chem. 19*:135.

33. Tan, W., Shi, Z. -Y. and Kopelman, R. (1992). The development of submicron optical fiber chemical sensor, *Anal. Chem. 64*: 2985–2990.

34. Tan, W., Shi, Z. -Y., Smith, S., Birnbaum, D., and Kopelman, R. (1992). Submicrometer intracellular chemical optical fiber sensors, *Science 258*: 778–781.

35. Nau, H., and Scott, W. J. (1987). Teratogenicity of valproic acid and related substances in the mouse: Drug accumulation and $pH_i$ in the embryo during organogenesis and structure–activity considerations, *Mech. Models Toxicol. Arch. Toxicol., Suppl. 11*:128.

36. Kopelman, R., Tan, W., and Shi, Z. -Y. (1992). Nanometer optical fiber pH sensor, *SPIE, Int. Soc. Op. Eng. 1796*:24.

37. Tan, W., Thorsrud, B., Harris, C. and Kopelman, R. (1993) Organogenesis and teratogenic insults: Real time pH measurements in the intact rat conceptus using ultramicrofiberoptic sensors, (to be published).

38. Quate, C. F. (1986). Vacuum tunneling: A new technique for microscopy, *Phys. Today*, August, p. 26.

39. Toledo-Crow, R., Chen, Y., and Vaez-Iravani, M. (1992). An atomic force regulated near field scanning optical microscope, *SPIE, Int. Soc. Opt. Eng. 1639*:44.

40. van Hulst, N. F., Moers, M. H. P., Noordman, O. F. J., Faulkner, T., Segerink, F. B., van dert Werf, K. O., de Grooth, B. G., and Bolger, B. (1992). Operation of a scanning near field optical field microscope in reflection in combination with a scanning force microscope, *SPIE, INT. Soc. Opt. Eng. 1639*:36.

41. Birnbaum, D., Kook, S. K. and and Kopelman, R. (1993). Near-field Scanning Optical Spectroscopy: Spatially resolved spectra of micro-crystals and nano-aggregates in doped polymers, *J. Phys. Chem.*, (in press).

42. Smith, S., Monson, E., Merrit, G., Tan, W., Birnbaum, D., Shi, Z-Y., Thorsrud, B. A., Harris, C., Grahn, H. T., Ploog, K., Merlin, R., Orr, B., Langmore, J. and Kopelman, R. (1993). Tip/Sample Interactions, Contrast and Near-Field Microscopy of Biological Samples, in Scanning Probe Microscopies, *SPIE*. The International Society for Optical Engineering, Vol. 1858,13.

43. Kopelman, R., Tan, W., Shi, Z-Y. and Birnbaum, D. (1993). Near Field Optical and Exciton Imaging, Spectroscopy and Chemical Sensors, ed. D. Pohl, NATO Workshop on Near Field Optics, Besancon, France (in press).

44. Tan, W. and Kopelman, R. (1993). Energy Transfer, Nanometer Crystals and Optical Nano-probes, in Dynamics in Small Confining Systems, edited by J. M. Drake, J. Klafter and R. Kopelman, *Proceedings of Materials Research Society*, Pittsburgh, PA (in press).

# 8

# Photoacoustic and Photothermal Imaging

**Joan F. Power**   *McGill University, Montreal, Quebec, Canada*

## I.  INTRODUCTION

Over the past decade, photoacoustic and photothermal phenomena have generated a class of scanning probe techniques that have been used in an extremely
wide variety of applications [1–5]. If a beam of intensity-modulated optical radiation is focused onto the surface of a solid sample, nonradiative decay processes following light absorption will generate in the sample a spatially damped
oscillating temperature field known as a thermal wave [5]. Because the damping
distance of thermal waves is very short range in most materials and because this
distance is readily controlled by varying the modulation frequency of the optical
beam, subsurface or buried features in a thin sample may be visualized through
their interaction with thermal waves. Photoacoustic and photothermal techniques that use thermal waves to probe materials have thus been employed in a
wide range of problems involving the recovery of depth-resolved information
from thin samples [6–8]. The scale of depth profiling available from photothermal methods ranges from a few hundred nanometers to a few millimeters [6],
depending on the thermal properties of the sample. This depth range is not accessible to other spectroscopic techniques on a nondestructive basis. Applications have been reported in biology and biophysics, nondestructive evaluation,
semiconductor and thin film imaging, polymer analysis, and materials science
and include the imaging of ion implantations in semiconductors [10], the inspection of welds and metallurgical bonds [11], the evaluation and detection of subsurface defects in paints [12,13] and coatings [14–17], and the evaluation of
thermal anisotropies in polymer films and injection-molded plastics [18]. A

**255**

number of major reviews and monographs detailing developments in photothermal technology have been reported over the last 5 years, and new applications continue to be reported [1–4,6,7].

## II. THERMAL WAVE GENERATION AND DETECTION

First, we clarify the meaning of the terms "photoacoustic" and "photothermal" as used in this chapter. *Photothermal* is the term applied to methods in which spectroscopic information is carried by a thermal wave. *Photoacoustic* measurements specifically involve the generation of sound as the result of light absorption. Sound generation occurs via (1) adiabatic expansion of the sample immediately following excitation or (2) the heating and cooling of a fluid layer adjacent to the sample via conduction from the heated sample. Photothermal techniques use heat conduction, primarily, as a carrier of spectroscopic and thermal subsurface information. Radiative heat transfer is exploited by some techniques (photothermal radiometry), but the relative heat loss introduced by this mechanism is very small. Convective heat transfer occurs on too slow a time scale to be useful.

The heat conduction equation is the basic mathematical relationship that describes the propagation of thermal waves in matter:

$$\nabla^2 T = \frac{1}{\alpha} \frac{\partial T}{\partial t} - \dot{q}(z,t) \tag{1}$$

where $\alpha$ is the thermal diffusivity of the sample, $T$ is the temperature, and $\dot{q}(z,t)$ is a spatially distributed heat flux source. In the sample, a heat source is present due to light absorption, but in adjacent layers such as the gas phase, no source is present: $\dot{q}(z,t) = 0$, and all heating of the layers occurs via conduction from the sample boundaries. Equation (1) clearly neglects the contribution of acoustic waves, which are generated at very early times past application of the heat flux source.

The spatial dependence of the temperature profiles established in the sample depends on the time dependence of the heat flux source $\dot{q}(z,t)$. If the heat flux source is periodically modulated, the temperature profile measured at all points in the sample has a sinusoidal time dependence and a spatial dependendence of the form shown in Fig. 1. In this example, the thermal wave is generated in the medium by periodic heating at the surface ($z = 0$ in Fig. 1).

Depending on the instantaneous phase of the heating cycle, $\phi = \omega t$, where $\omega$ is the angular frequency, the temperature profile shows relative heating or cooling with distance in a thin layer near the surface. The thickness of this region of significant temperature change is effectively given by the so-called thermal diffusion length $\mu$; in the sample, $\mu = (2\alpha/\omega)^{1/2}$. The key feature of the

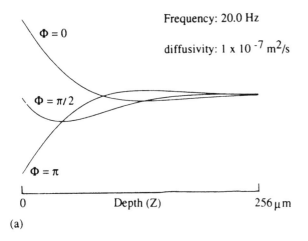

(a)

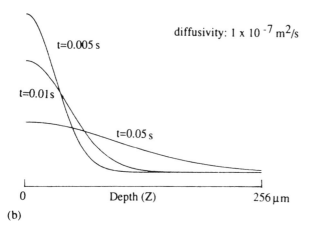

(b)

**Figure 1** Spatial dependence of (a) thermal wave propagation in the frequency domain and (b) corresponding impulse response.

frequency domain thermal wave is that it is a critically damped phenomenon. The subsurface penetration depth of such a wave does not significantly exceed a distance of $2\pi\mu$, which is controlled by varying the modulation frequency.

Time domain thermal waves may also be produced in solids by transient heating. A commonly encountered excitation waveform is the impulse or short pulse. Figure 1b shows the theoretical subsurface temperature response expected when the front surface of the sample is heated by an instantaneous unit pulse at time $t = 0$. The temperature profile consists of a spatial Gaussian whose peak width $\mu(t)$ varies as a function of time: $\mu = \sqrt{4\alpha t}$. This quantity is the time domain analog of the thermal diffusion length, giving the effective depth of penetration of the thermal wave as a function of time past excitation. The thermal transit time $\tau_p$ is the time required for heat conduction from a buried line source (as in Fig. 1b) to the sample surface, given by $\tau_p = d^2/4\alpha$, where $d$ is the depth of the absorber. The thermal transit time will drop rapidly for short distances and in good conductors, where $\alpha$ is large.

A number of instrumental configurations exist for the generation and detection of thermal waves. Figure 2a summarizes some of the most commonly reported examples.

In principle, a wide frequency range of electromagnetic radiation from the radiofrequency (rf) to x-rays is adaptable to photoacoustic/photothermal measurement. Particle beams may also serve as sources. It is assumed here that nonradiative decay processes occur instantaneously relative to the time scale of heat conduction. This is a reasonable assumption for many materials in which no significant luminescence or photochemical reactions occur. Table 1 summarizes the performance characteristics of the various measurement schemes.

Heat generation in the sample produces a number of physical effects. The surface of the sample may expand or buckle [19], and the surface reflectance may change through a weak dependence on temperature [20]. The reflectance change may be monitored by a low intensity probe beam aligned collinear with the excitation beam. The surface displacement of the sample may be detected by aligning the probe beam at the inflection point of the thermal bump. These detection modes, photothermoreflectance (PTR) and photothermoelastic effect (PTE) detection [20,21], are extremely wide bandwidth detection techniques, used commercially for nondestructive evaluation of semiconductors.

Additional modes of detection are furnished by heat conduction from the sample surface into the gas phase. Such heat flow produces a thermal lens (if the beam is focused) or a "mirage," effectively, a one-dimensional thermal lens (if the beam is defocused) in the gas phase above the sample [22]. Surface mirages may be monitored through the deflection of a probe beam, which is aligned parallel to the surface, termed "mirage effect" or "photothermal deflection" spectrometry (PDS).

In PDS the measured signal is proportional to the deflection of the probe beam. The beam deflection may occur in a direction normal (N-PDS) or parallel to the sample surface (transverse or T-PDS). The signal in N-PDS is proportional to the spatial gradient of the refractive index, and therefore of the temperature above the sample surface. The T-PDS signal measures transverse heat flow and is therefore sensitive to thermal anisotropies in the sample. In most cases, the PDS signal contributions are due to thermal variations in the gas phase refractive index, but modulated displacements of the sample surface may contribute a thermoelastic component in samples with low optical absorbance or in rubbery materials that deform significantly with heating.

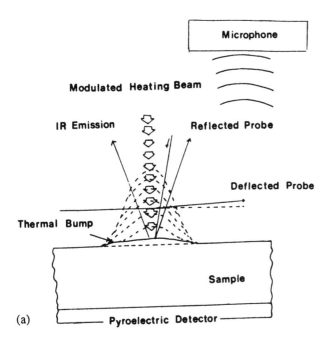

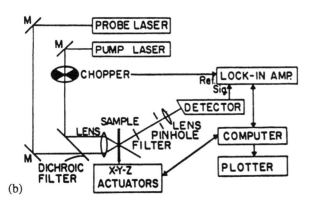

**Figure 2** (a) Schematic of processes occurring in a sample heated by a modulated beam of radiation. (b) Optical geometry used by the high resolution crossed-beam thermal lens microscope. (From Ref. 33.) Reproduced with permission from Pergamon Press PLC.

**Table 1**  Photoacoustic and Photothermal Detection Methods

| Method | Method signal | Sensitivity $(K/Hz^{1/2})$[a] | Minimum response time or maximum bandwidth (approximate) |
|---|---|---|---|
| Gas microphone photoacoustic spectroscopy | Acoustic pressure change | $10^{-5}$ | 10 kHz |
| Mirage effect/ optical beam deflection | Beam deflection (by thermooptical gradient) | $10^{-7}$ (medium, geometry dependent) | 500 kHz |
| Infrared photothermal radiometry | Infrared blackbody emission (integrated) | $10^{-4}$ | Several megahertz (common detectors) |
| Photothermo reflectance | Temperature-induced change in surface reflectivity | 1 | >1 GHz |
| Photothermoelastic effect detection | Temperature-driven surface displacement | $<10^{-12}$ (interferometer detection) | >GHz |
| Photopyroelectric effect detection | Average temperature change in thin film sensor | No figures available | <1 ns (dependent on sample thickness) |

[a]Figures obtained from Boccara and Fournier, p. 486 in Ref. 2.

Heat conduction into the gas layer from the sample surface also causes the modulated expansion of a boundary layer of gas. The periodic rarefactions produce a sound wave, which may be detected with a sensitive microphone. This "gas microphone photoacoustic spectroscopy" was the earliest reported photothermal wave technique [5]. Of all the thermal wave based models, it is probably the most familiar to chemists. Commercial photoacoustic cells are available for several Fourier transform infrared (FT-IR) systems.

Heat that propagates below the sample surface produces infrared emission that may be collected by lenses and focused onto a sensitive infrared detector. This infrared photothermal radiometry (PT-IR) is used in the remote inspection of materials [23,24] because it is completely contactless, and the detection system is relatively insensitive to alignment instabilities. Bandwidths of several megahertz or greater are available.

Heat transmission by the sample may be monitored by contacting a thin film pyroelectric sensor to the sample's rear surface and monitoring the temperature changes in the pyroelectric [25–27]. This mode of detection, "photopyroelectric effect spectrometry," furnishes a general technique for measuring the optical spectrum, thickness, or thermal diffusivity of a thin film. It has been used in a number of recent depth profiling applications of the optical [28–30] properties of thin samples.

Thin film polyvinylidene fluoride pyroelectric sensors are sensitive and have rise times of nanoseconds or better. The bandwidth of photopyroelectric effect spectrometry actually depends on the thickness of the sample contacted to the temperature sensor, in addition to the sensor's electrical bandwidth. While the sensitivity and simplicity of this technique are unrivaled, a key weakness is that contact of the sample to the sensor is required.

Another mode of detection is provided through the generation of elastic waves in a solid as a result of heating. These elastic waves propagate at speeds determined by the Lamé constants of the material and may be detected at the sample's rear surface by a piezoelectric transducer or via interferometry [31]. If the sample is homogeneous with respect to its elastic properties, image and/or depth profile information will be dominated by thermal wave generation in the sample. While semiconductors may obey this assumption to a good approximation [19], elastic inhomogeneities are present in most materials and may contribute to the image or depth profile [32]. Such images are difficult to interpret quantitatively, and image contrast mechanisms have been under considerable debate. We consider cases in which the image contrast is explainable on the basis of heat conduction mechanisms only.

The detection mechanisms diagrammed in Fig. 2a are all adaptable to optically single-ended measurements. They all may be used to study opaque materials. Depth profile information is obtained through the time or frequency dependence of the photothermal signal rather than through spatial localization of the heating beam. If single-ended detection may be sacrificed, the configuration of Fig. 2b, crossed-beam thermal lens effect microscopy [33,34], may be used. A focused pump beam generates a highly localized thermal gradient. A low intensity probe beam aligned at 90° to the pump beam, is defocused by this thermal lens. The detection volume may be made as small as 150 attoliters (= $150 \times 10^{-18}$L) [34]. Alternatively, the thermal gradient may be used to deflect a probe beam aligned at a shallow angle to the heating beam (not shown) [35]. A depth profile is recovered by scanning the beam intersection region over the depth of the sample.

Two classical excitation and signal processing strategies have been used with photothermal data, namely pointwise single-frequency measurements and impulse methods of detection.

The single-frequency method excites the sample with a radiation beam, which is sinusoidally modulated at a fixed frequency. The magnitude and phase of the sample's photothermal response is recorded at the modulation frequency using a lock-in amplifier. If the measurement is repeated at different frequencies, a depth profile of the sample's subsurface properties may be recovered. The advantages include large dynamic range, rejection of photothermal nonlinearities, and rejection of wideband noise. But, multifrequency measurements are time-consuming, with this method.

In principle, the pulsed method of detection superimposes all the excitation Fourier components in phase at zero time and therefore excites the sample with all the modulation frequencies within the sample's response bandwidth simultaneously [36]. The impulse method is faster than single-frequency measurements and has high resolution [37]. The main disadvantage is high peak power, which may cause optical or thermal damage to the specimen under test. Other disadvantages include poor measurement dynamic range and a susceptibility to nonlinearities and wide band noise.

More recent approaches [37–40] involve the use of wide band excitation and signal recovery techniques. These approaches involve the excitation of the photothermal system with waveforms, which, like the impulse, deliver uniform energy to the photothermal system over its response bandwidth, to give impulse (time domain) and frequency response information simultaneously via correlation and spectral analysis. They possess many of the advantages of both the time and frequency domain methods, including speed, low peak power, and high resolution. Their main drawbacks at present include complexity and expense of the specialized measurement apparatus.

## III. PRINCIPLES OF THERMAL WAVE DEPTH PROFILING

Thermal waves may be used to image objects of microscopic, mesoscopic, or macroscopic dimensions [41,42]. Objects that are thermally microscopic possess transverse length dimensions of the same order as the thermal diffusion length. Macroscopic objects, on the other hand, have transverse dimensions many orders of magnitude greater than $\mu$. In the macroscopic case, one-dimensional thermal waves may be used to interrogate the subsurface properties of the sample. The analysis of images generated using such waves is considerably simpler than the corresponding three-dimensional case.

A key factor that affects the propagation of thermal waves in a nonhomogeneous sample is the interfacial thermal efflux ratio $b_{ij}$ [5,19]. This parameter determines whether thermal waves are absorbed or reflected at an interface between two media (indexed $i$ and $j$) where the thermal properties change. The thermal efflux ratio $b_{ij}$ is obtained as the ratio of the thermal efflux values in two adjacent media $b_{ij} = (\kappa_i/\alpha_j^{1/2})/(\kappa_j\alpha_i^{1/2})$, where $\kappa$ is the thermal conductivity and $\alpha$ is the thermal diffusivity.

The effects of thermal reflection and absorption at an interface are easily seen in the time domain [43]. A simple three-layer model of Fig. 3a illustrates these points. The sample is assumed to be irradiated with an impulse heat flux source whose transverse dimensions are large compared to the thickness. An infinitely thick backing layer (3) is contacted to the sample. As the properties of this backing layer change, the reflection or absorption of thermal waves occurs at the rear interface, producing a measurable change in the temperature–time profile measured at the air–sample interface.

The extent of absorption or reflection at a thermal interface may be evaluated quantitatively by means of the thermal reflection coefficient:

$$\zeta = \frac{1 - b_{ij}}{1 + b_{ij}} \tag{2}$$

This parameter has a value in the range $-1 \leq \zeta \leq 1$. If $\zeta = 0$, thermal matching is maintained across an interface. On the other hand, for $\zeta = +1$, thermal waves are fully reflected at the interface, while if $\zeta = -1$ they are fully absorbed.

If the backing layer is highly insulating relative to the sample ($b_{32} \gg 1$), heat flux across this interface approaches zero, and the temperature profile is fully reflected at the rear interface of the sample, causing an increase in the front surface temperature to be observed, over and above the value expected for the case of an infinitely thick sample. The reflected temperature profile, returned to the front surface, encounters the gas–solid interface (where $b_{21} \gg 1$ for a wide range of solids and gases).

Relatively soon after the application of the heating pulse to the sample, the impulse response traces show no thickness dependence because negligible energy reaches the backing interface in this time range. At later times, the contribution of energy returned to the surface causes the surface temperature to increase above the response of the semi-infinite sample. As the sample thickness decreases (at constant diffusivity), this deviation occurs at increasingly early time. At sufficiently long times past excitation, the surface temperature decays to a steady state background level (rather than to zero). This feature arises as a result of the thermal wave trapping effect and demonstrates that in thin layers, a significant number of multiple reflections between surfaces offset the effect of thermal damping in time.

Absorption is observed in this system if the backing is an excellent conductor relative to the sample ($b_{32} \ll 1$). Thermal energy reaching the rear surface of the sample is preferentially conducted into the backing layer. As the sample thickness decreases, damping contributes to the impulse response at increasingly earlier times.

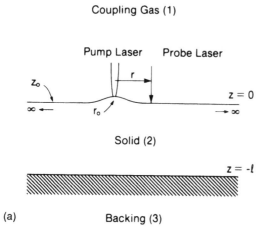

Coupling Gas (1)

Solid (2)

(a)                    Backing (3)

**Figure 3** (a) Three-layer heat conduction model for surface temperature impulse response. (b) Theoretical predictions of the effect of sample thickness on the one-dimensional surface temperature of a sample with perfectly insulative backing. Sample thicknesses were 22, 10, and 7 $\mu$m for curves (1), (2), and (3), respectively; $\alpha = 1 \times 10^{-7}$m$^2$/s. (c) Theoretical predictions of the effect of sample thickness on the one-dimensional surface temperature of sample with perfectly conductive backing. Sample thicknesses were 100, 5, 3, and 1 $\mu$m from curves (1), (2), (3), and (4), respectively.

Very complex patterns of interfacial reflections are possible in samples composed of multiple layers. Currently available theoretical models may be applied to systems in which the variation of thermal and/or optical properties is either discrete or continuous. Samples composed of discrete layers frequently use a forward modeling approach, while samples containing a continuous subsurface variation of thermal/optical properties use an inverse approach.

In the forward approach, the heat conduction equation is solved for a sample of arbitrary layer structure and thermal parameters. The temperature profile is compared with the experimental response and iteratively refined. This approach assumes some prior knowledge of the sample structure; otherwise, the time required to explore the model space becomes prohibitively high. Furthermore, there is no way to determine the number of different models that could generate the same experimental response profile within experimental error. Forward modeling of heat conduction in one-dimensional (layered) structures has been thoroughly studied by [19,28,44,45].

The earlier theory of Ref. 44 is applicable to a frequency domain description of one-dimensional and simple three-dimensional thermal waves (in which the transverse variation of the thermal properties of the sample was slow compared to a thermal diffusion length) in multilayered media. Aamodt, MacLachlan-

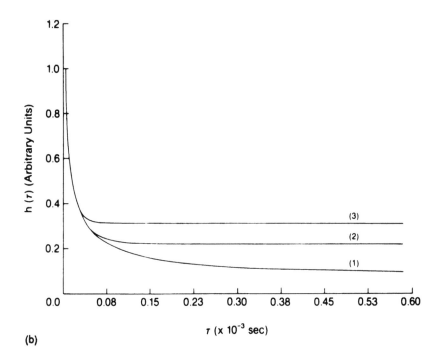

(b)

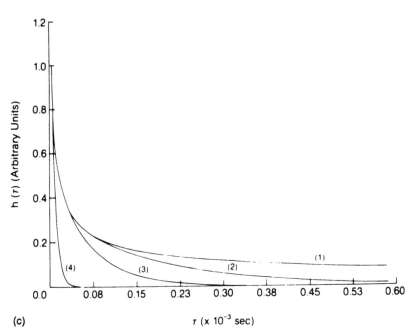

(c)

Spicer, and Murphy [45] have generalized the one-dimensional theoretical description of thermal waves to the time domain, for systems composed of arbitrary multilayers excited with different types of transient waveform.

Additional approaches to theoretical modeling include numerical simulations. Finite difference and finite element simulations have traditionally provided means of modeling heat conduction in complex structures [46]. Several authors have used numerical modeling to interpret photothermal signal generation in one-dimensional structures [47–49]. Although they do not yield the insight provided by a mathematical theory, numerical models remain the preferred methods for describing more complex three-dimensional structures, where the variation in thermal properties is rapid compared to a thermal diffusion length [50].

In systems that exhibit a continuous variation of thermal properties with depth or spatial position, inverse problem theory provides a means of interpreting experiments. Many inverse problems are ill-posed however, and a reliable recovery of the depth dependence of thermal properties is not always ensured at realistic levels of experimental noise. Recently, Mandelis et al. reported an inversion technique based on an earlier Hamilton–Jacobi [51] model of thermal wave propagation in condensed media [52,53] and have successfully modeled magnetic-field-dependent spatial inhomogeneities in the thermal properties of liquid crystals.

A problem of more interest to spectroscopists is the recovery of a depth profile of the optical properties of a thin layer. Several different classes of samples exist, in which the optical absorption coefficient varies with depth, while the thermal efflux of the sample is nearly constant. Examples include tissue specimens (where the sample properties are dominated by those of water) [54,55], polymer laminates of thin layers of similar thermophysical properties [28–30], and thin polymer films containing nonuniform concentrations of absorbing impurities [56]. In some diffusion problems, spatial concentration profiles of strongly absorbing impurities may be established in a thin substrate without significantly affecting its thermal properties.

Impulse photothermal measurements provide excellent demonstrations of optical depth profiling in multilayers [57]. The three-layer structure of Fig. 4 was imaged using gas microphone photoacoustic spectroscopy (PAS). The specimen was irradiated with a wide band modulated waveform, and an approximation to the impulse response was recovered at several time delays past excitation. A focused beam scanned across the surface yielded the images of Fig. 4. The photoacoustic method in this case records a signal proportional to surface temperature. The time delay for emergence of buried features in the recorded thermal wave image is consistent with the subsurface depth of the absorber. At 0.21 ms only surface features contribute to the image. But at 2.3 ms delay the thermal contribution from the middle layer arrives at the front surface. Finally,

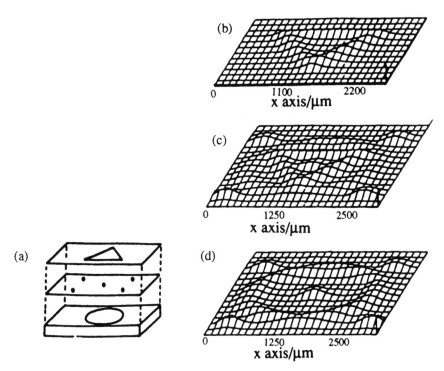

**Figure 4** Correlation photoacoustic spectroscopic (CPAS) imaging of a three-layer sample from Sugitani and Uejima [57]. Thermal wave images were recovered at time delays of 0.21 ms (top trace), 2.3 ms (middle trace), and 3.7 ms past excitation at $t = 0$. (Reproduced with permission.)

at the longest time delay, the rear surface contribution is detected and the front surface feature has decayed.

In frequency domain depth profiling, the phase of a photothermal signal is especially sensitive to the distance and distribution of chromophores in the sample layer [30]. This is true in general, because the photothermal phase measures the effective distance traveled by a thermal wave within a half-period of the modulation. Thermal delays that are a fraction of the modulation period will produce a measurable phase shift in the output. Unlike the impulse response, the phase measurement is independent of the absolute amount of energy delivered to the system.

Phase versus sample thickness relationships are summarized in Table 2 for some simple geometries. These expressions show how the absolute value of the surface temperature's phase varies with sample thickness and/or thermal diffusivity. In these expressions, thermal reflections at the various surfaces are

**Table 2**  Phase Versus Distance Relationships for Some Simple Geometries

(a)

$$|\varphi(\omega)| = 1 \left(\frac{\omega}{2\alpha}\right)^{1/2} + \frac{\pi}{4}$$

$$\delta l \ll l$$

$$\omega \gg \left(\frac{2}{9}\right) \frac{\alpha}{l^2}$$

(b)

$$|\varphi(\omega)| = d \left(\frac{\omega}{2\alpha}\right)^{1/2}$$

$$\delta l \ll d$$

(c)

$$|\varphi(\omega)| = \left[l_1 \left(\frac{\omega}{2\alpha_1}\right)^{1/2} + l_2 \left(\frac{\omega}{2\alpha_2}\right)^{1/2}\right] + \frac{\pi}{4}$$

$$\delta l \ll l_1, l_2$$

(d)

$$|\varphi(\omega)| = \left[l \left(\frac{\omega}{2\alpha_1}\right)^{1/2} + \frac{\pi}{4}\right]$$

$$\delta l \ll 1$$

neglected. Light absorption is assumed to be occurring in layers that are negligibly thin compared with the thickness of the sample. Each relationship applies to the specific surface referred to in the diagram.

In the frequency domain, the phase, in addition to the modulation frequency, may be used to separate the spectral or thermal contributions of the individual layers in a multilayered sample. At fixed modulation frequency, a sample con-

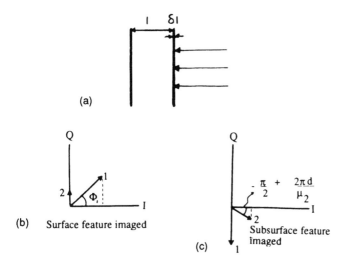

**Figure 5** Illustration of thermal wave image contrast enhancement by phase rotation method. A surface method of detection is assumed.

sisting of two thin absorber layers separated by a layer of transparent material will have a surface temperature given by the sum of the components of temperature contributed by the upper and lower layers. Both these contributions vary sinusoidally in time at the modulation frequency, although the respective magnitudes and phases will be different.

In steady state harmonic detection, a phase-sensitive measurement made with a lock-in amplifier recovers projections of the response that are in phase ($I$) and in quadrature ($Q$) ($+ 90°$) with a sinusoidal reference (Fig. 5). If a sufficient phase offset is applied, a desired spectral feature (e.g., layer 2) may be driven in quadrature with the reference and therefore aligned 90° out of phase with the reference. A phase-sensitive measurement made in phase will show zero contribution from the second layer. The recovered signal consists of the upper layer contribution only, in phase with the reference. In a multilayered structure, it is possible to apply this procedure sequentially to all layers in the structure and to orthogonalize each component out of the measurement in sequence. Finally, it makes no difference whether the phase rotation is carried out mathematically or instrumentally: the procedures are equivalent in principle.

An example of how phase detuning affects a scanned thermal wave image was reported by Kirkbright and Miller [58]. Figure 6 is a diagram of a test feature inlaid on 70 $\mu$m polyester. The annular ring feature was etched on the upper surface, while the central disk was located on the lower surface of the film. These absorbing layers were considered to be thin relative to the polyester spacer.

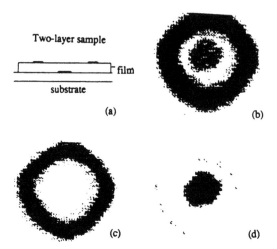

**Figure 6**   (a) Phase-selective imaging of an annular test pattern inlaid on a polyester substrate. (From Ref. 58.) (b) Magnitude image. (c) Image recorded with phase optimization of front surface image feature. (d) Image recorded with phase optimization of rear surface feature.

Thermal wave images of these samples obtained by PAS at 22 Hz and are shown in Fig. 6b–d.

The magnitude image (without phase optimization) (Fig. 6b) shows a strong weighting of the upper surface relative to the lower surface feature. Only the tail of the thermal wave from the lower feature reaches the front surface. By addition of a sufficient phase offset to the image data, the lower surface contribution is removed. The resulting image (Fig. 6c) shows no contribution of the central circular feature. In Fig. 6d, the phase offset was tuned so that only the rear surface contributed to the image.

The phase detuning method has also been used to resolve the optical absorption spectra of individual layers in multilayered samples [59,60]. Figure 7 illustrates the phase rotation method used with gas microphone photoacoustic spectroscopy to resolve the absorption spectra of two separate layers in a composite sample [60], pigmented polymethyl methacrylate layers fused together by a mild heat treatment. The top layer was 10 $\mu$m thick and doped with methylene blue, while the substrate was 61 $\mu$m thick and contained $\beta$-carotene. In Fig. 7a no phase rotation was applied. The absorption spectra of both layers appear in the output: the $\beta$-carotene band appears at 460 nm and a peak due to methylene blue is observed at 660 nm. Applying a phase rotation drove the signal contribution from the lower layer onto the in-phase axis in Fig. 7c. The spectrum of methylene blue recovered in quadrature clearly shows the absence of any con-

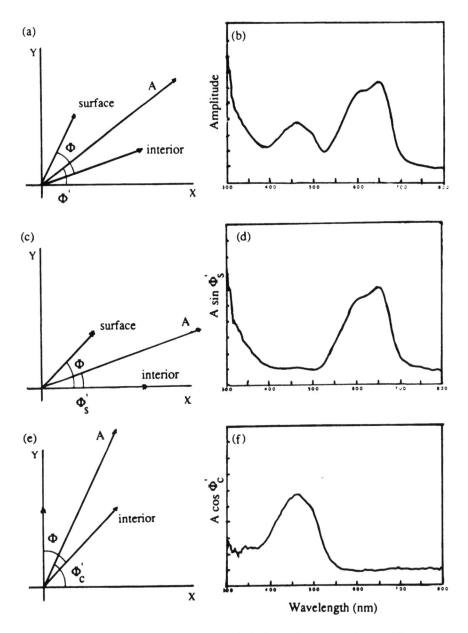

**Figure 7** Recovery of depth-resolved optical absorption spectra from a bilayer test sample by phase rotation method from Ref. 60: (a) phasor diagram and (b) experimental photoacoustic response with no phase optimization at 26 Hz; (c) axis rotation applied to resolve surface layer contribution; (d) PAS spectrum of surface layer containing methylene blue; (e) axis rotation applied to resolve lower layer spectrum; (f) PAS spectrum of β-carotene contained in lower layer. See text for details of the sample structure. (Reproduced with permission from Pergamon Press PLC.)

tribution from the $\beta$-carotene peak in the lower layer. A phase offset applied to drive the surface component in quadrature with the reference (Fig. 7e) recovered the methylene blue spectrum of the upper layer (Fig. 7f).

Phase rotation is often carried out at constant frequency. As the modulation frequency changes, the relative lengths and phases of the individual phasors will change. At high frequency, only the contribution of the topmost layers is significant. The lower layers in the structure may be resolved by reducing the frequency range of the measurement and repeating the phase rotation procedure.

While many depth profiling studies have been carried out in the visible and UV regions, some of the best applications of photothermal depth profiling are potentially available in the mid-infrared, where spectroscopic signals are rich in molecular structural information. In the "step-scan" FT-IR method [61,62], the Michelson interferometer beam is independently chopper modulated to enable a complete recovery of the photothermal frequency response at each position of the movable mirror. The phase and magnitude interferograms recovered at all mirror positions are processed to yield depth profile information as a function of wavelength. Procedures for modifying commercial instrumentation have been reported by Palmer et al. [63].

Although many depth profile studies have been carried out using FT-IR in the "constant scan rate" mode of operation [64,65], the modulation frequency of each component varies with the infrared frequency. This mode of operation has an obvious disadvantage of ambiguity of data interpretation, because the thermal diffusion length is a function of the wavelength. The limited range of mirror velocities is also a serious problem for depth profiling.

## IV. THERMAL WAVE MICROSCOPY

In thermal wave microscopy, the dimensions of the features imaged approach the thermal diffusion length. A one-dimensional analysis is no longer applicable. The analogies between thermal and optical scalar waves have been invoked by various authors. Like optical waves, three-dimensional thermal waves experience diffraction, interference [66], and reflection.

Thermal wave microscopy, like light microscopy, may be conducted in the near or far field [41,67,68]. Since the analogy between thermal and optical waves is clearest in the frequency domain, we discuss only harmonically driven thermal waves. It is assumed, for simplicity, that the sample is optically opaque. In far-field thermal wave microscopy, the resolution between adjacent features is set by the thermal diffusion length. The transverse separation between adjacent thermal features cannot be resolved unless it is greater than the thermal diffusion length. In the vertical direction, the photothermal signal becomes insensitive to features that lie significantly below a thermal diffusion length of the surface.

It is also possible to spatially resolve highly localized image features using near-field thermal imaging [67,68]. The excitation beam is tightly focused and modulated at low frequency. The spatial separation of the image features is orders of magnitude less than the thermal diffusion length. The thermal wave amplitude becomes proportional to the inverse distance of the detector (or probe spot) from the image feature. There is no significant phase lag with depth (or distance) because all depths are small compared with the thermal wavelength, and depth discrimination is largely lost.

In far-field imaging, by contrast, features on the order of a thermal diffusion length below the surface exhibit a significant phase lag with depth. An increase in the modulation frequency rejects buried image features at depths significantly greater than a thermal diffusion length. In both near- and far-field detection modes, the maximum achievable spatial resolution is set by optical diffraction of the heating beam.

Magnitude and phase thermal wave images as measured in the far field are significantly different. In Fig. 8, photothermoelastic effect detection was used to recover images of an integrated circuit metallization on a silicon substrate [69]. Such images may be interpreted by purely thermal contrast mechanisms, because of the approximate elastic homogeneity of semiconductors and because the PTE signal is detected at close proximity to the heating source [32]. The magnitude image (Fig. 8a) often resembles the optical micrograph of the pad structure, since it is sensitive to variations in the optical reflectivity of the surface. Also, near-field subsurface defects may contribute to the magnitude image. The phase image (Fig. 8b) is insensitive to surface reflectivity, contains purely thermal information, and is much less sensitive to near-field image features. It therefore provides the superior thermal depth profiling probe.

Most classes of thermal wave microscope use raster scanning in which the excitation source is localized, while the detection is either localized or broadfield. In general, detection will be of the area type (as in gas microphone PAS detection), line type (as in photothermal beam deflection), or point type (as in photothermoreflectance, photothermoelastic effect, and PT-IR spectrometry). The symmetry of the detection scheme may limit the nature of the image information available [41].

In area detection, the measured temperature is averaged over the sample surface, preserving planar symmetry. The theoretical interpretation is considerably simplified. But this mode of detection is insensitive to certain symmetrical defects such as closed vertical cracks. The limited modulation bandwidth of the gas microphone method ($< 10$ kHz) severely limits the depth profile resolution of that technique.

Methods such as transverse photothermal deflection spectrometry use a line probe geometry that removes the planar symmetry and enhances the sensitivity to anisotropies in heat flow. Closed vertical cracks can be imaged. The

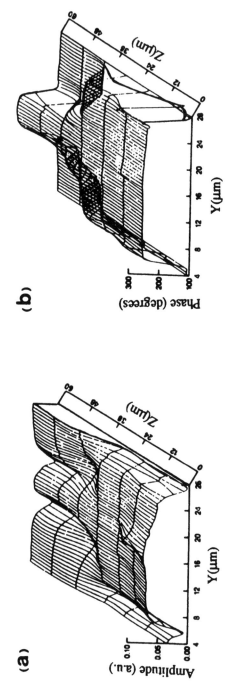

**Figure 8** Thermal wave images of a micrometer-sized metal pad on a silicon substrate, recorded using photothermoelastic thermal wave imaging in Ref. 69: (a) magnitude image and (b) phase image. Data were recorded at a modulation frequency of 1 MHz. (Reproduced with permission.)

theoretical description of images is complicated by the cylindrical geometry. In practice, the line probe method has the additional problems of alignment sensitivity and instability. Because the signal is sensitive to the vertical offset of the probe, images may show a dependence on the local surface topography. Furthermore, the method requires a flat or curved sample surface to permit sufficiently close alignment of the probe beam. In practice, probe beam offsets less than about 10 μm are not practical, [70]. The time delay required for heat conduction from the sample surface to the probe limits the bandwidth of the technique, and therefore, the depth resolution, to a few hundred kilohertz.

Methods such as photothermoreflectance, photothermal infrared radiometry, and photothermoelastic effect spectrometry use point source excitation and detection. None of these methods requires a gas phase medium for detection, and all are adaptable to high vacuum. In PT-IR magnitude images (and signal-to-noise ratios) are sensitive to the infrared emissivity of the surface, while in PTE and PTR methods, such images are dependent on the surface reflectivity.

In the case of PTR and PTE imaging, the spatial resolution is set by the spot size of the focused exciting and probe beams. A spatial resolution of about 1 μm is achievable with visible wavelength radiation. Electron or ion beams may be used in higher resolution applications and will allow direct subsurface penetration to a spatial resolution determined by the beam's subsurface "bloom" [32,41]. Subsurface scatterers may be excited directly and detected via the thermal energy arriving at the surface. The resolution is higher than would be obtained using a surface source, where far-field thermal wave diffraction occurs over the round-trip distance from the surface to the scattering feature.

Finally, the nature of the thermal wave image information depends critically on the length scale for which the image is recovered. At the highest resolution level, thermal waves are capable of resolving structures smaller than individual grains (on a length scale of a few hundred nanometers). At the other extreme, (length scales > 1 mm), local variations in the bulk sample properties are observed. The intermediate "mesoscopic" length scale ranging from about 10 to 500 μm is especially interesting for the thermal wave imaging of polycrystalline samples because the structure of percolation networks is elucidated on this scale [42]. These networks are composed of structural units exhibiting ordered behavior over a characteristic correlation distance (which is usually determined by the presence of large voids). The structure of these networks is of primary importance in determining the bulk thermophysical properties of many materials such as glasses, gels, and polymers. An analysis using fractal geometry may be appropriate to many problems in this area [42].

A primary limitation of many thermal wave imaging techniques is the time consumed by scanning. Point-by-point scanning is slow, and if thermal wave techniques are to be used in routine inspection applications, some means of rapid image processing is essential. Solutions include the use of spatial transform techniques [71–74], parallel video processing of photothermal signals [75–77], and rapid scanning methods [78].

Parallel thermal wave image processing techniques have been developed using infrared video cameras [75–77]. These methods have been successfully adapted to the rapid inspection of large industrial samples such as engine components and sheet metal panels. The depth profile range is much greater than for the usual laboratory methods, because the modulation frequency range is typically much lower. The rapid radiometric detection of near surface defects may also be achieved using a "flying spot" scanning approach [78].

Thermal waves have also been used to perform tomography by the mirage effect [79], the photopyroelectric effect [80], and PT-IR spectrometry [81,82]. In Ref. 80, the classical approach to tomography was used, in which a single-frequency, modulated point excitation beam was scanned across the sample surface and a thermal wave line scan was recovered at each point of the heating beam. The three-dimensional image structure was then recovered. A more recent approach used parallel (video) impulse excitation and detection [81]. Thermal wave tomograms were recovered from the peak values of a set of differential thermal impulse transients recorded on a pixel-by-pixel basis. The tomogram generation reported in Ref. 82, was carried out in real time and used to generate three-dimensional images of subsurface features in samples of plastics, as well as impact damage in graphite–epoxy composites.

## V. APPLICATIONS OF THERMAL WAVE IMAGING AND MICROSCOPY

This section reviews some individual applications in the areas of nondestructive evaluation (NDE) and the analysis of polymers and biophysical materials. Our discussion cannot be comprehensive: the volume of material in areas such as semiconductor imaging and NDE is beyond the scope of this work. Emphasis is given, instead, to applications that illustrate the principles of spectroscopic or thermal depth profiling in representative samples of practical interest.

### A. Metallurgical Samples

Thermal waves exhibit the greatest depth profiling range in metals because of their high thermal diffusivities. Thermal diffusion lengths of the order of several millimeters are possible. Thermal wave imaging may be used to inspect sheet metal specimens and welds and to detect subsurface flaws in metals. Thermal wave microscopy is capable of spatially resolving the grain structures of alloys in the vicinity of welds [83]. Strong grain–boundary contrast is observed with techniques that use thermoelastic contrast [32]. This visualization of microstructure may be useful in evaluating the failure mode of structures such as welding seams.

Defects such as cracks or slots introduce an air–metal interface that strongly reflects thermal waves. The scattering of thermal waves due to some of these features may be computed accurately from theory [41]. Depending on the geometry of a flaw and the symmetry of the photothermal detection, a feature may not be detectable in a recovered image. Closed vertical cracks, as we have already seen, contribute too small a thermal mass for gas microphone PAS. Thermal reflection at an air–metal interface produces a phase shift in the surface photothermal response. Phase images are especially sensitive to the effects of thermal discontinuities introduced by cracks and defects. Changes in subsurface thermal efflux may be present as a result of thermal processing, hardening, or deformation of the material. The photothermal phase will therefore be sensitive to these features.

PT-IR detection is frequently chosen for these studies. The noncontact, remote nature of the detection makes PT-IR adaptable to many geometries. Parallel image processing allows rapid inspection of a variety of industrial samples.

Inspection of welding seam structures [11] is a good example of practical PT-IR inspection. Welding seams may be imaged using either backscattering or thermal transmission mode detection. In backscattering mode, the image is dominated by features located within a thermal diffusion length of the surface, while in transmission mode detection the scanned phase signal resolves an image related to the average thickness of the seam. Some line scans of welding seams recovered by the PT-IR method [11] shown in Fig. 9. Phase images in both transmission and backscattering mode give good agreement with a thickness model based on an analysis similar to that of Table 2 (entry c). In other words, thermal reflections at the seam/substrate can be neglected, simplifying the analysis.

Thermal wave signals are found to be affected by the thermal and mechanical treatment of samples, which may induce phase transitions in the material or affect its grain structure. Luukkala and Askerov [84] provide a striking demonstration. Polished samples of aluminum and brass were struck with a wedge and deformed mechanically, then repolished to a flat surface finish. Thermal wave line scans of the samples were recorded before and after deformation. The wedge structure produced by deformation was still detected even after repolishing in both front surface and transmission mode images. In aluminum, subsequent annealing could render the deformation undetectable. Because the sample was polycrystalline, plastic deformation of the sample caused an increase in the number of grain boundaries, which persisted in the material after repolishing, but not after annealing. Similar results were observed by Busse et al. on thermally treated and mechanically deformed steel [11]. In the latter case, a residual phase image persisted after annealing, because of the high sensitivity of the phase signal to even small thermal delays.

The investigation of many metallurgical problems indicate that surface treatments and local microstructures exert significant effects on the surface and

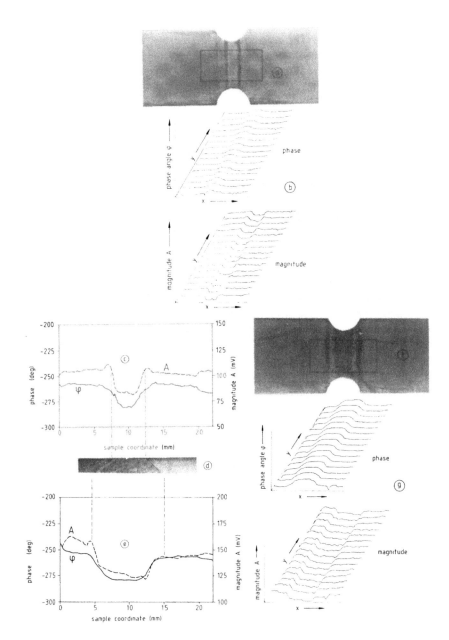

**Figure 9** Photothermal infrared radiometric imaging of welding seams performed in backscattering mode by Busse et al. [11]: (b, c) magnitude and phase images of front surface seam; (e, g) magnitude and phase images of rear surface. (Reproduced with permission.)

interfacial properties of metals. The bulk effect may be seen as a change in the measured thermal contact resistance between interfaces. Over time, surfaces may become fouled with dirt, causing thermal "short circuits" and consequent reduced image contrasts for crack structures. On the other hand, oxide formation at surfaces may result in a significant increase in the interfacial thermal contact resistance and increased thermal delays.

It is of interest in metals processing to determine near-surface hardening pro-files. In steel samples, hardening causes the thermal conductivity to vary by a factor of 2 relative to the bulk, as a result of the presence of a carbon-enriched hardening layer near the surface [85]. Depth profile recovery has been addressed through inverse problem theory, but the problem is ill-posed, and the resulting depth profile is artifact-prone and unreliable in the presence of realistic exper-imental noise. A more recent inversion method based on the Hamilton–Jacobi formulation is well-posed and may offer greater potential for recovering reliable depth profiles [53].

In the examples discussed above, thermal wave imaging has provided useful depth-resolved nondestructive information in a variety of practical situations. However, as pointed out by Busse [86], thermal wave based methods must com-pete in this area with other established techniques based on electrical conduc-tivity or magnetic properties. Their full impact may be felt in evaluation of nonmetals, where few alternatives exist.

## B.  Semiconductors

Thermal wave microscopy is used for inspection of defects in silicon and for monitoring the quality of ion implantations. Because of the large number of pro-cessing stages required in the production of integrated circuits, it is desirable to detect defects or impurities at early stages.

The thermoacoustic (TA) microscope has traditionally been the most com-monly used instrument for the thermal wave imaging of semiconductors [32]. Commercial TA microscopes operate at frequencies of up to 2 MHz, so that high resolution images are obtained. TA images exhibit both elastic and thermal con-trast and are capable of resolving subsurface cracks, grain boundary structures, and defects with high contrast. Many of these features are invisible in scanning electron microscopy. The primary drawback of the TA microscope is that it re-quires contact between the transducer and the specimen under study. More re-cent techniques such as photothermoreflectance and photothermoelastic effect spectrometry provide contactless thermal wave image detection at frequencies up to 10 MHz [19,20].

Microscopes using PTE and PTR detection are marketed under the trade name Thermaprobe [87]. These microscopes resolve images that are thermal, not thermoacoustic, and thus the elastic contrast mechanism associated with TA

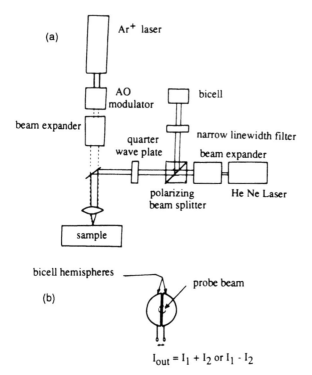

**Figure 10** (a) Block diagram of modulated photothermoreflectance–photothermoelastic effect spectrometer; AO, acousto-optic (modulator). (b) Bicell detector [7]. (From Ref. 6. Reproduced with permission from Pergamon Press PLC.)

detection is lost. The schematic of a typical PTR/PTE microscope is shown in Fig. 10. Excitation is obtained from an intensity-modulated argon ion laser. The probe beam is aligned with the pump beam, directed through a microscope objective, and focused onto the surface of the sample as a spot about 1 $\mu$m in diameter. In PTR detection, the alignment of the pump and probe beams is collinear, whereas in PTE detection, the beams are slightly displaced and arrive at the sample surface separated by a few micrometers. Because the PTE/PTR method is contactless, it may, in principle, be used under the high vacuum conditions required by the use of electron beams [32].

Thermal wave imaging may be used to evaluate devices in both active (with current flowing) and inactive modes [69]. We discuss only some relatively simple examples of inactive mode semiconductor imaging. The richness of the non-radiative physical phenomena occurring in semiconductor materials places a discussion of many contrast mechanisms outside the scope of this review. The reader is referred to a recent monograph on the subject for more details [10].

An example of inspection of an integrated circuit is shown in Fig. 11. The feature imaged was a metal–silicon interface on a metal oxide semiconductor field effect transistor (MOSFET) device. The thickness of the metallization was about 2 μm. Because of the very large thermal diffusivity of silicon, and the thinness of the feature, megahertz frequencies are required for depth profiling. The magnitude image is sensitive to optical reflectivity and generally resembles a conventional optical micrograph. Also, the contribution of near-field thermal features to the magnitude image cannot be discounted. The phase image, on the other hand, gives a true far-field thermal image, whose contrast depends on the modulation frequency. While much of the detail in the magnitude image appears to be frequency independent, below 100 kHz the phase contrast is completely lost.

## C. Paints and Coatings

Thermal wave imaging has been applied to many problems involving the evaluation of paints and coatings. Photothermal radiometry has been the method of choice, probably because it offer contactless detection and alignment insensitivity. In many applications, the objective is the detection of spalling or delamination of the coating layer. Adhesive bond strength may be evaluated via the surface temperature response [88]. More sophisticated applications may require the monitoring of thermal property variations in several coating layers. In paint weathering, changes in the spectrum or spatial distribution of the pigments may also be important.

Coating delamination is frequently modeled as the development of an air layer between the coating and substrate [45,48]. The thermal properties of many paints, the common coating, are similar to those of polymers. If the substrate is a metal surface, then the thermal reflection coefficient will be −1 at the coating–substrate interface, indicating strong absorption of thermal waves by an excellent conductor. If a thick air gap is present under the paint layer, the reflection coefficient changes from −1 to +1. A scanned image of the surface based on impulse data at long times past the application of the pulse shows the delaminated regions appearing as "hot spots," while the bonded areas appear "cool" (see Fig. 3). If the thermal diffusivity of the paint is known, it is possible to use the impulse response to obtain the paint layer thickness.

A more difficult problem is the measurement of paint thickness in samples that are well bonded to a polymer substrate. In these situations, the thermal efflux ratio at the interface is so close to unity that effects on the impulse response are minimal. In the frequency domain, the equivalent thickness probe is the phase, which is also small. Busse and Vergne [14] were able to monitor thicknesses of polyurethane-based paint on a polyurethane substrate using PT-IR. A high phase resolution and stability were essential because phase variation of only about 1° was produced for a 20 μm layer. Near-surface microstructures in the paint layer cause phase fluctuations of the order of 0.25°, but the paint layer

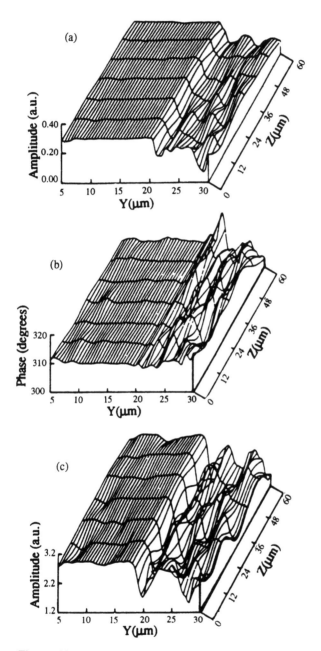

**Figure 11** Thermal wave images of metal–silicon test feature recorded in Ref. 69 at different modulation frequencies. Traces (a, b) 1 MHz; (c, d) 50 kHz; (e, f) 1 kHz. (Reproduced with permission.)

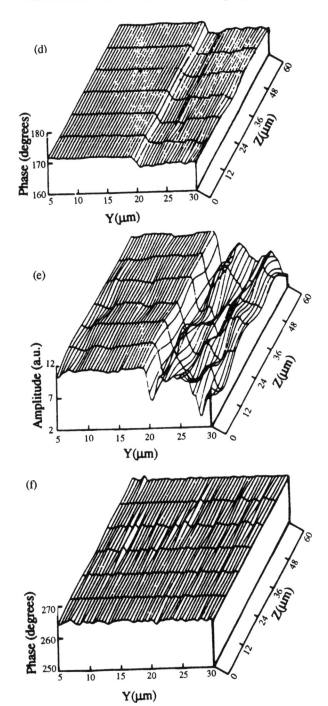

thickness dependence was still detectable. It is therefore possible to probe the thermal effects of substrate preparations and contaminants at the bond interface using this method.

For very thick paint layers, the response to pulsed or modulated heating may be too slow ($> 1$ s, $f < 1$ Hz) for conventional lock-in or transient techniques. Detection methods based on parallel (video) PT-IR imaging give good performance [75–77].

Another area of application of photothermal imaging is evaluation of paint weathering. Traditional weathering tests are very time-consuming, requiring up to 2 year of environmental sample exposure to observe appreciable deterioration. Imhof et al. [12] used pulsed infrared radiometry to study the effects of exposure of $TiO_2$-based paints in a weatherometer. As the paint layer was exposed, over a period of up to 100 hours, changes in the early time decay of the impulse response occurred. These were consistent with changes in the surface properties of the paint layer on a depth scale of about 10 μm. They correlated more specifically with spectral changes (yellowing) observed in the surface pigments over time. Since the gloss of the paint layer was not strongly affected in the course of the test, an explanation based on changes in the thermal diffusivity appeared less likely.

The foregoing work underlines some of the complexities encountered in evaluation of paints. As a paint specimen weathers, the surface temperature impulse response may be affected by spectral changes in the paint layer, thermal changes, shrinkage, and increases in light scattering as surface erosion of resins occur. This complexity suggests that thermal waving imaging techniques provide the maximum degree of insight when interpreted in concert with other probes, such as reflectance and FT-IR spectrometry.

There has been considerable interest in the use of thermal wave based techniques for the evaluation of thermal barrier and plasma spray coatings. Many conventional analytical techniques perform poorly on the latter because of their complex microstructure. Ultrasonic inspection methods, for example, fail because of high levels of sound scattering by the matrix. The conventional methods of evaluation are visual inspection and eddy current thickness measurements. One-dimensional thermal wave imaging measures the bulk diffusivity of the sample, which averages over the local variations in microstructure [88].

A number of authors have investigated these coatings [15,16,48,88,89]. Experimental methods include both transient and continuous wave detection. The coatings may be assumed to be opaque or partially translucent. The thermal contact resistance between the coating layer and the substrate is related to the adhesive bond strength [88]. The exact association between these two parameters, however, must be determined by experiment for a particular coating–substrate combination. An alternative theoretical approach demonstrates that a disbond present in a two-layer system behaves as an additional layer of low efflux material. By varying the thickness of the additional layer, the behavior of the disbond may be experimentally simulated.

The $N$ layer theory developed by Murphy et al. [45] has been used to model heat flow in multilayer coatings excited by a step heating source. This theory has been successfully applied to interpret the experimental signals observed for two- and three-layer thermal barrier coatings. Such studies are expected to gain in importance as engineered coatings comprised of multiple layers are developed.

## D. Polymers

Thermal wave imaging has broad applications in polymer analysis. Structured thin film specimens having a variation in the optical absorption coefficient with depth may be studied in both the visible and the infrared regions. Thermal anisotropy enables visualization of fiber orientation in a variety of materials including graphite-reinforced matrices. Dynamic processes such as diffusion in thin films are readily studied by using thermal wave depth profiling. Phase transitions in materials are probed through changes in thermal diffusivity and thermal efflux. However, considerable work using photothermal methods for the thermal characterization of bulk samples has been reviewed elsewhere [90] and is not discussed here.

Optical depth profiling studies, discussed earlier in detail, have been demonstrated in the UV and visible regions on a variety of thin film laminates [29,30,60,91–94]. Depth profiling has been used to nondestructively evaluate manufactured samples such as food wrappers [92] and the laminates used in credit cards [93]. The technique would be most useful in the mid-infrared but is hampered by the constant scan rate design used by most commercial FT-IR interferometers. A number of limited-resolution depth profile studies have appeared, including bilayers of polyethylene terephthalate and polyvinylidene fluoride [95], dyed textile fibers [96], bilayers of epoxy resin on polypropylene [97], and metallized cellulose films [98]. Recent work in the infrared depth profiling of polymer laminates has been reported by Dittmar et al. using the step-scan technique and a phase rotation method [99].

In highly anisotropic systems such as fibers or in stretched polycrystalline polymers, a preferred axis for heat flow lies along the direction of the fiber orientation. Measurements by Busse and Eyrer demonstrated these fiber orientation effects in stretched samples of semicrystalline polyethylene [18]. The induced thermal anisotropy caused the thermal diffusivity to increase along the stretch axis, relative to the perpendicular axis. The average value of the thermal diffusivity also changed, consistent with void formation under stretch as well as molecular alignment effects.

Thermal wave imaging methods may also be applied to the spatial mapping of thermal anisotropy in a backscattering mode measurement using continuous wave PT-IR. The excitation beam and the probe spot are scanned transversely along the sample surface at an offset distance of a few hundred micrometers. As the scanning direction of the probe spot is changed, in an anisotropic and optically opaque material, the measured thermal diffusivity varies, producing variations in the photothermal signal dependence [100].

Thermal waves have also been used to evaluate composite materials for delamination and fiber orientation. Again, photothermal radiometry has been the method of choice. In experiments by Busse et al. [100], PT-IR was used to measure thermal anisotropy in a carbon fiber reinforced matrix. Defects were induced in materials by electrical heating along an axis parallel to the fiber orientation. This thermal damage could be readily localized in phase angle scans of the material measured by transmission PT-IR.

Cielo used pulsed PT-IR to evaluate delaminations in graphite–epoxy composites [48]. The experimental impulse response data were compared to numerical models expected for matrices with different fiber orientations and distributions of contact points. Delaminations of a randomly oriented matrix could be readily identified and distinguished from those of a uniform fiber matrix.

Another area of considerable general interest to polymer science is in the study of diffusion in thin films. When a membrane is wetted by a solvent, for example, changes in the thermophysical properties (and thickness) of the membrane occur as the solvent diffuses through the material. These changes may, in principle, be probed using thermal wave depth profiling. A simpler case is the diffusion of an absorbing impurity through a solvent-saturated membrane. Here, the low concentration absorber does not perturb the thermophysical properties of the membrane while inducing a measurable absorption profile across the film. Optical depth profiling of the concentration gradient in the membrane provides a method for directly visualizing the diffusion process.

Depth profiling reported by Miller and coworkers [101,102] used gas microphone photoacoustic detection to study migration of a colored dye through a hydrophobic membrane. Some examples of their measurements are shown in Fig. 12. The dye was initially contained in a reservoir underneath the membrane specimen. The experiment used the photoacoustic impulse response to study the upward migration of the dye through the film. The peak delay in the impulse response was used as a principal means of evaluating the distance traveled by the dye (Fig. 12b).

Good agreement was found between experiment and a model based on Fickian diffusion (Fig. 12c), although small systematic deviations in the peak delay dependence were observed. These deviations, which may be due to effects of the microphone transfer function or other factors in the experiment, prevent

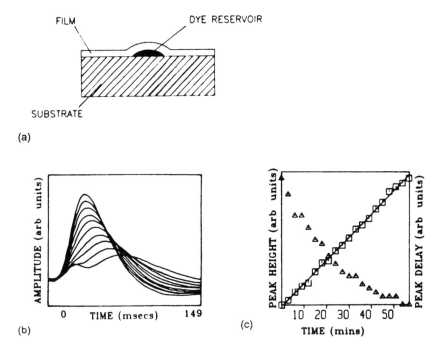

**Figure 12** Correlation (impulse response) photoacoustic study of low concentration ink diffusion in a thin film, performed by Miller [111]. (a) Geometry of diffusion cell. (b) Impulse response data obtained from a typical dye diffusion experiment, recorded at 3-minute intervals. (c) Comparison of impulse response peak delay (triangles) and peak height (squares) variation with time to a model based on Fickian diffusion. (Reproduced with permission from Pergamon Press PLC.)

extraction of absolute diffusion coefficients. With a clarification of this latter point, however, the method just described shows excellent prospects for future application.

## E. Biophysical Applications

Thermal wave depth profiling has found application in a variety of samples of biological origin because it can resolve optical absorption spectra in the complex layered systems characteristic of many tissues and biological samples. This depth profile capability has more recently become available with optical techniques such as coherent Raman spectroscopy [103] or confocal scanning microscopy [104]. In the former case, however, the measured signal does not vary linearly with excitation intensity, while in the latter, artifacts may be encountered in media where the refractive index changes rapidly over highly localized spatial regions, as in samples that exhibit significant spatial variations in the concentration of ionic species. Thermal wave imaging, as implemented in most of the methods discussed below, is not sensitive to such variations in refractive index.

In static mode, thermal wave depth profiling may provide a map of the chromophore distribution in a complex system. A sample may be exposed in dynamic mode to a variety of chemical and physical stimuli such as toxic substances, dyes, drugs, or temperature changes. The resulting chromophore distribution may be monitored nondestructively over time. Gas microphone photoacoustic spectroscopy has made significant contributions to photosynthesis research as a result of its capability for detecting oxygen evolution from leaf specimens in situ [60,105]. Photothermal techniques are useful for studying orientational effects in chlorophyll, since this molecule is very sensitive to aggregational and matrix effects [106]. Other areas of interest addressed by photothermal techniques include studies of human tissues, especially skin and the transport of pigments and topically applied compounds through the skin [107,108].

Thermal wave imaging is well adapted to the study of leaves. The leaf structure observed for a number of plant species consists of at least two photothermally resolvable layers, which have been accessed mainly by gas microphone photoacoustic methods. Some species, such as green radish and maize, do not appear to exhibit this well-defined morphology, however, and the distribution of pigments appears to be less sharply defined [109]. These layers consist of a waxy cuticle overlayer, which absorbs mainly in the near UV, and the palisade parenchyma, which contains the leaf chloroplasts [105,110,111]. This lower layer therefore contains chlorophyll and related pigments, which give leaf specimens their distinctive color. The waxy cuticle overlayer may be up to 50 μm thick depending on the species [112] and functions as a UV-blocking window. The pigment distribution in the lower layer is nonuniform and changes as the leaf yellows.

In the depth profiling of leaf specimens, the techniques of choice have been phase-resolved spectroscopy and impulse response PAS. A number of different plant species have been examined, including coffee and soybean leaves [113], lettuce [112], and maize and green radish [109]. Changes in the relative intensities of chlorophyll pigments in intact leaves may be followed as a species matures. Kirkbright et al. used impulse response PAS to examine this process in pubescent and mature laburnum leaves [112]. Several authors have used PAS depth profiling to monitor the pigment distribution changes associated with the yellowing of leaves [105,110]. As yellowing progresses, production of the pigment anthocyanin occurs, with the growth of an absorption band at 525 nm. A depth profile of the pigment distribution in sugar maple leaves [105] is shown in Fig. 13.

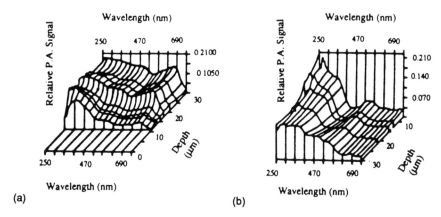

**Figure 13** Chromophore depth profiles in sugar maple leaves recovered by N'Soukpoe-Kossi and Leblanc [105] using photoacoustic spectroscopy. The study shows development of the anthocyanin absorption band as the leaf yellows. (a) Green leaf with chlorophyll absorption bands near 470 and 680 nm. (b) Yellowing leaf showing anthecyanin absorption at 525 nm. (Reproduced with permission from Elsevier Science Publishers BV.)

Gas microphone PAS has been used to evaluate herbicide and acid rain damage to leaf specimens [112]. Nery et al. used phase-resolved spectroscopy to study spectral changes in the bilayer structure of the leaf brought on by paraquat treatment [113]. Evidence was found for bleaching and loss of pigment material over time, consistent with the action of paraquat. The effects of shrinkage of the leaf specimen over time due to dehydration were observed and related to optically measured changes in the thickness of the leaf specimen.

Miller [111] demonstrated that part of the shrinkage effect probably could be ascribed to the natural drying of the leaf specimen over the several hours of the study. The centroid of the PAS impulse response was used as a measure of the shape of $h(t)$ and followed over time in an attempt to interpret changes in the distribution of the pigments. The results could not be determined unambiguously, however. Bleaching effects in the parenchyma were suggested, along with evidence for an increase in the depth of visible absorbing photosynthetic pigments. It was also demonstrated that changes in the impulse response centroid became significant only after light exposure of the herbicide-treated samples. This was consistent with the known toxic action of paraquat.

The PAS impulse response, while sensitive to the spatial distribution of the chromophores, may be difficult to interpret in complex systems because a number of thermo-optic parameters change simultaneously. Consequently, a thermal wave depth profile recovered in such a problem is best interpreted in conjunction with evidence obtained from several other more established analytical techniques.

Other phytochemical samples have been studied using thermal wave depth profiling. A fairly complex layered structure is the lichen *Acarospora schleichieri*, spectrally resolved by Moore et al., using phase-resolved photoacoustic spectroscopy [60]. Figure 14 gives a summary of the lichen structure and the phase-resolved spectra for the various layers. At the higher modulation frequency (150 Hz), the thermal diffusion length was too short to penetrate the algal layer. The spectrum of the upper layer containing rhizocarpic acid could be cleanly resolved from the cytochrome layer. At sufficiently low frequency, the thermal diffusion length could be made long enough to penetrate relatively deeply into the algal layer and recover the spectrum of the chloroplasts.

Specimens of animal tissue have been depth profiled using photoacoustic spectroscopy. Leblanc et al. used PAS to profile the chromophore distribution in bovine retinal tissue [54], with a micrometer resolution. Figure 15 shows the map for the ocular tissue, consisting of the neural retina–photoreceptor layer and the retinal pigment epithelium (RPE). This map was recovered by depth profiling the retina–photoreceptor and the RPE–choroid layers individually. The samples were sectioned as shown in the inset and studied in a specially designed low temperature cell. The neural retina is seen as a highly transparent region, which admits light into the lower tissues. The photoreceptor layer clearly shows the absorption band of the visual pigment bathorhodopsin, as well as traces of the visual pigment retinal, which is present as an oxidative impurity. Finally, light exiting the photoreceptor encounters the retinal pigment epithelium and choroid, where high concentrations of melanin and hemoglobin act as an effective light stop.

The study of skin is another area of general interest for thermal wave depth profiling applications. Human skin has thickness and thermal properties that make it readily amenable to study in the modulation frequency range of gas microphone photoacoustic spectrometry. Applications include monitoring of moisture content of skin, drug delivery, and pigment distribution, as well as the evaluation of topically applied cosmetic treatments [9].

Recently Imhof [108] used transient infrared photothermal radiometry to follow the penetration of a sunscreen into the skin layer as a function of time. Earlier studies had used gas microphone photoacoustic spectroscopy to study similar processes [107]. The versatility and alignment insensitivity of the PT-IR method made it a much more convenient measurement method. The impulse method also gave an improved visualization of the dynamics of the process. It was possible to identify the time at which the sunscreen no longer appeared as a discrete phase because of a probable loss of volatiles. A residence time for the product in the skin could be estimated. Measurements of this type have promising applications in the study of drug delivery via the skin.

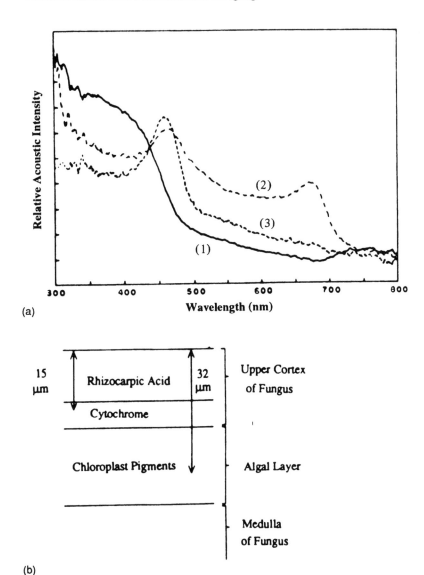

(a)

(b)

**Figure 14** Photoacoustic depth profile study performed by Moore et al. [60] on a sample of the lichen *A schleicheri*. (a) Depth-resolved spectra recorded at 150 Hz using the phase rotation methods: (1) spectral contributions of rhizocarpic acid; (2) bulk cytochrome absorption; (3) chloroplast absorption. The latter spectrum was recorded at 32 Hz. (b) Schematic diagram of the layer structure of the lichen. (Reproduced with permission from Pergamon Press PLC.)

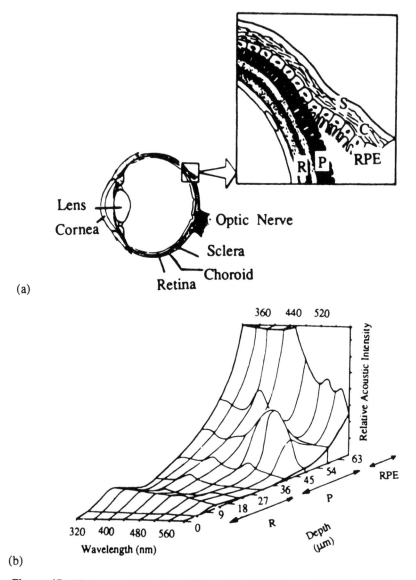

(a)

(b)

**Figure 15** Photoacoustic depth profiling study of LeBlanc et al. [54] on bovine ocular tissue. (a) Schematic drawing of vertebrate ocular tissue showing separation of the retina from the choroid and location of the neural retina (R), photoreceptor cells (P), and retinal pigment epithelium (PRE). (b) Depth distribution of chromophores in the neural retina, photoreceptor, and retinal pigment epithelium.

If tissue depth profiling is required on a microscopic scale, the crossed-beam photothermal refraction microscopy technique developed by Dovichi provides an excellent means of image recovery [33]. This method combines scanning microscopy with point laser heating to give unambiguous localization of subsurface image features on a depth scale as low as hundreds of nanometers. The spatial resolution, about 500 nm, is close to the optical diffraction limit in the transverse plane. By translating the sample through the intersection point of the two beams along the depth axis, a depth profile is recovered. Very weakly absorbing features may be detected. The dynamic range is outstanding: about nine orders of magnitude. Images of optical absorption and thermal diffusivity may be recovered by monitoring the phase and magnitude response, respectively. In some applications, such as the infection of tissues with strains of tuberculosis bacteria, the phase signal may be relatively featureless while the magnitude image contains considerable structural detail. Calcified tissues, on the other hand, are expected to exhibit significant diffusivity variations so that the phase image would be rich in structural detail [34].

With lock-in detection, thermal diffusivity variations on the order of 1% are readily detectable using the scanning photothermal microscope. With more conventional photothermal techniques, such small variations in diffusivity probably would not be detectable: the phase channel would be dominated by the depth dependence of the absorbing features in the sample.

Figure 16 shows some typical absorptivity and thermal diffusivity images recovered with the crossed-beam thermal lens microscope for a specimen of stem tissue of the species *Tilia americana*. Because of the sensitivity and depth profile capabilities of this microscope, the need to microtome samples is virtually eliminated. The requirement for staining the samples is significantly reduced relative to conventional microscopy. While the scanning time for an image was originally rather long, parallel generation and probing of the thermal lens may decrease image acquisition times from several hours to about one minute [114].

The crossed-beam thermal lens microscope exhibits several key advantages compared to single-ended photothermal measurement (PAS, PT-IR, etc.). The most significant advantages are the clear resolution of image components due to thermal diffusivity and optical absorptivity variations and the absolute nature of the image localization.

The disadvantages of the crossed-beam thermal lens microscope are the requirement for optical transmission of the probe beam and the scanned nature of the depth profile recovery. The technique is not suitable for highly absorbing samples. Relatively low modulation frequencies must be used, causing a fairly long recovery time for the recording of a high resolution depth profile. Crossed-beam thermal lens depth profiling may not be suitable for the observation of kinetic or dynamic phenomena. Finally, because of the thermo-optical detection mechanism, the recorded magnitude images may be sensitive to

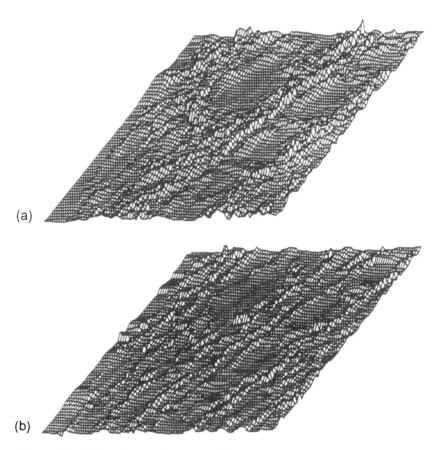

(a)

(b)

**Figure 16** High resolution photothermal images reported by Burgi and Dovichi [34] using the crossed-beam thermal lens microscope. The sample was a specimen of the plant species *Tilia americana*. Amplitude (a) and phase (b) images were recorded with a 1 μm pixel spacing. (Reproduced with permission.)

spatial variations in *dn/dT*, the temperature coefficient of refractive index, as well as the sample absorptivity.

No single method combines all the properties of an ideal imaging technique. The absolute localization of image features provided by the crossed-beam thermal lens technique is a valuable features because it reduces a great deal of ambiguity in the image interpretation. However, the crossed-beam method requires that the sample be optically transmitting. Thermal wave images generated by single-ended surface techniques, on the other hand, exhibit more complicated contrast mechanisms and yield insight less unambiguously. In optically opaque or highly absorbing samples, however, the latter are capable of generating depth-resolved images of materials that may not be obtainable by any other method.

## VI. SUMMARY AND FUTURE DIRECTIONS

Photoacoustic and photothermal imaging techniques have developed rapidly and are now reaching maturity. The process of interpreting one-dimensional images is now well advanced. Further tomographic reconstruction algorithms based on one-dimensional analysis will be forthcoming. Continuous profiles of subsurface thermal efflux may currently be recovered via inversion techniques, and the potential exists for applying thermal wave imaging more broadly to the study of phase transitions in thin films or near surfaces. Spectroscopic depth profiling has been developed as an analytical tool for the study of layered materials. Future work will be required to evaluate the information content of photothermal depth profiles in the presence of significant levels of experimental noise. A detailed experimental comparison of surface-detected thermal wave imaging techniques with localized methods such as crossed-beam thermal lens and confocal scanning microscopy should provide more information about the relative advantages and merits of the various measurement approaches. Finally, processing, acquisition, and detection schemes that enable the rapid recovery of thermal wave images will provide significant further advancement, since the majority of current thermal wave imaging methods are scanning techniques, in which image processing in real time usually is not achievable.

## REFERENCES

1. Hess, P., and Pelzl, J., Eds. (1988). *Photoacoustic and Photothermal Phenomena*, Springer-Verlag, Berlin.
2. Murphy, J. C., Maclachlan-Spicer, J. W. Aamodt, L. C., and Royce, B. S. H., Eds. (1990). *Photoacoustic and Photothermal Phenomena*, Vol II, Springer-Verlag, Berlin.
3. Hutchins, D. A., and Tam, A. C., Eds. (1986). *IEEE Trans. Ultrason. Ferroelec. Freq. Control, UFFC-33*, (5) (special issue devoted to photoacoustic and photothermal sciences).
4. Bertrand, L., Cielo, P., Leblanc, R., Montchalin, J. P., and Mongeau, B., Eds. (1986). *Proceedings of the Fourth International Topical Meeting on Photoacoustic, Thermal and Related Phenomena, Esterel, Quebec (1985), Can. J. Phys. 64* (9).
5. Rosencwaig, A. (1980). *Photoacoustics and Photoacoustic Spectroscopy*, John Wiley & Sons, New York.
6. Power, J. F. (1989). Thermal wave imaging in the physico-chemical analysis and evaluation of solid phase samples, *Prog. Anal. Spectrosc. 12*:453.
7. (1991). *Proceedings of the Seventh Topical International Meeting on Photoacoustic and Photothermal Phenomena (STIMPAP)*, Doorwerth, the Netherlands, Aug. 26–30.

8. Hess, P., Ed. (1989). *Photoacoustic, Photothermal and Photochemical Processes at Surfaces and in Thin Films*, Springer-Verlag, Berlin.
9. Miller, R. M. (1989). Spectroscopic depth profiling using thermal waves, in *Photoacoustic, Photothermal and Photochemical Processes at Surfaces and in Thin Films* (Hess, P., Ed.), Springer-Verlag, Berlin, p. 171.
10. Mandelis, A., Ed. (1987). *Photoacoustic and Thermal Wave Phenomena in Semi-Conductors*, North-Holland, New York.
11. Busse, G., Rief B., and Eyrer P. (1986). Photothermal non-destructive evaluation of welding seams, *Can. J. Phys.* 64:1195.
12. Imhof, R. E., Birch, D. J. S., Thornley, F. R., Gilchrist, J. R., and Strivens, T. A. (1985). Opto-thermal monitoring of paint degradation, *J. Phys. D*, 18:L103.
13. Busse, G., Vergne D., and Wetzel, B. (1988). Photothermal nondestructive inspection of paint and coatings, in *Photoacoustic and Photothermal Phenomena* (Hess, P., and Pelzl, J., Eds.), Springer-Verlag, Berlin, p. 427.
14. Busse, G., and Vergne, D. (1989). Remote inspection of coatings with thermal wave radiometry, *Infrared Phys.* 29:839.
15. Morris, J. D., Almond, D. P., and Patel, P. M. (1990). Modelling of Sample Translucency, Substrate Absorption and Interface Defects for Thermal Wave Infrared Inspection of Thermal Barrier Coatings, in *Photoacoustic and Photothermal Phenomena*, Vol. II (Murphy, J. C., Maclachlan-Spicer, J. W., Aamodt, L. C., and Royce, B. S. H., Eds.), Springer-Verlag, Berlin, p. 71.
16. Lau, S. K., Almond, D. P., and Patel, P. M. (1990). Transient thermal wave techniques for the characterisation of surface coatings, in *Photoacoustic and Photothermal Phenomena*, Vol. II (Murphy, J. C., Maclachlan-Spicer, J. W., Aamodt, L. C., and Royce, B. H. S., Eds.), Springer-Verlag, Berlin, p. 74.
17. Maclachlan, J. W., Aamodt, L. C., and Murphy, J. C. (1989). Time resolved infrared radiometric imaging of coatings, in *Review of Progress in Quantitative Non-Destructive Evaluation* (Thompson, D. O., and Chimenti, D. E., Eds.), Plenum Press, New York, p. 1297.
18. Busse, G., and Eyrer, P. (1983). Thermal wave remote and non-destructive inspection of polymers, *Appl. Phys. Lett.* 43:355.
19. Opsal, J., and Rosencwaig, A. (1982). Thermal wave depth profiling: Theory, *J. Appl. Phys.* 53:4240.
20. Rosencwaig, A., Opsal, J., and Willenborg, D. (1985). Detection of thermal waves through optical reflectance, *Appl. Phys. Lett.* 4:1013.
21. Opsal, J., Rosencwaig, A., and Willenborg, D. (1983). Thermal-wave detection and thin-film thickness measurements with laser beam deflection, *Appl. Opt.* 22:3169.
22. Boccara, A. C., Fourier, D., and Badoz, J. (1980). Thermooptical spectroscopy: Detection by the mirage effect, *Appl. Phys. Lett.* 36:130.
23. Tam, A. C., and Sullivan, B. (1983). Remote sensing applications of thermal wave radiometry, *Appl. Phys. Lett.* 43:333.
24. Imhof, R. E., Birch, D. J. S., Thornley, F. R., Gilchrist, J. R., and Strivens, T. A. (1984). New opto-thermal radiometry technique using wavelength selective detection, *J. Phys. E; Sci. Instrum.* 17:521.

25. Coufal, H. (1983). Photothermal spectroscopy using a pyroelectric thin-film detector, *Appl. Phys. Lett. 42*:33.
26. Mandelis, A. (1984). Frequency-domain photopyroelectric spectroscopy of condensed phases (PPES): A new, simple and powerful spectroscopic technique, *Chem. Phys. Lett. 108*:388.
27. Coufal, H. (1986). Pyroelectric detection of radiation-induced thermal wave phenomena, *IEEE Trans. Ultrason., Ferroelec. Freq. Control, UFFC-33*:507.
28. Power, J. F. (1991). Impulse photopyroelectric depth profiling of multilayers. I: Theory, *Appl. Spectrosc. 45*:1240.
29. Power, J. F. (1991). Impulse photopyroelectric depth profiling of multilayers. II: Experiments with amplitude/phase modulated wideband spectrometry, *Appl. Spectrosc. 45*:1252.
30. Prystay, M. C. and Power, J. F. (1993). Spatial depth profiling of chromophores in thin polymer films using photopyroelectric spectroscopy, *Polym. Eng. Sci. 33*: 43.
31. Monchalin, J. P. (1986). Optical detection of ultrasound, *IEEE Trans. Ultrason. Ferroelec. Freq. Control, UFFC-33*:485.
32. Rosencwaig, A., and Opsal, J. (1986). Thermal wave imaging with thermoacoustic detection, *IEEE Trans. Ultrason. Ferroelec. Freq. Control*, UFFC-33:516.
33. Dovichi, N. J. (1988). Thermo-optical spectroscopy for trace microchemical analysis, *Prog. Anal. Spectrosc. 11*:179.
34. Burgi, D. S., and Dovichi, N. J. (1987). Submicrometer resolution images of absorbance and thermal diffusivity by the photothermal microscope, *Appl. Opt. 26*:4665.
35. Huang, X., and Chen, W. (1990). Measurement of the absorption distribution of optical thin films by scanning photothermal microscopy, in *Photoacoustic and Photothermal Phenomena*, Vol. II (Murphy, J. C., Maclachlan-Spicer, J. W., Aamodt, L. C., and Royce, B. S. H., Eds.), Springer-Verlag, Berlin, p. 133.
36. Bendat, J. S., and Piersol, A. G. (1980). *Engineering Applications of Correlation and Spectral Analysis*, Wiley-Interscience, New York.
37. Power, J. F. (1990). Amplitude and phase modulated (AM-PM) wideband photothermal spectrometry. I: Theory. II: Experiment, *Rev. Sci. Instrum. 61*:90.
38. Kato, K., Ishino, S., and Sugitani, Y. (1980). Correlation photoacoustics, *Chem. Lett. Chem. Soc. Jap. 7*:783.
39. Mandelis, A. (1986). Time delay domain and pseudo-random-noise photoacoustic and photothermal wave processes: A review of the state of the art, *IEEE Trans. Ultrason. Ferroelec. Freq. Control, UFFC-33*:590.
40. Kirkbright, G. F., and Miller, R. M. (1983). Cross-correlation techniques for signal recovery in thermal wave imaging, *Anal. Chem. 55*:502.
41. Thomas, R. L., Favro, L. D., and Kuo, P. K. (1986). Thermal wave imaging for non-destructive evaluation, *Can J. Phys. 64*:1234.
42. Boccara, A. C., and Fournier, D. (1989). Heat diffusion and random media, in *Photoacoustic, Photothermal and Photochemical Processes at Surfaces and in Thin Films* (Hess, P., Ed.), Springer-Verlag, Berlin, p. 303.
43. Mandelis, A., and Power, J. F. (1988). Frequency-modulated impulse response photothermal detection through optical reflectance. I: Theory, *Appl. Opt. 27*:3397.

44. Iravani, M. V., and Wickramasinghe, H. K. (1985). Scattering matrix approach to thermal wave propagation in layered structures, *J. Appl. Phys. 58*:122.
45. Aamodt, L. C., Maclachlan-Spicer, J. W., and Murphy, J. C. (1990). Analysis of characteristic thermal transit times for time-resolved infrared radiometry studies of multilayered coatings, *J. Appl. Phys. 68*:6087.
46. Patankar, S. V. (1975). *Numerical Heat Transfer and Fluid Flow*, Clarendon Press, Oxford.
47. Miller, R. M. (1986). Digital simulation of photoacoustic impulse responses, *Can. J. Phys. 64*:1049.
48. Cielo, P. (1984). Pulsed photothermal evaluation of layered materials, *J. Appl. Phys. 56*:230.
49. Lau, S. K., Almond, D. P., and Patel, P. M. (1990). Numerical analysis of the transient thermal response of subsurface defects, in *Photoacoustic and Photothermal Phenomena*, Vol. II (Murphy, J. C., Maclachlan-Spicer, J. W., Aamodt, L. C., and Royce, B. S. H., Eds.), Springer-Verlag, Berlin, p. 78.
50. Lau, S. K., Almond, D. P., and Patel, P. M. (1991). A quantitative analysis of pulsed video thermographic imaging of subsurface defects, in *Proceedings of the Seventh International Topical Meeting on Photoacoustic and Photothermal Phenomena* (STIMPAP), Aug. 26–30, Doorwerth, the Netherlands, *Conference Digest*, p. 503.
51. Mandelis, A. (1985). Hamilton–Jacobi formulation and quantum theory of thermal wave propagation in the solid state, *J. Math. Phys. 26*:2676.
52. Mandelis, A., Schoubs, E., Peralta, S. B., and Thoen, J. (1991). Quantitative photoacoustic depth profiling of magnetic field-induced thermal diffusivity inhomogeneity in the liquid crystal octylcyanobiphenyl (8CB), *J. Appl. Phys. 70*: 1771.
53. Mandelis, A., Peralta, S. B., and Thoen, J. (1991). Photoacoustic frequency-domain depth profiling of continuously inhomogeneous condensed phases. Theory and simulations for the inverse problem, *J. Appl. Phys. 70*:1761.
54. Boucher, F., Leblanc, R. M., Savage, S., and Beaulieu, B. (1986). Depth-resolved chromophore analysis of bovine retina and pigment epithelium by photoacoustic spectroscopy, *Appl. Opt. 25*:515.
55. Touloukian, Y. S., Powell, R. W., Ho, C. Y., and Nicolaou, M. C. (1973). *Thermophysical Properties of Matter*, Vol. 10, IFI Plenum, New York, pp. 622–648.
56. Harata, A., and Sawada, T. (1990). Photoacoustic quantitative depth profiling of an ion pair adsorbed polyvinylchloride film using an approximation of the inverse Laplace transform, in *Photoacoustic and Photothermal Phenomena*, Vol. II (Murphy, J. C., Maclachlan-Spicer, J. W., Aamodt, L. C., and Royce, B. S. H., Eds.), Springer-Verlag, Berlin, p. 64.
57. Sugitani, Y., and Uejima, A. (1985). Depth resolved thermal wave imaging of layered samples by correlation photoacoustics, in *Proceedings of the 4th International Topical Meeting on Photoacoustic, Thermal and Related Phenomena*, Esterel, Quebec, Canada, pp. TUD2.1–TUD2.4.
58. Kirkbright, G. F., and Miller, R. M. (1982). Thermal wave imaging of optically thin films using the photoacoustic effect, *Analyst, 107*:798.

59. Cesar, C. L., Vargas, H. Pelzl, J., and Miranda, L. C. M., Phase-resolved photo-acoustic microscopy: Application to ferromagnetic and layered samples, *J. Appl. Phys*, *55*:3460.
60. O'Hara, E. P., Tom, R. D., and Moore, T. A. (1983). Determination of the in-vivo absorption and photosynthetic properties of the lichen *Acarospora schleichieri* using photoacoustic spectroscopy, *Photochem. Photobiol.* *38*:709.
61. Debarre, D., Boccara, A. C., and Fournier, D. (1981). High luminosity visible and near-infrared Fourier transform photoacoustic spectrometer, *Appl. Opt.* *20*:4281.
62. Manning, C. J., and Palmer, R. A. (1989). Design principles and instrumentation for Step-Scan FT-IR, *SPIE Proc. 1145*:540.
63. Manning, C. J., Palmer, R. A., and Chao, J. L. (1991). Step-scan Fourier-transform infrared spectrometer, *Rev. Sci. Instrum. 62*:1219.
64. Rockley, M. G. (1979). Fourier transform infrared photoacoustic spectroscopy of polystyrene films, *Chem. Phys. Lett. 68*:455.
65. Vidrine, D. W. (1980). Photoacoustic Fourier transform infrared spectroscopy of solid samples, *Appl. Spectrosc. 34*:314.
66. Mandelis, A. (1989). Theory of photothermal-wave diffraction and interference in condensed media, *J. Opt. Soc. Am. A6*:298.
67. Inglehart, J., Lin, M. J., Favro, L. D., Kuo, K., and Thomas, R. L. (1983). Spatial resolution of thermal wave microscopes, *IEEE Ultrason. Symp.* 668.
68. Inglehart, L. J., Grice, K. R., Favro, L. D., Kuo, P. K., and Thomas, R. L. (1983). Spatial resolution of thermal wave microscopes, *Appl. Phys. Lett.* *43*:446.
69. Mandelis, A., Williams, A., and Siu, E. K. (1988). Photothermal wave imaging of metal oxide semi-conductor field effect transistor structures, *J. Appl. Phys. 63*:92.
70. Mandelis, A., Borm, L. M. L., and Tiessinga, J. (1986). Frequency modulated (FM) time delay domain photoacoustic and photothermal wave spectroscopies, technique, instrumentation and detection. II: Mirage effect spectrometer design and performance, *Rev. Sci. Instrum. 57*:622.
71. Coufal, H., Moller, U., and Schneider, S. (1982). Photoacoustic imaging using the Hadamard transform technique, *Appl. Opt. 21*:2339.
72. Coufal, H., Moller, U., and Schneider, S. (1982). Photoacoustic imaging using a Fourier transform technique, *Appl. Opt. 21*:2339.
73. Fotiou, F. K., and Morris, M. D. (1986). Hadamard transform photothermal deflection imaging, *Appl. Spectrosc. 40*:704.
74. Treado, P. J., and Morris, M. D. (1988). Modulation transfer function evaluation of a Hadamard transform photothermal deflection imaging system with beam condensing optics, *Appl. Spectrosc. 42*:1487.
75. Favro, L. D., Ahmed, T., Jin, H. J., Chen, P., Kuo, P. K., and Thomas, R. L. (1990). Real time asynchronous/synchronous lock-in thermal wave imaging with an IR video camera, in *Photoacoustic and Photothermal Phenomena*, Vol. II (Murphy, J. C., Maclachlan-Spicer, J. W., Aamodt, L. C., and Royce, B. S. H., Eds.), Springer-Verlag, Berlin, p. 490.
76. Ahmed, T., Jin, H. J., Chen, P., Kuo, P. K., Favro, L. D., and Thomas, R. L. (1990). Real time thermal wave imaging of plasma sprayed coatings and adhesive

bonds using a box-car video technique, in *Photoacoustic and Photothermal Phenomena*, Vol. II (Murphy, J. C., Maclachlan-Spicer, J. W., Aamodt, L. C., and Royce, B. S. H., Eds.), Springer-Verlag, Berlin, p. 30.

77. Lau, S. K., Almond, D. A., Patel, P. M., Corbett, J., and Quigley, M. B. C. (1990). Analysis of transient thermal inspection, *Proc. SPIE, 132*:178.

78. Yang, Y. Q., Kuo, P. K., Favro, L. D., and Thomas, R. L. (1990). A novel "flying spot" infrared camera for imaging very fast thermal-wave phenomena, in *Photoacoustic and Photothermal Phenomena*, Vol. II (Murphy, J. C., Maclachlan-Spicer, J. W., Aamodt, L. C., and Royce, B. S. H., Eds.), Springer-Verlag, Berlin, p. 24.

79. Fournier, D., Lepoutre, F., and Boccara, A. C. (1983). Tomographic approach for photothermal imaging using the mirage effect, *J. Phys. C (Paris) C6-44, suppl. 10*:479.

80. Munidasa, M., and Mandelis, A. (1991). Photopyroelectric thermal wave spatial and depth-resolved imaging with ray optic tomographic reconstruction, in *Proceedings of the Seventh International Topical Meeting on Photoacoustic and Photothermal Phenomena* (STIMPAP), Aug. 26–30, Doorwerth, the Netherlands, *Conference Digest*, p. 288.

81. Vavilov, V. P., Ahmed, T., Dzhin, Kh.D., Thomas, R. L., and Favro, L. D. (1990). Experimental thermal tomography of solids in pulsed, one sided heating, *Sov. J. Nondestr. Test. 26*:882.

82. Favro, L. D., Jin, H. J., Kuo, P. K., Thomas, R. L., and Wang, Y. X. (1991). Real time thermal wave tomography, in *Proceedings of the Seventh International Topical Meeting on Photoacoustic and Photothermal Phenomena* (STIMPAP), Aug. 26–30, Doorwerth, the Netherlands, *Conference Digest*, p. 497.

83. Rosencwaig, A. (1983). Applications of thermal-wave physics to semiconductor materials analysis, *J. Phys. (Paris) 44* (C6):436.

84. Luukkala, M., and Askerov, S. G. (1980). Detection of plastic deformation in metals with a photoacoustic microscope, *Electron. Lett. 16*:84.

85. Vidberg, H. J., Jaarinen, J., and Riska, D. O. (1986). Inverse determination of the thermal conductivity profile in steel from the thermal-wave surface data, *Can. J. Phys. 64*:1178.

86. Busse, G. (1989). Nondestructive evaluation with thermal waves, in *Photoacoustic, Photothermal and Photochemical Processes at Surfaces and in Thin Films* (Hess, P., Ed.), Springer-Verlag, Berlin, p. 211.

87. Thermaprobe, from Thermawave Inc., Fremont CA 94539, U.S.A.

88. Jaarinen, J., Hartikainen, J., and Luukkala, M. (1989). Quantitative thermal wave characterisation of coating adhesion defects, in *Review of Progress in Quantitative Non-Destructive Evaluation* (Thompson, D. O., and Chimenti, D. E., Eds.), Plenum Press, New York, p. 1311.

89. Imhof, I., Thornley, F. R., Gilchrist, J. R., and Birch D. J. S. (1986). Optothermal study of thermally insulating films on thermally conducting substrates, *J. Phys. D: Appl. Phys. 19*:1829.

90. Coufal, H. (1991). Photothermal methods for the measurement of thermal properties of thin polymer films, *Polym. Eng. Sci. 31*:92.

91. Schweitzer, M. A., and Power, J. F. (1991). Unpublished results.

92. Cummins, P. G., Johnson, S. A., Miller, R. M., Allan, A., and Watkins, A. (1990). Impulse response measurements using both photoacoustic and photothermal deflection methods, in *Photoacoustic and Photothermal Phenomena*, Vol. II (Murphy, J. C., Maclachlan-Spicer, J. W., Aamodt, L. C., and Royce, B. S. H., Eds.), Springer-Verlag, Berlin, p. 99.

93. Imhof, R. E., Whitters, C. J., and Birch, D. J. S. (1990). Time-domain optothermal spectro-radiometry, in *Photoacoustic and Photothermal Phenomena*, Vol. II (Murphy, J. C., Maclachlan-Spicer, J. W., Aamodt, L. C., and Royce, B. S. H., Eds.), Springer-Verlag, Berlin, p. 46.

94. Kirkbright, G. F., Miller, R. M., Spillane, D. E. M., and Sugitani, Y. (1984). Depth resolved spectroscopic analysis of solid samples using photoacoustic spectroscopy, *Anal. Chem. 56*:2043.

95. Urban, M. W., and Koenig, J. (1986). Depth-profiling studies of double-layer PVF2-on-PET films by Fourier transform infrared photoacoustic spectroscopy, *Appl. Spectrosc. 40*:994.

96. Yang, C. Q., and Fateley, W. G. (1987). Fourier-transform infrared photoacoustic spectroscopy evaluated for near-surface characterisation of polymeric materials, *Anal. Chim. Acta, 194*:303.

97. Teramae, N., and Tanaka, S. (1985). Subsurface layer detection by Fourier transform infrared photoacoustic spectroscopy, *Appl. Spectrosc. 39*:797.

98. Varlashkin, P. G., and Low, M. J. D. (1986). IR spectral depth profiling using Fourier transform photothermal beam deflection, *Infrared Phys. 26*:171.

99. Dittmar, R. M., Chao, J. L., and Palmer, R. A. (1991). Photoacoustic depth profiling of polymer laminates by step-scan Fourier transform infrared spectroscopy, *Appl. Spectrosc. 45*:1104.

100. Busse, G., Rief, B., and Eyrer, P. (1987). Non-destructive evaluation of polymers with optically generated thermal waves, *Polym. Compos. 8*:283.

101. Miller, R. M., Surtees, G. R., Tye, C. T., and Vickery, I. P. (1986). The study of time dependent phenomena in materials by impulse response photoacoustic spectroscopy, *Can. J. Phys. 64*:1146.

102. Miller, R. M., Surtees, G. R., and Tye, C. T. (1989). Study of dynamic processes by impulse response photoacoustic spectroscopy, *Analyst, 114*:547.

103. Duncan, M. D., Reintjes, J., and Manucci, T. J. (1982). Scanning coherent anti-Stokes Raman microscope, *Opt. Lett. 7*:350.

104. Boyde, A. (1985). Stereoscopic images in confocal (tandem scanning) microscopy, *Science, 230*:1270.

105. N'Soukpoe-Kossi, C. N., and Leblanc, R. M. (1990). Application of photoacoustic spectroscopy in photosynthesis research, *Photosynth. Res. 24*:69.

106. Desormeaux, A., and Leblanc, R. M. (1985). Electronic and photoacoustic spectroscopies of chlorophyll in monolayer and multilayer arrays, *Thin Solid Films, 132*:91.

107. Giese, K., Nicolaus, S., Sennhenn, B., and Kolmel, K. (1986). Photoacoustic in-vivo study of the penetration of sunscreen into human skin, *Can J. Phys. 64*:1139.

108. Imhof, R. E. (1990). Optothermal monitoring of sunscreens in skin, *Phys. Med. Biol. 35*:95.

109. Nagel, E. M., and Lichtenthaler, H. K. (1988). Photoacoustic spectra of green leaves and white leaves treated with the bleaching herbicide norfluorazon, in *Photoacoustic and Photothermal Phenomena* (Hess, P., and Pelzl, J., Eds.), Springer-Verlag, Berlin, p. 552.

110. Helander, P., Lundstrom, I., and McQueen, D. (1984). Photoacoustic study of layered samples, *J. Appl. Phys. 52*:1146.

111. Miller, R. M. (1988). Applications of impulse response photoacoustic spectroscopy to the characterisation of dynamic physiochemical systems, *Spectrochim. Acta, 43B*:687.

112. Kirkbright, G. F., Miller, R. M., Spillane, D. E., and Vickery, I. P. (1984). Impulse response photoacoustic spectroscopy of biological samples, *Analyst. 109*:1443.

113. Nery, J. W., Pessoa, O., Vargas, H., de A. M. Reis, F., Gabrielli, A. C., Miranda, L. C. M., and Vinha, C. A. (1987). Photoacoustic spectroscopy for depth profile analysis and herbicide monitoring in leaves, *Analyst, 112*:1487.

114. Dovichi, N. J., and Weimer, W. A. (1988). Multichannel thermal lens measurements for Absolute absorbance determination with pulsed laser excitation, *Anal. Chem. 60*:662.

115. Boccara, A. C., and Fournier, D. (1990). Photothermal investigations: Why, when and How? in *Photoacoustic and Photothermal Phenomena*, Vol. II (Murphy, J. C., Maclachlan-Spicer, J. W., Aamodt, L. C., and Royce, B. S. H., Eds.), Springer-Verlag, Berlin, p. 486.

# 9

# X-Ray Emission Imaging

**Barry M. Gordon and Keith W. Jones**  *Brookhaven National Laboratory, Upton, New York*

## I.  INTRODUCTION

Shortly after the discovery of x-rays, Barkla [1] established that elements emitted groups of high and low energy x-rays under electron bombardment. In 1913 Moseley [2], in a systematic study of this phenomenon, showed a monotonically increasing relationship between the energy of these $K$- and $L$-shell x-rays and atomic number of the targets, as shown in Fig. 1. The detection and measurement of the characteristic x-rays became the basis of a technique for elemental analysis. The subject of this chapter is the application of the excitation of a sample target and the imaging of the elemental distributions of the target at small spatial resolutions using the characteristic fluorescence x-rays.

While the technique of x-ray fluorescence analysis was established with Moseley's discovery and was applied very early, the sensitivity of the technique has steadily improved with technological advances, and its applicability has widened. In particular the introduction of nuclear techniques has led to the attainment of small spatial resolutions and low detection limits. The influence of the nuclear sciences is shown in the sometimes misleading use of nuclear units to describe the atomic process of ionization, such as quoting cross section values in barns. One liberty taken in this chapter is the specification of x-ray energy $E$ in kiloelectronvolts (keV), not frequency, wavenumbers, or wavelength units. The conversion is $E = 12.3985/\lambda$, where $\lambda$ is wavelength (Å).

X-ray fluorescence analysis and imaging, in particular, have been applied in a wide variety of fields including the biomedical, geochemical, and material sciences. The technique is usually nondestructive and can be performed in air or inert atmospheres. For the most part, the technique has been limited to elemental

**303**

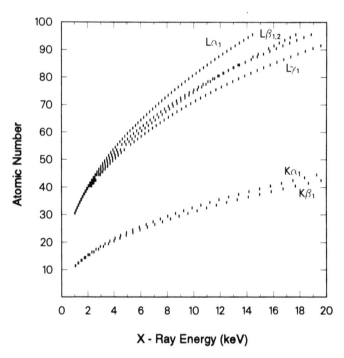

**Figure 1** Moseley's law: fluorescence energies increase monotonically with atomic number. The width of the lines is 150 eV, approximating peak overlaps obtained with solid state detectors. The K lines for higher atomic numbers (not shown) at higher energies do not overlap L lines of existing elements.

distributions without regard to chemical speciation. The introduction of synchrotron radiation sources makes possible the study of chemical speciation through the study of x-ray absorption spectra with narrow energy bandwidth radiation.

X-ray fluorescence (XRF) techniques consist basically of two operations. The first is the excitation of core electrons of the atom to produce holes in the $K$ or $L$ shells. Relaxation of the atom occurs by filling the produced vacancy with higher orbital electrons within the atom. The transition can be radiative, resulting in fluorescence x-ray emission, or the transition can result in emission of a cascade of Auger electrons. The probability of a $K$-shell fluorescence x-ray being emitted—that is, the fluorescence yield $\omega_k$—can be expressed as a function of atomic number $Z$ as follows:

$$\omega_K(Z) = \frac{Z^4}{a + bZ^4} \tag{1}$$

where $a$ and $b$ are constants such that $\omega_K$ varies from near 0 at low $Z$ to 1 at high $Z$ [3]. The L-shell yield, $\omega_L$, varies from 0.1 for middle $Z$ elements to 0.5 at high $Z$. The second part of the technique is the detection and quantification of the fluorescence radiations characteristic of the elements being determined. The application of Moseley's law, and the energy regime used, allow for few artifacts. The parameters of the method are well understood, and yields can be calculated from first principles. Nevertheless, as with most analytical techniques, great reliance is placed on standard reference materials.

## A. Excitation Methods

The energy required to ionize K- and L-shell electrons for fluorescence detection ranges from a few to more than 100 keV. This amount of energy is generally provided by particle and photon beams, the most efficient of which maximize the amount of energy that goes into core electron excitation. For imaging at high spatial resolution, the beams must be focused into a small spot. The beam source is described by the quantity "brilliance," which is defined as the fluence of particles per second per unit area per unit solid angle. The beam can be described by ellipses in phase space representing position and momentum for each dimension perpendicular to the beam propagation direction. Liouville's theorem states that the elliptical areas in phase space do not change while traversing the focusing system. Therefore any focusing of the beam will result in a corresponding broadening of the angular dispersion.

Both x-ray tubes and radioactive sources are energy sources that are useful in bulk analysis and industrial applications. These sources are not accelerator based and can be used in any laboratory and in many field situations. The brilliance of these sources is generally too small to be useful for imaging at trace levels. However, powerful rotating anode x-ray tubes are useful in some tomographic applications. Some success has been achieved with rotating anode x-ray tubes for elemental imaging [4,5].

Electron microprobes are used extensively for imaging distributions of the major and minor elements in a target. Electron projectiles, which are very light, emit a large bremsstrahlung background as energy is lost, thereby limiting the technique in the most favorable cases to detection limits of 100 ppm. It is not possible to focus electrons to the usual 10 nm resolution when bombarding relatively thick targets for x-ray emission imaging because electrons are easily scattered. The scattering will result in an interaction volume comparable with the range of the electron beam.

The electron probe microanalyzer (EPMA) was initially developed in 1951 by Castaing [6]. Since that time extensive work has been done to develop its capabilities. The current state of the field has been reviewed by Newbury et al. [6] and in greater detail by Goldstein et al. [7] and again Newbury et al. [8]. The

EPMA is the most used of all fluorescence probes because of its high spatial resolution and its availability as a commercial product. The EPMA is the best engineered system and possesses the most advanced analytical software of all the instruments described here.

Protons accelerated in the 2–4 MeV range are excellent energy sources for x-ray fluorescence analysis. This energy range is in some ways a compromise among several factors. A large number of such accelerators were used as sources for nuclear physics studies and have been converted to proton-induced x-ray emission (PIXE) systems. The initial microPIXE beam was produced at Harwell shortly after the introduction of the PIXE method [9]. In 1980 there were 20 proton microprobes, and this number has grown substantially [10].

For the trace elements of greatest interest (i.e., through the first-row transition elements), K x-ray production cross sections are at a maximum from 3 to 20 MeV protons. The most important background source in proton bombardment in this range is secondary electron bremsstrahlung. This background has, in terms of the electron mass and the proton mass and energy, an end point approximately equal to $4M_eE_p/m_p$, which represents the maximum energy a proton can transfer to a free electron in a collision. A better signal-to-noise ratio is obtained with 3 MeV protons with a smaller bremsstrahlung background rather than with 20 MeV protons, which are at the cross section maximum but with a greater background. Proton cross sections are highest for low $Z$ elements, which are the most abundant elements, and lowest for the heavier elements of lower abundance. Since the signal one observes is the product of the cross section and the concentration, proton bombardment sensitivity is skewed toward the light elements. The opposite trend is seen with x-ray bombardment.

The newest method for producing characteristic x-rays is synchrotron radiation (SR). SR is electromagnetic radiation emitted by an electron beam stored in a synchrotron ring. As shown in Fig. 2, an electron beam orbiting in a circular path emits dipole radiation in the radiofrequency (rf) region of the electromagnetic spectrum. However, at gigaelectronvolt energies, the electrons are relativistic, and transformation of the frequency of the emitted radiation to the laboratory frame of reference results in the emission of a continuous "white light" spectrum highly collimated in the forward direction.

The electrons are kept at a constant energy by replacing the energy emitted with energy picked up in rf cavities in the orbit. For a 2.5 GeV storage ring such as the National Synchrotron Light Source (NSLS) at Brookhaven National Laboratory, useful intensities extend to 40 keV at the bending magnets. The white light spectrum is many orders of magnitude more intense than that of a rotating anode tube.

Another property that makes SR a good excitation source is collimation in the forward direction. The half-angle vertical dispersion is about $1/\gamma$, where $\gamma$ is the ratio of the electron energy to the rest mass energy, in radians. At the NSLS this

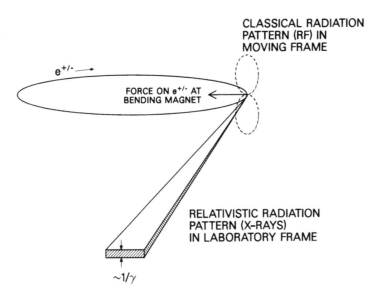

CLASSICAL RADIATION
PATTERN (RF) IN
MOVING FRAME

$e^{+/-}$

FORCE ON $e^{+/-}$ AT
BENDING MAGNET

RELATIVISTIC RADIATION
PATTERN (X-RAYS)
IN LABORATORY FRAME

$\sim 1/\gamma$

**Figure 2** Perspective view showing the radiation pattern emitted by relativistic positrons or electrons in a circular orbit of a storage ring. The radiation is emitted in the plane of the ring with a vertical angular dispersion of $1/\gamma$, in radians.

angle is 0.2 mrad. The light is highly polarized in the plane of the ring, resulting in reduced scattering for observation of targets at 90°. The $4\sigma$ size of the NSLS electron beam giving rise to the SR is 1.5 mm wide × 0.5 mm high. This factor, combined with the high fluence and the small angular dispersion, results in a high brilliance source.

Besides allowing production of collimated beams, the high brilliance of SR makes it possible to obtain small beam spots by focusing the x-rays. Ordinary refractive optical elements cannot be used to focus x-rays because of their short absorption lengths and indices of refraction near unity. Under certain conditions, however, curved mirrors can be used to focus x-rays. Several crystalline materials can be used for Bragg reflection to focus and select monochromatic beams of varying energy bandwidth. While the use of "white light" provides the greatest number of incident photons, monochromatic beams have the advantage of lower scattering backgrounds under the fluorescence peaks.

A narrow energy bandwidth can be used to scan through the absorption edge of an element and thereby obtain information about its chemical state. This technique is called x-ray absorption near edge structure (XANES). The position of the absorption edge is characteristic of the oxidation state of the element and the electronegativity of the bonding atoms. The structure of the absorption edge reflects the bonding, and often, the symmetry of the bonding of the element.

Images of different species of the same element can be obtained by excitation with different narrow bandwidth energies within the range of the absorption edge. SR bombardments can be carried out in air or in a helium atmosphere, allowing, for example, analysis of wet and even live cells. On the other hand, while proton beams have been used in air, they are usually confined to vacuum chambers. With SR, samples need not be coated before bombardment and no charging of the target is observed, as occurs with charged particle bombardment.

## B. Detection Methods

In the incident x-ray fluorescence imaging of photons, ions, or electrons, energy is focused into a single picture element (pixel). Images are formed by raster scanning. Ions and electrons, which are charged particles, can be scanned over a target either by moving the beam with steering elements in the accelerator beamline, or by mechanically scanning the target across a stationary beam. A photon beam, on the other hand, cannot be magnetically or electrically scanned, and so the target must be translated. Target translation is slower than beam scanning. For x-ray fluorescence imaging, resolution is currently in the submicrometer range for protons and about one micrometer for photons.

Proton microprobes are focused after collimation and do not generally have beam-defining pinholes before the target. Photon microprobes using Kirkpatrick–Baez [11] and Wolter [12] focusing have object slits before the focusing optics. In the case of unfocused collimated photon beams, the pinhole must be close to the target to prevent the angular divergence from degrading the resolution. This is particularly true if a focusing mirror is used to increase the fluence, because the divergence is also increased.

Element imaging requires energy measurement of the emitted characteristic x-rays. The natural line widths of characteristic x-rays are in the range of less than 0.1 eV for low $Z$ K lines to over 1 eV for the Mo $K_\alpha$ line [13]. The best energy resolution is obtained with crystal spectrometers and is usually limited by the angular divergence of the accepted radiation or the rocking curves of the analyzing crystal. The efficiency of early detectors was low, and their use was limited to major element analysis using less brilliant sources. The introduction of proton excitation and efficient solid state Si(Li) detectors (SSD) from the nuclear sciences caused renewed interest in x-ray fluorescence analysis and imaging [14]. The SSD energy resolution, typically about 140 eV for Mn $K_\alpha$, provides for good element discrimination, although overlaps sometimes obscure low concentrations in the presence of high concentrations of interfering elements. The effect is diagrammed in Fig. 1, where the bars are 150 eV wide to show possible overlaps between lines of different elements.

The SSD, when the energy resolution is optimized, is limited to a few thousand counts per second by the dead time incurred in pulse processing. Dead time limits our ability to take advantage of the highest beam intensities now available or planned for the near future. Crystal spectrometers, which provide higher resolution with lower efficiency, may become very useful once again with the development of more intense sources. Segmented SSDs, with as many as 100 elements on a common cryostat, are being developed [15]. They are intended to be much more efficient and will increase throughput, but at the cost of lower energy resolution.

## II. COMPARISON OF X-RAY EMISSION ANALYSIS TECHNIQUES

X-ray emission is induced by the interaction of matter with a variety of particle radiations, including photons. A rigorous expression for $N_i$, the number of characteristic x-rays of element $i$ produced per second, is complicated because it takes into account the production of x-rays along the path of the excitation radiation and the attenuation of the fluorescence x-rays on their path out of the target [16]. The limiting expression for production rate of a given fluorescence x-ray in a thin target, where attenuation and of x-rays and energy loss of excitation particles are negligible, is:

$$N_i = I n_i \sigma_i x \tag{2}$$

where $I$ is the fluence, the number of incident excitation particles per second, $n_i$ the number of target atoms per unit volume (i.e., concentration), $\sigma_i$ the production cross section of the interaction leading to fluorescence, and $x$ the thickness of the target. With Eq. (2) and a target containing elemental standards, a working sensitivity curve for the sensitivity (counts/s/ppm/beam current) versus $Z$ can be established for a given geometry and sample matrix. Measurement of peak areas and the beam intensity yields the elemental concentrations. The National Institute for Science and Technology (NIST) has issued a multielemental glass film standard that is certified uniform at micrometer resolution [17]. Less fragile and less costly standards, shown to be uniform at 10 µm, have also been used [18].

The K-shell and L-shell x-ray spectra of the elements are simple and well understood. The analysis of multielement spectra is complicated by interferences between lines of neighboring elements as well as by overlap of L-shell fluorescence of high $Z$ elements with K-shell fluorescence of low $Z$ elements. Further complications result from the presence of scattering backgrounds, escape peaks, and sum peaks. The degree of difficulty of the analysis depends on the resolution of the detector system.

The principal excitation radiations—electrons, protons, and photons—differ significantly in their interaction with matter and have different strengths for trace element analysis. Figure 3 summarizes x-ray production cross sections for electrons, protons, and photons [19]. Detector response is the product of the production cross section and the elemental abundance. Fluorescence cross sections generally increase with $Z$ for photons, but decrease with $Z$ for charged particles. Therefore the response with photons will be more uniform across the periodic table, whereas the use of charged particles will emphasize low $Z$ elements and discriminate against middle $Z$ elements.

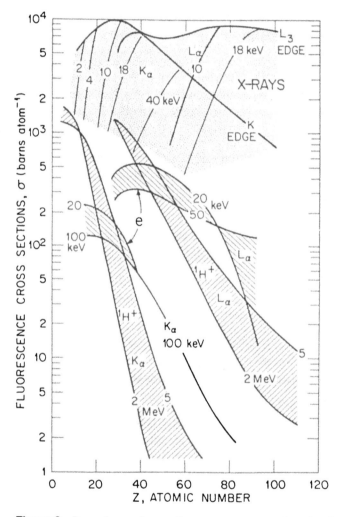

**Figure 3** Dependence of x-ray fluorescence cross sections for electrons, protons, and x-rays, on atomic number. (Reprinted with permission from Ref. 19.)

## A.  Production by Protons

The cross section for producing an inner core vacancy by protons is at a maximum when the proton velocity is comparable to the core electron velocity. The usefulness of electrons as projectiles is limited by the large bremsstrahlung backgrounds resulting from the slowing down in the target. Proton bremsstrahlung, however, is negligible because of the large proton mass. The major source of background is the bremsstrahlung arising from secondary electrons slowed in the target and is concentrated at the low energy end of the spectrum where the cross sections are highest. The maximum of the x-ray production cross section as a function of proton energy for the first-row transition elements is around 10 MeV [20]. The maximum sensitivity, however, is in the 2.5–4.0 MeV range, because of the bremsstrahlung background.

Figure 4 shows a typical PIXE spectrum obtained with a Si(Li)SSD. As noted, the cross sections favor greater efficiency for the low $Z$ elements. This is compensated to some degree by using a thick absorber with a small hole over the detector to limit the low $Z$ count rate. K-shell excitation of high $Z$ elements with low energy protons is impractical because cross sections are small. Cyclotron-generated proton probes up to an energy of 40 MeV take advantage of higher cross sections and have been used at a resolution of 10 μm [21]. The depth of penetration for protons is illustrated in Fig. 5 for a carbon target of unit density. The cross section falls off rapidly with energy. Electrons would penetrate only a few micrometers into the target and would be scattered into a spherical reaction volume.

## B.  Production by X-Rays

The maximum cross-section for the production of x-ray emission is obtained when the photon energy just exceeds the binding energy of the core electron. Above the maximum, the cross section falls off with the 2.7 power of the energy. The radiations observed in the target, in addition to the desired fluorescence lines of the target atoms, are Rayleigh (elastic) scattering and Compton (inelastic) scattering. These radiations do not directly interfere with the fluorescence peaks, but their high intensities tend to saturate SSDs, making them less efficient. An x-ray tube generates monochromatic fluorescence radiation characteristic of the tube's target. Synchrotron radiation has properties, discussed previously, which aid in overcoming deficiencies of the x-ray tube as an excitation source. The most important is the availability of much higher intensities, which allows the use of collimated continuum radiation, or "white light," in a microprobe configuration.

Spectra of an NIST standard reference material (SRM) obtained with a white light source (Fig. 6a) and with a monochromatic source (Fig. 6b) indicate that with monochromatic excitation, the scattering peaks are clearly visible and

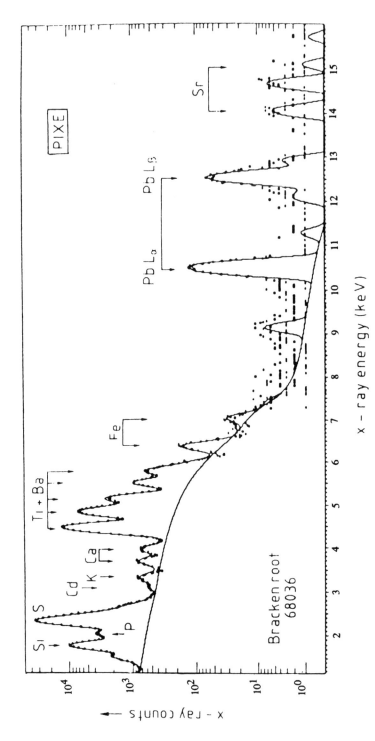

**Figure 4** Integrated PIXE fluorescence spectrum of a bracken root. There is a large bremsstrahlung background at lower energies. For an image of the root, see Fig. 15. (Reprinted with permission from Ref. 46.)

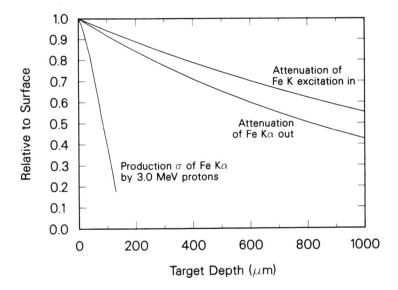

**Figure 5** The depth of field for protons and SR for the analysis of iron in carbon by x-ray fluorescence. Electron excitation samples only the first few micrometers.

interference is limited to the region of the Compton scattering. The backgrounds under the fluorescence peaks are uniformly low and are due to incomplete charge collection in the SSD. The white light spectrum shows a broad and relatively high background, which is the envelope of the scattering peaks from all energies of the excitation spectrum. The NSLS bending magnet source can be used with unfocused collimation to micrometer dimensions without saturating the detector. A focused white light beam would require a crystal spectrometer to avoid dead time problems. The broad continuous SR spectrum permits the selection of any desired energy and so can differentiate extreme overlapping situations. A good example is the overlapping of Pb $L_\alpha$ and As $K_\alpha$ peaks at 10.54 keV. A 13.0 keV beam will excite the As $K_\alpha$ fluorescence but will be below the $L_{III}$ edge of Pb and only the As $K_\alpha$ peak will appear.

As shown in Fig. 5, the absorption depth of x-rays is great and does not allow any depth resolution in samples. However, the relatively long absorption lengths permit the application of fluorescence microtomography techniques with resolution governed by the size of the beam.

## C. Heat Deposition

Because projectiles have very different efficiencies of conversion of lost energy into core vacancy production, and because scanning microprobes require high

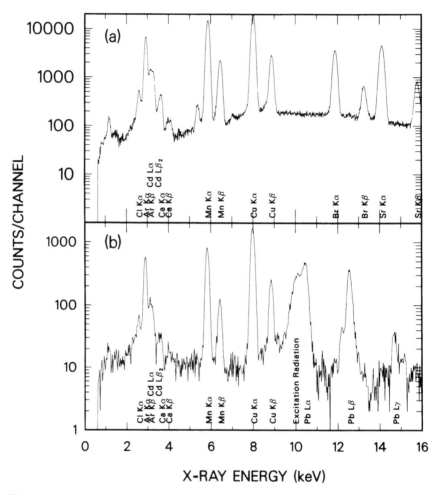

**Figure 6** X-ray emission spectra of NIST SRM 3171 aqueous solution deposited on a Millipore filter and obtained by excitation with (a) continuous SR spectrum and (b) monochromatic 10 keV SR from LBL Kirkpatrick–Baez optics at the NSLS. The Pb peaks in (b) originate from scattering of lead shielding.

intensities in small beam spots, heat dissipation can be a problem in targets that are not good heat conductors. Sparks [22] has compared the energy deposited by different projectiles, at constant detection level. Protons deposit less energy than electrons by one to two orders of magnitude, and monochromatic x-rays deposit less energy than protons by a few orders of magnitude. For a thick biomedical target 10μm thick, a flux of 100 pA of 3 MeV protons ($6 \times 10^8$ particles/s/μm$^2$

will deposit $10^{-5}$W/$\mu$m$^2$, whereas $10^9$ photons/s/$\mu$m$^2$ of 10 keV x-rays will deposit $3 \times 10^{-9}$W/$\mu$m$^2$ As will be seen, proton microprobes can alleviate this situation by rapidly scanning the beam spot. Rapid scanning is not available with an x-ray microprobe. Although it is not now a major problem with x-ray microprobes, sample heating will become more serious as more intense imaging sources are developed at the new light sources.

## III. FOCUSING OF X-RAYS AND PROTON BEAMS

The sensitivity of x-ray fluorescence depends on the deposition of intense particle beams on small areas of the targets. The brilliance of a source is inversely related to the emittance, defined in the phase space of position and angular dispersion as the area of an ellipse circumscribing the beam for each direction perpendicular to the propagation direction. The ellipses represent inclusion of $4\sigma$ of the beam distribution. The areas of the ellipses do not change upon passage through an optical element, but the orientations do change. As the beam is focused at the target position, the ellipses shrink along the position axes and elongate along the angular dispersion axes.

### A. Proton Beams

The initial rapid progress of the PIXE method was aided by the availability of a large number of aging charged particle accelerators, mostly 2–4 MeV van de Graaff machines, whose original purpose was the study of nuclear physics. Because of the early success of PIXE, new and more stable accelerators have been constructed and installed in dedicated PIXE facilities.

Figure 7 is a schematic diagram of a microPIXE facility. The proton beam is accelerated to the desired energy and is passed through a sector bending magnet, which causes focusing in the horizontal direction. The emittance of the beam is established by the object aperture. This aperture must be carefully constructed to prevent scattering, because placing a beam-defining pinhole in front of the target is not desirable. The beam is focused by a set of quadrupole magnets, each of which focuses in one direction and defocuses in the orthogonal direction. Various combinations have been used to correct for aberrations [23]. The proton beam can also be deflected by programmed electrostatic and magnetic fields to scan a stationary target. Repeated scanning can reduce the heating that accompanies proton irradiation by reducing the time spent at any one pixel. Furthermore, scanning the beam is much faster than mechanically moving sample stages, and so more time is available for data collection.

The maximum signal-to-noise ratio is obtained at a backscatter angle, making it advantageous to place the target at 90° to the beam. With no pinhole in front of the target, other detectors can be placed at backscatter angles to monitor the

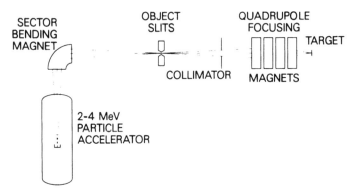

**Figure 7** Schematic diagram of a typical proton microprobe beamline. The beam spot size is fixed by the size of the object slit and the demagnification of the focusing system. The components for scanning the beam are not shown.

interactions. These detectors can determine major element content by Rutherford backscattering (RBS) or can view the sample by secondary electron emission. This contrasts with SR irradiations, where the requirement of keeping the detector at a right angle to the beam means hitting the target at 45°. Most irradiations are carried out in vacuum, because of the energy loss of protons in matter. There are microprobes that do use an external beam formed by extraction through a foil or pinhole with different pumping.

## B. Focusing of X-Rays

Focusing high energy x-rays is more difficult than focusing lower energy radiation because the absorption depths are short and the index of refraction is slightly less than unity. Reflection of x-rays is most easily accomplished with crystals, as with Bragg reflection. The relative energy bandwidth is approximated by the expression

$$\frac{\Delta E}{E} = -\cot \theta \, \Delta \theta$$

obtained by differentiating Bragg's law where $\theta$ is the reflection angle and $\Delta \theta$ is the angular dispersion of the beam. It is preferable to have as wide a bandwidth as possible consistent with the maintenance of monochromatic conditions for the scattered radiation under the fluorescence peaks.

Reflections from perfect crystals yield bandwidths of the order of 0.01%. The commonly used pyrolytic graphite is a mosaic crystal, with a spread of 0.4°, or 3.7% bandwidth at 10 keV. A recent advance is the development of synthetic multilayer structures; these alternating deposits of heavy and light elements produce pseudolayers with large "$d$" spacings. Bragg reflection from such structures at incidence angles of about 20 mrad produces bandwidths up to 10%.

The total reflection of x-rays can be accomplished by reflection at smooth surfaces at very small incidence angles. The index of refraction for x-rays is approximated as $1 - \delta$, where $\delta$ is $10^{-5}$ to $10^{-6}$ [24]. As an x-ray beam enters a mirror surface, the path bends away from the normal only slightly, as shown in Fig. 8. Rays a, b, and c are of the same energy. At some critical angle $i$, the ray will be totally reflected, as represented by $c$. For a given surface, determined by the electron density of the material, there is a high energy cutoff above which x-rays will not be reflected. That energy is proportional to $1/\sin i$. This arrangement can be used to advantage as a high energy filter.

All the reflectors discussed have been produced in bent shapes to achieve focusing as well as monochromation. The sizes of the optical components are dependent on the angles of incidence, the need to accept as much as possible of the 0.2 mrad vertical angular dispersion, and the 1 mrad horizontal acceptance. The necessary beam transport systems and safety considerations require that optical components be situated no closer than a few meters from the source. For a

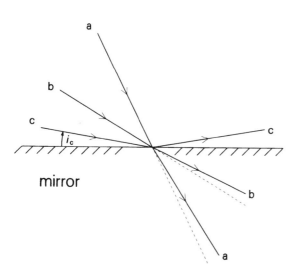

**Figure 8** The path of x-rays at a mirror surface refracts slightly away from the normal. Rays a, b, and c have the same angle. At the critical angle, indicated by ray c, the ray will not enter the mirror and will totally reflect. At this angle, x-rays of higher energy will still be absorbed, and the mirror behaves as a filter of higher energy rays.

crystal at 8 m and a Bragg angle of 30°, the crystal would have to be greater than $3.2 \times 8$ mm$^2$. However, a mirror with an incidence angle of 5 mrad would have to be 320 mm long to intercept the vertical dispersion of the beam.

## 1. Sources

The brilliance of SR sources allows efficient focusing of x-rays for use in imaging trace elements at micrometer resolutions. The brilliances of existing SR sources, represented by the NSLS, and new third-generation SR sources now under construction, are shown in Fig. 9. Most current sources use the bending magnet radiation, since it is the most readily available. The third-generation storage rings now under construction will have many straight sections in the orbit to accommodate wigglers. Wigglers are periodic magnetic structures in the ring straight sections which force the beam to follow a sinusoidal path, in turn causing the emission of SR. The wiggler radiation is characterized by a deflection parameter $K$, given by:

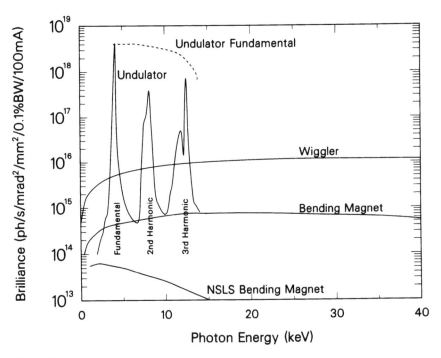

**Figure 9**   The brilliance of radiations planned at the APS and representative of third-generation storage rings in the 6–8 GeV range. The dashed line indicates the range of fundamental energies accessible by changing the gap between magnetic poles. The brilliance of the NSLS bending magnets is shown.

$$K = \frac{eB_0\lambda_u}{2\pi mc} = 0.934\lambda_u B_0 \qquad (3)$$

where $\lambda_u$ is the magnet period (cm) and $B_0$ is the magnetic field in teslas. For $K \gg 1$, as in the case of superconducting wigglers, the radiations at each pole add. The radiation spectrum is characteristic of that magnetic field, and the intensity is multiplied by the number of poles.

When $K \leq 1$ the radiations from each period interfere and add coherently to produce a fundamental energy and higher harmonics of very high brilliance. These magnets are usually permanent alloy magnets and are called undulators. Figure 10 represents an undulator and the resulting angular dispersion of the light. The energy of the fundamental for a given ring and magnet structure is a function of the field strength, which is regulated by a variable gap. The brilliance curve for the undulator in Fig. 10 shows the fundamental and higher harmonics for a given gap distance between pole faces. The dashed line represents the envelope of the fundamental peaks as the gap increases and the energy of the fundamental increases. With these sources, development of much more sensitive x-ray microphobes will be possible in the near future, but the problems associated with beam-induced sample heating and damage will be formidable. An unfocused collimated 1 μm beam of 10 keV undulator radiation at the Advanced Photon Source (APS) will have a fluence of $3 \times 10^9$ photons/s [25]. A focused beam will have a flux of $1 \times 10^{14}$ photons/s/mm² [26].

## 2. Focusing Optics

The SR sources currently available and those planned for the future have a Gaussian intensity distribution and generally have a 4σ diameter of less than 1 mm. To demagnify these sources it is necessary to define a beam area with a collimator, either as an object slit upstream of the focusing element or one downstream immediately in front of the target. The use of slits in front of the target limits detector and TV monitor access to the target. Slits are best suited to

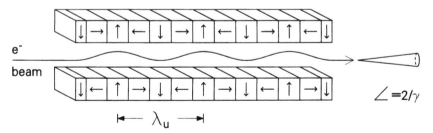

**Figure 10**  The magnetic lattice of an undulator designed to yield a high intensity source strongly peaked in the forward direction.

systems of low demagnification or to collimated unfocused beams with small angular divergence. A highly demagnified beam would necessarily have a high angular divergence and therefore would diffuse the beam between the slit and the target. A pinhole or slit smaller than a micrometer will produce noticeable diffraction effects. The passage of 10 keV photons through a 1μm pinhole 1 cm in front of the target results in a broadening of about 1 μm.

Conical section mirrors operated at grazing angles for total reflection are used to demagnify SR sources. The easiest shape to fabricate for both vertical and horizontal focusing is a spherical mirror. However, because the incidence angle is only a few milliradians, extreme astigmatism is introduced. This can be corrected by using two such mirrors, perpendicular to each other, as in a Kirkpatrick–Baez system [11]. An ellipsiodal mirror can be used to focus a source located at one focal point of the mirror at the second focal point. At small glancing angles an aberration, coma, results from the difference in the demagnifications from the front end of the mirror and the back end. The demagnification is the ratio of the path lengths to and from the mirror. This aberration can be corrected by reflection off two conic sections, as in the Wolter system [12].

Mirrors focus all energies of the SR spectrum up to the high energy cutoff. Although low energy absorbers are used to filter the excitation beam, it is often more advantageous to use crystals to provide monochromatic excitation radiation with varying energy bandwidths. We have described these crystals as monochromating elements, but it is important to realize that these monochromator crystals can also be shaped into a variety of focusing elements.

Two new developments in x-ray focusing will probably play a role in the future development of x-ray microprobe (XRM) systems. The first, the Fresnel zone plate, has been used as a focusing element in x-ray microscopy at energies below 1 keV [27]. The plate is a structure made up of alternating opaque and transparent concentric rings of decreasing widths toward the outside. Zone plates have successfully produced spatial resolutions of about 0.1 μm at energies less than 1 keV. These zone plates have thicknesses that are not opaque to radiations above 1 keV. However, thick zone plates are now being produced by deposition of alternating layers of heavy and light elements on a gold wire substrate [28]. The wire is then sliced to the required thickness to produce the zone plate. These zone plates are being tested in a number of laboratories. The second innovation is the focusing of x-rays by total reflection in tapered capillary tubes [29]. However, multiple reflections within the tube increase the angular divergence of the photon beam and therefore require the tip of the capillary structure to be close to the target.

## 3. X-Ray Microprobe Systems

Most of the types of focusing optics discussed have been applied to XRM systems at various SR sources. The first XRM was constructed by Horowitz and

Howell at the Cambridge Electron Accelerator (CEA) [30]. It used a quartz focusing mirror and a 2 μm pinhole. It suffered from low beam intensities, and its development was halted by the closing of CEA. Nevertheless, it contained all the elements of the microprobes now being developed. The next prominent effort in using a microprobe with x-ray emission detection was the attempt to find superheavy elements in superhalo structures of mica [31]. This system at the Stanford Synchrotron Radiation Laboratory (SSRL) was used for that specific experiment and was not developed further.

Several approaches to imaging microprobes are illustrated in Fig. 11. The system at the NSLS (Fig. 11a) uses an 8:1 ellipsoidal mirror with a Pt coating. The target is on an $X$-$Y$-$Z$-$\theta$ stage that has a 1 μm reproducibility. The surface of the ellipsoidal mirror has imperfections greater than the original mirror specifications but it does increase the beam intensity by a factor of 30. The mirror can be used in conjunction with a monochromator. At present, however, unfocused white light is being used with collimators as small as 2 μm. A collimator is a pair of crossed slits 2.5 cm apart and made from polished tantalum edges separated by thin film spacers. The target is at 45° to the beam and is viewed with a 30 mm$^2$ Si(Li)SSD at 90° to the beam. The sample is observed with a Nikon SMZ-10 binocular zoom microscope equipped with a high resolution, closed-circuit TV camera. The field of view is 2–10 mm, and the optical resolution is about 10 μm.

A group at the Center for X-Ray Optics at Lawrence Berkeley Laboratory (LBL) has developed the Kirkpatrick–Baez system illustrated in Fig. 11b [32]. The crossed spherical mirrors are surfaced with multilayer coatings, which, at a Bragg reflection angle calculated to yield 10 keV radiation, provides an ideal bandwidth of 1 keV. The system has been studied at the NSLS X-26C beamline. The size of the mirrors and the small incidence angle limit the angular divergence accepted. The focused spot is about 5 μm in diameter and does not require a collimator in front of the target. The energy of the microprobe is tunable between 6 and 14 keV.

The XRF analysis group at the Photon Factory (PF) in Japan has been innovative in the development of different approaches to x-ray imaging. The Wolter focusing system is shown in Fig. 11c. It consists of a condenser mirror and a tubular focusing mirror [33]. Each mirror is made of Pt-coated ellipsoidal and hyperboloidal surfaces to eliminate coma. The mirrors are designed for a 7 mrad incidence angle resulting in a high energy cutoff of 10 keV. With demagnification factors of 20 and 13 for the condenser and focusing mirrors, respectively, and a 60 μm pinhole at $F_1$, a beam spot of 4.6 μm was predicted. In fact, the throughput of the second mirror is only 2%, and the final beam spot is 1.6 × 34 μm$^2$.

The Daresbury Synchrotron Radiation Source (SRS) has been used by a group from the University of Warwick and the Free University of Amsterdam as

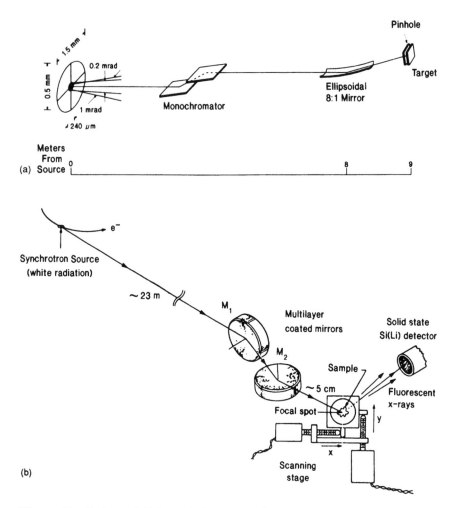

**Figure 11** Designs of high resolution x-ray microscopes: (a) ellipsoidal mirror at NSLS, (b) LBL Kirkpatrick–Baez spherical multilayer surfaces, (c) Wolter focusing system used at the Photon Factory, and (d) curved crystal system used at Daresbury SRS. For details, see text. (Reprinted with permission from Ref. 53.)

the source for a XRM using an Si(111) crystal bent to an ellipsoidal surface. The XRM shown in Fig. 11d achieved a spot size of $10 \times 20 \ \mu m^2$ but is limited in flux because of the narrow bandwidth reflecting from a crystal. This XRM lacks tunability but does have the advantage of being monochromatic.

These examples are representative of the different microprobe designs in use or planned. Table 1 summarizes performance and design data for most of the microprobes now in operation.

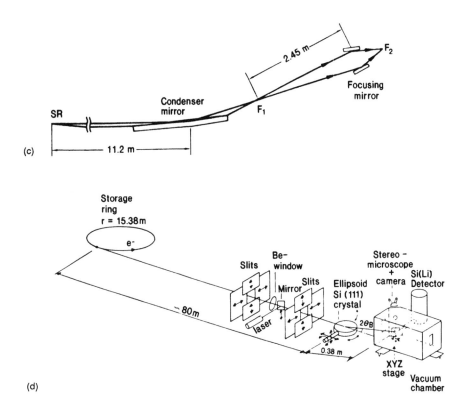

(c)

(d)

## IV.  QUANTITATION AND DETECTION LIMITS

The various components of the x-ray fluorescence signal are all calculable from first principles, theoretically eliminating the need for reference standards. For convenience, however, quantitative measurements are usually made by calibration against standards known to be homogeneous. Trace multielement standards have been used at resolutions of 10 $\mu$m [34,35].

The high sensitivity of the x-ray emission technique depends on detecting single photons and is therefore limited by counting statistics. Almost all XRMs to date have used SSDs for their combination of energy resolution and efficiency. The width of the x-ray lines as determined by SSDs is about 150 eV or greater, largely because of the nature of energy deposition in the detector. The integrated count rate for a given fluence of excitation particles determines the sensitivity of the method for a given experimental apparatus. The ability to measure a small peak in the background radiation establishes the detection limit, generally called the minimum detection limit (MDL).

A recent comparison of detection levels using proton, x-ray, and electron excitation microprobes on the same glass microsphere standard has been reported

**Table 1** Characteristics of SR Microprobes

| Storage ring | Excitation energy (keV) | Beam optics | Spatial size | Photon intensity (ph/s/100 mA) | Ref. |
|---|---|---|---|---|---|
| DORIS | 10–20 ($\Delta E = .44$–1.7) | Graphite Monochromator | 10 $\mu$m | | 54 |
| NSLS (X-26A) | 4–30 | Crossed slits | $2 \times 2\mu$m$^2$ | $1.8 \times 10^8/\mu$m$^2$ ("white") | 55 |
| NSLS (X-26C) | 6–14 ($\Delta E/E \approx 10\%$) | Kirkpatrick–Baez multilayer (LBL) | $6 \times 6\mu$m$^2$ | $5 \times 10^7/\mu$m$^2$ | 32 |
| PF (BL4A) | $\leq 10$ | Wolter | $1.6 \times 34\mu$m$^2$ | $4 \times 10^6$ (8 keV) $8 \times 10^8$ ("white") | 33 |
| PF (BL4A) | $\leq 12$ | Si(111) monochromator, ellipsoidal mirror | 200$\mu$m | $1 \times 10^9$(10 keV) $2 \times 10^9$ (8 keV) | 56 |
| PF (BL4A) | | Zone plate (50 $\mu$m thick) | $\sim 3 \times 10\mu$m$^2$ | $2 \times 10^4$ (8 keV) | 28 |
| PF (BL8C) | 6.2 | Si(111) monochromator ellipsoidal Kirkpatrick–Baez | $4.2 \times 5.5\mu$m$^2$ | | 57 |
| SRS | 15 | Ellipsoidal Si(111) | $10 \times 20\mu$m$^2$ 15$\mu$m | $10^6/\mu$m$^2$ $10^9$ | 58 |
| VEPP-3 | 4–40 | Graphite monochromator | 60$\mu$m 30$\mu$m | (monochromator) ("white") | 59 |

[36]. The microPIXE beam was $5 \times 5$ $\mu$m$^2$ in size and was limited to a current density of 25 pA/mm$^2$ because of charging effects. The NSLS-XRM results were obtained with a $5 \times 8$ $\mu$m$^2$ beam spot and a white excitation spectrum with a maximum fluence at 8 keV of $10^4$ photons/s/mm$^2$/eV/100 mA. The SRS-XRM used a $10 \times 15$ $\mu$m$^2$ focused spot with 15 keV x-rays and a 0.3 keV bandwidth. The electron microprobe operated at 25 kV with a 1 nA beam. The results are summarized in Fig. 12. The microPIXE run exhibited the best signal-to-background ratio for Mn and Fe but suffered from a deficiency of total counts and therefore required a longer count interval to obtain a significant number of events. The microPIXE method is better suited to thin biological matrices at smaller spatial resolutions. The best MDL values were obtained with the NSLS X26 XRM.

The MDL levels discussed below are for single-point analysis where irradiation times may range from 1 to 10 minutes. The scanning process to make distribution maps may allow only a second or two per pixel to complete a scan in a reasonable time. The MDL for the scanning mode would be correspondingly greater.

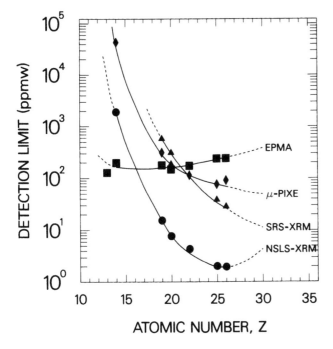

**Figure 12**  Minimum detection limits obtained for four microprobes on NIST SRM K309 glass microspheres of 20–30 μm diameter and 100 second count intervals. (From Ref. 36.)

## A.  Charged Particles

The MDL using electron bombardment is approximately 100 ppm. The size of the interaction volume is about 1–3 μm in all but the thinnest specimens because of scattering of the light projectile. The MDL for protons is lowest for the first-row transition elements and is about 0.5 ppm. For lower Z, the broad bremsstrahlung spectrum interferes, and absorption and low fluorescence yields also contribute to a high MDL. At higher Z, the cross sections for x-ray fluorescence production decrease rapidly. A similar picture holds in the same fluorescence energy range for L-shell excitation for heavier elements.

The quantitation of fluorescence analysis requires a knowledge of the pixel's major element components. PIXE XRMs can take advantage of the proton bombardment by simultaneously measuring backscattered protons. These particles are particularly sensitive to light elements, which are not observable by fluorescence [37]. In the absence of backscattered protons, or in runs where Faraday cup readings of beam intensities are not possible, portions of the bremsstrahlung spectrum may be used to normalize the measurements.

## B.  Synchrotron Radiation

In anticipation of new dedicated facilities, there have been a number of calculations of MDLs using SR excitation [38,39]. The results illustrate the great advantage of tunability of the excitation energy. Assuming an ideal system at the NSLS, an ellipsoidal 8:1 focusing mirror preceded by a multilayer monochromator of $\Delta E/E = 0.02$, one could attain a fluence of $3 \times 10^9$ photons/$\mu m^2$ in a 30 beam spot. These assumptions lead to MDLs of 50 ppb using an SSD in a 1-minute run. It is apparent that because of saturation of the SSD, one could not take advantage of the available beam. In contrast, assuming a wavelength dispersive system (WDS), such as a scanning crystal spectrometer, and a 5-second dwell time per element, the MDLs would be reduced to a range of 2–20 ppb for all elements in an organic matrix.

Another WDS that may play an important role in the future is a stationary analyzer crystal, which disperses the fluorescence to be measured in a position-sensitive proportional counter. Such a detector has been used at the PF and achieved a 12 eV resolution at the Cu $K_\alpha$ line [40]. This detector has the advantage of being positioned to shield the active volume from the excitation and scattered radiation, thereby making it possible to use the much more intense sources being developed.

It is apparent from the reports of the XRMs at the existing SR sources that development of focusing systems is still in its infancy and that beam losses due to deficiencies in optical components are considerable. Properties of components will surely improve with manufacturing experience. Still, MDLs for bulk sample analysis will fall in the range of 50–100 ppb. For the XRM, MDLs are now in the 0.5–1 ppm range for beam spots of 8 $\mu m^2$, in agreement with calculations.

## V.  IMAGING AND SCANNING TECHNIQUES

Automated elemental mapping using x-ray fluorescence began with electron microprobes in which the position of the scanning electron beam of an oscilloscope tube was synchronized with the scanning electron beam of an electron microprobe [41]. If during the short time the microprobe beam impinged on a pixel, a fluorescence x-ray associated with a particular element was detected, the electron gun of the oscilloscope was activated, and the light signal was recorded at that pixel position. If the scan rate was adjusted to the concentration levels for that element, a meaningful spatial distribution of the element was obtained. At best, however, the distribution was qualitative. Multiple events at a pixel were not distinguished, and more quantitative results could be obtained only by adjustment of pixel size and scan rates. More quantitative elemental distributions were obtainable with the introduction of high speed computers with large memory systems.

## A.  Data Acquisition

The amount of data available from a modern microprobe is indeed formidable because spectra are collected with multiple detector systems at tens of thousands of pixels for each scan. Data acquisition systems used today can provide megabytes of memory to store the data and can analyze it in reasonable times. As discussed previously, microPIXE images can be scanned by moving the sample or by deflection of the beam, but only slow sample scanning is available to SR imaging. However, a beam can be scanned with frequencies of 1–10 kHz. In such cases an event entering a "ready" detection system is recorded by noting the energy together with the pixel coordinates. The data can be used to provide real-time images of elemental distributions as they build up in time by sorting for energies in preselected ranges representing the chosen fluorescence lines for each pixel. Generally, this procedure loses all spectral information. At least one facility, the Melbourne proton microprobe, saves all $(E,x,y)$ information for later use [42]. Although disk-intensive, the procedure allows recovery of fluorescence spectra of any region of the scanned area.

The demands of an XRM are similar to those of a proton microprobe, but scanning the sample requires dead time for mechanical stage translation. At the NSLS X-26A beamline, a fluorescence spectrum is accumulated at each pixel. A spectrum is illustrated in Fig. 13. Regions of interest (ROI) in energy representing elemental peak positions are established. The counts in the ROIs, shown by the hatched areas, are corrected for background by using the channels adjacent to the ROIs. Overlapping peaks can be resolved in inclusive ROIs with appropriate channel counts and simple algorithms. In this case only the correct counts for up to 32 ROIs are transferred to disk.

## B.  Presentation of Results

There are many ways to display elemental distributions of a sample. Preferred techniques are matters of individual taste. Examples of various techniques will be used to illustrate applications, although printing limitations restrict us to black and white (BW) reproductions.

The introduction of instrumentation for producing color hard copy has expanded the range of available techniques. The most dramatic is the use of pseudocolor images, where the elemental concentration of each pixel is represented by a color. The image is accompanied by a color bar scale showing the progression of colors through the intensity scale. A blackbody temperature color scale is also used. The scale progresses from black through red to yellow to white as concentrations increase [43]. The advantage is that this scale is intuitive for most scientists and requires little reference to a color bar scale.

The typical BW presentations are three-dimensional surface plots in which the elevation axis representing concentration of an element is plotted versus the

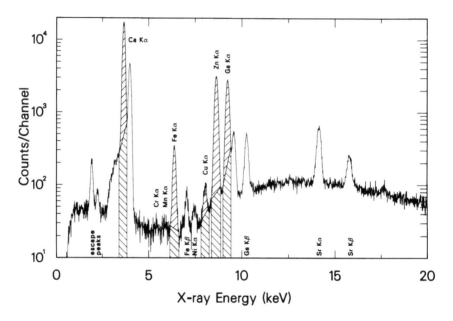

**Figure 13**   A fluorescence spectrum of a rat bone using "white light" SR excitation for a study of calcium resorption using a Ga therapeutic agent. The hatched areas represent the regions of interest used to select data to be collected for imaging scans. The broad background at higher energies reflects scattering from the continuous excitation spectrum.

pixel coordinates. The surface may be a mesh plot or a three-dimensional histogram. The histogram is easier to read when the statistics are poor and the scatter great. The three-dimensional surface can also be drawn as contour lines on a plane surface. Contour plots are sometimes the easiest to interpret because nothing is obscured by the hills and valleys of three-dimensional perspective plots. Care must be taken to choose contour intervals wide enough to avoid the effects of counting statistical noise, which may make the plot look like a jungle. The contour lines should be numbered at intervals, or at least have tick marks to show the slope of the surface in any region. While these plotting routines are generally available, others can easily be constructed. For example, a simple scheme is to draw at each pixel a black square whose area is proportional to concentration, normalized to the highest concentration pixel.

Line plots of any row or column can be extracted from the stored raw data. Also ratios of element concentrations can be plotted for images or line plots. Where available, color can be used to correlate concentrations of two or three elements in two dimensions. Primary colors are used to represent the elements and a triangular mixed color scale is used to show the overlaps.

## C. Applications

The following examples illustrate the wide range of applications of x-ray emission imaging. They have been chosen to show the variety of studies published and the range of spatial resolutions available from the different microprobes discussed here.

Foster and Saubermann [44] have developed EPMA imaging techniques including the examination of frozen hydrated specimens and the use of a personal computer for data acquisition, analysis, and display. Measurements are made in an energy-dispersive spectroscopy (EDS) mode. An elegant example of data obtained using this methodology is given by LoPachin et al. [45], who studied the distribution of Na, P, Cl, K, and Ca in sections of rat peripheral nerve axons and dorsal root ganglion cell bodies. These investigators were able to show that trace elemental concentrations varied among the anatomical structures studied. Figure 14 shows mapping of the elemental distributions in the sciatic nerve in frozen hydrated and dehydrated states.

The study of heavy metal uptake by bracken (*Pteridium aquilinum*) illustrates the application of x-ray fluorescence for trace analysis with high spatial resolution [46]. These plants are mycorrhiza, associations of plant roots and fungi. These are the only plant life found in a waste dump area of Krakow, Poland. The study, performed at the Oxford Proton Microprobe (PMP), examined the hypothesis that the bracken was permitted to grow as a result of metal immobilization by its mycorrhizae. Images of a 60 μm longitudinal section with 0.5 μm resolution are shown for five elements in Fig. 15. The secondary-electron map shows a void in the structure. The Ba concentration in the metal-rich areas is 17%, whereas that of the Cd is 530 ppm and those of Pb and Ca are about 1%. The RBS results showed the presence of $BaSO_4$ crystals. This result is surprising because the Ba concentration is lower than that of the other metals found in the waste dump.

A 2 MeV He$^+$ beam at the Melbourne microprobe was used to study the quality of small GaAs epitaxial regions in GaAs/Si heterostructures grown by molecular beam epitaxy [47]. The cross-sectional view of a structure containing trenches 15, 30, 50, and 100 μm$^2$ is shown in Fig. 16. Figure 17 shows both RBS and x-ray fluorescence images of different size trenches in which the beam was aligned with the GaAs epitaxial regions and also at random angles with the epitaxial regions. The contrast in the RBS images reflects the lower backscattering yields from the regions of epitaxial GaAs compared to yields from the polycrystalline GaAs on the oxide surrounding the trenches. The contrast of the RBS images with deeper penetration decreases because of the dechanneling process as the beam penetrates the sample. The Ga and As K x-ray fluorescence images are poor because of the low cross sections. The Si $K_\alpha$ x-rays should be strongly absorbed by the GaAs overlayer. The PIXE images of the Si indicate

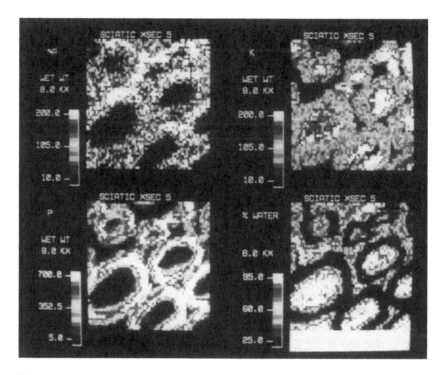

**Figure 14** Digital x-ray image (64 × 64 pixel matrix) of distributions of (clockwise from upper left) Na, K, and P in the hydrated state for several myelinated axons from a distal sciatic nerve section. The sample thickness is 0.2 μm, and the spatial resolution is 0.25 μm. Elemental concentrations are in millimoles per kilogram of wet weight. The original image is in pseudocolor. (Reprinted with permission from Ref. 45.)

that considerable diffusion into the GaAs has occurred and that more diffusion has taken place in the polycrystalline phase than in the epitaxial phase, possibly because of diffusion along grain boundaries.

Conventional theories of the formation of feldspar phenocrysts from large silica magma chambers predict a uniform distribution of trace elements in the solids. Figure 18 shows the distributions of some elements in the Bishop Tuff, California [48]. It is seen that the feldspar phenocrysts, although very uniform in major elements such as potassium, show $\geq$ 20:1 zonation in trace elements such as barium. Because these new observations are inconsistent with current theories, they will change our understanding of how these silica magma chambers evolve. The presence of Sr $K_\alpha$ x-rays in the lower right-hand corner is a result of the x-ray beam striking the target at 45° and the long range of the Sr fluorescence.

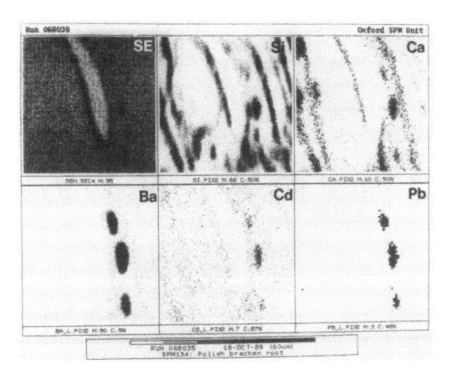

**Figure 15** Secondary-electron (SE), Si, Ca, Ba, Cd, and Pb maps from a longitudinal section of bracken root containing mycorrhizal fungus. The section is 60 μm on a side, and the resolution is about 0.5 μm (Reprinted with permission from Ref. 46.)

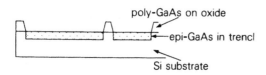

**Figure 16** The configuration of GaAs/Si heterostructures. (Reprinted with permission from Ref. 47.)

Lead is a major public health problem in many countries. It causes learning retardation in children and possibly cardiovascular and kidney problems in adults [49]. Most of the lead in the body is stored in the skeleton, and if released into the blood, it can cause organ damage over long period of time. In vivo measurements of lead are made using K- and L-shell excitation of x-ray fluorescence. The L x-rays are sensitive to the surface concentration, while the K x-rays give information on lead distribution throughout the bone. To make comparisons

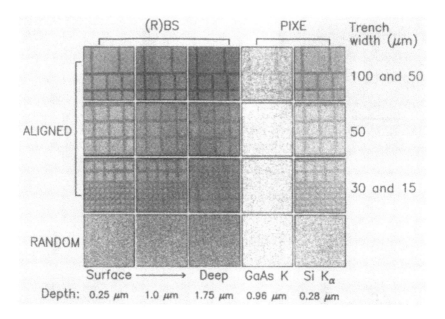

**Figure 17** RBS and PIXE images of GaAs/Si heterostructures with the crystal axis aligned with and at random orientations to the analysis beam. The depths for the RBS images are the mean depths from which the backscattered particles originated. The depths for the PIXE images are the calculated mean x-ray production depths. (From Ref. 47.)

between these methods, the Brookhaven group has studied the spatial distribution of lead in bone, using the NSLS x-ray microprobe. The sample was a thin section of bone from an amputation specimen of a humerus from a 9-year-old boy, who had not been exposed to lead. Maps of the distributions of Ca, Cu, Zn, and Pb over the surface region of the bone were made. The mapped region includes the periosteum, where bone is developing, and a portion of the fully mineralized region of the bone. The maps are shown in shaded surface format in Fig. 19. It can be seen that the metals are highly concentrated at the surface of the bone in the periosteal region outside the region of full calcification. They also show that direct comparison of the two in vivo fluorescence approaches is, in general, difficult.

## VI. CHEMICAL SPECIATION IMAGING

The ability to produce and use very narrow energy bandwidths permits the extension of x-ray fluorescence techniques to chemical speciation imaging of trace elements by microXANES. The spectrum obtained by scanning through the ab-

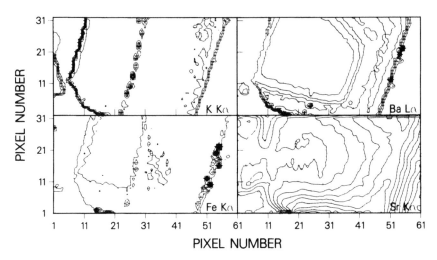

**Figure 18** Contour plot of elemental distributions in Bishop Tuff feldspar. The contour intervals for K, Ba, Fe, and Sr are 2000, 500, 1000, and 100 counts/pixel, respectively. The beam size was $50 \times 50 \ \mu m^2$. The tick marks point in the direction of fewer counts (i.e., downhill). (Reprinted with permission from Ref. 48.)

sorption edge of an element contains information about the oxidation state of the element and the nature of the bonding of the element. For example, the oxidized species of an element will provide less shielding of the core electrons and so will require more energy to ionize, thus a shifting of the absorption edge to higher energy. The electronegativity of ligands and the ionicity of the bonds also contribute to the absorption edge energy. Other features in the XANES region can be indicative of bonding and symmetries.

Most speciation experiments consist of comparing the XANES spectrum of an unknown to those of model compounds covering the range of oxidation states and types of bonding. The range beyond about 40 eV from the edge is termed the EXAFS (extended x-ray absorption fine structure) region and reflects the positions and scattering powers of the neighboring atoms, but tells little about speciation.

Figure 20 illustrates spectra of chromium compounds, used in study of chromium oxidation states in lunar olivine samples [50]. The $K_2CrO_4$ spectrum shows a prominent pre-edge peak characteristic of tetrahedrally coordinated transition elements with empty $3d$ orbitals. The results indicate the prevalence of Cr(II) in lunar olivines, in contrast to terrestial olivines, in which Cr(III) is the major oxidation state.

A group at the PF has imaged chemical species of iron in a Japanese implement from the Yayoi period (300–200 BC to AD 200–300) [51]. Figure 21 is a photograph of the area analyzed. Figure 22 shows spectra of areas of different

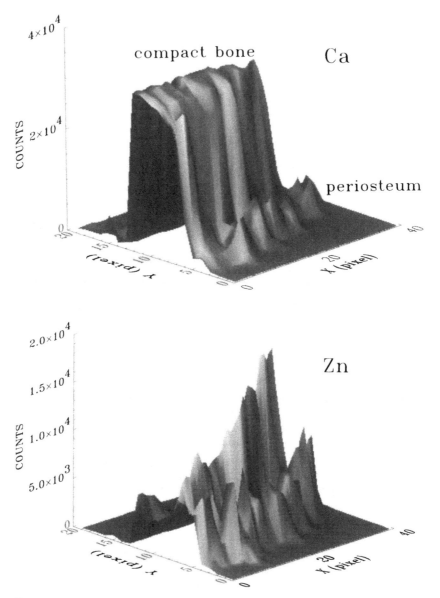

**Figure 19**   Surface distributions of Ca, Sr, Zn, and Pb in a 635 μm thick section of bone taken from the humerus of a 9-year-old boy. The pixel size was 50 μm in both $x$ and $y$ directions. The lead concentration is highest in the region outside the fully mineralized bone. (Reprinted with permission from Ref. 49.)

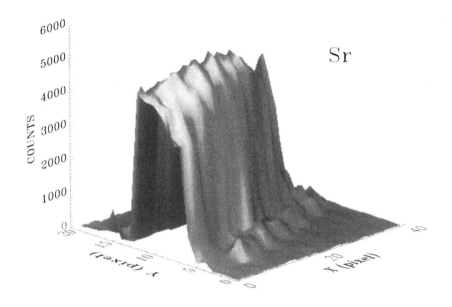

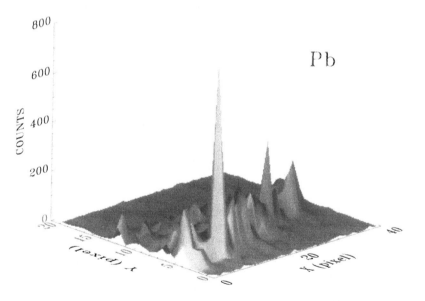

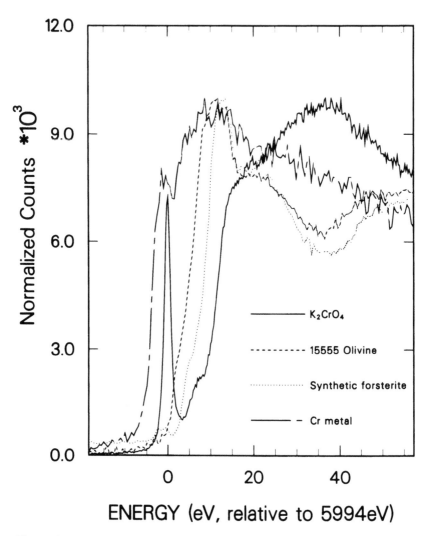

**Figure 20**  XANES spectra taken at the NSLS X-26A microprobe at a spatial resolution of 200 × 200 μm$^2$ for a study of chromium oxidation states in lunar olivine. The spectrum is compared to model compounds $K_2CrO_4$, Cr metal, and $Cr_2O_3$ in synthetic forsterite. (Reprinted with permission from Ref. 50.)

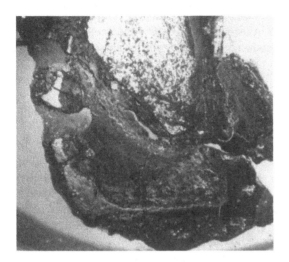

**Figure 21** Polished sample of Yayoi period corroded iron implement, showing zoning of rusts.

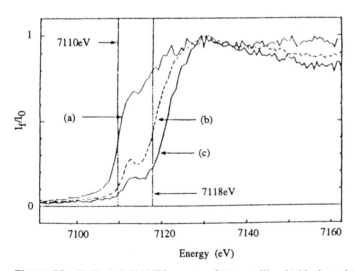

**Figure 22** Fe K-shell XANES spectra of (a) metallic, (b) black, and (c) reddish brown parts of the object in Fig. 21. (Reprinted with permission from Ref. 51.)

colors of metallic iron and rusts. Comparisons with model compounds show the curves a, b, and c to be spectra of metallic iron, $Fe_3O_4$, and goethite (FeOOH), respectively. The scans were produced with a 200 $\mu m^2$ beam spot over an area of 60 × 47 pixels. The scans in Fig. 23 a, b, and c were excited with 7110, 7118, and 7410 eV, respectively. The distributions in Fig. 23 d, e, and f were obtained by appropriate subtraction of a, b, and c according to the relative cross sections at each energy, as shown in Fig. 22. The distributions correspond to metallic iron, $Fe_3O_4$, and iron (III) oxide, respectively.

The detection limits for point-by-point speciation of first-row transition elements are approximately 100 ppm for beam sizes of about 30 $\mu m$, if a focusing mirror is used. This sensitivity will improve greatly with the use of undulators at the third-generation storage rings. It appears that SR is the only technique for imaging chemical species at trace and minor levels by x-ray fluorescence. Speciation has been performed by observing chemical shifts of x-ray fluorescence lines using crystal spectrometers [52]. Because of the low efficiency of such spectrometers, the technique is used currently only for major constituents.

## VII. CONCLUSIONS

This chapter has reviewed the principal excitation sources for XRF imaging, which are electron, proton, and photon beams. Although the principles for using these projectiles are similar, the characteristics of each particle vary widely, and each technique serves different purposes. One should remember that interesting problems exist at all levels of sensitivity and at all resolutions. There is need for specialized techniques available only at a few centralized facilities, as well as techniques locally available to many laboratories. For example, scanning electron microprobes (SEM) are less sensitive than the probes described here, and imaging is limited to major and minor constituents. The spatial resolution is limited to a few micrometers, because of the ease with which electrons scatter. However, SEMs are commercially available and are readily accessible to researchers at their home institutions.

The proton microprobe at present provides the best spatial resolution for XRF imaging and has relatively good sensitivity for trace elements. It is limited to a large extent by the heating and radiation damage caused by the high proton fluences of PMPs. This limitation, particularly for biomedical samples, decreases the value of further development of high intensity ion sources or the use of crystal spectrometers for detecting the fluorescence lines with better resolution than solid state detectors. PMPs are still associated with nuclear physics laboratories, and their use generally requires collaboration with resident staff members. New tandem accelerators are commercially available, and the number of these sources dedicated to PIXE and PMPs is growing.

The x-ray microprobe is the most recent form of the technique and, as such, will gain most by further development. Because of diffraction, the current system

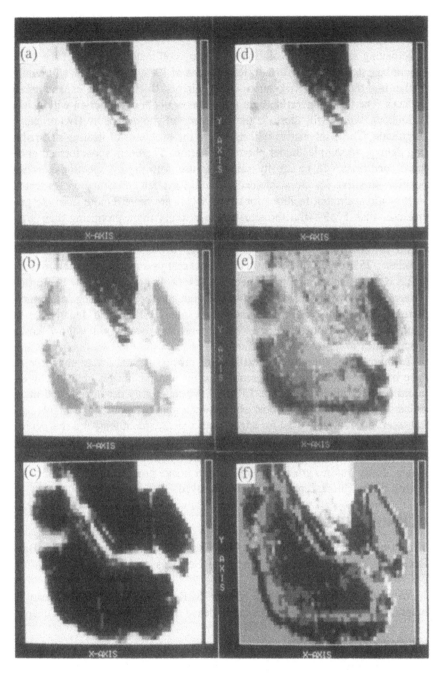

**Figure 23** Fluorescence images of the Yayoi period implement with energies (a) 7110 eV, (b) 7118 eV, and (c) 7410 eV. The pixel resolution is 200 μm². The subtracted images (d, e, and f) show the distributions of metallic iron, Fe₃O₄, and iron(III) oxide, respectively. (Reprinted with permission from Ref. 51.)

of achieving micrometer resolution by placing collimators in front of the target cannot be extended much further. Resolutions of 1000 Å and below will require further improvement of Kirkpatrick–Baez mirrors, thick zone plates, and mirror surfaces. The third-generation light sources now under construction will include undulators, which will increase the brilliance of the source by five orders of magnitude. These intensities will introduce the problems of heating and radiation damage to samples under photon excitation. The beam spots formed under these conditions will be ideally suited for use with crystal spectrometers and position-sensitive detectors shielded to avoid Rayleigh and Compton scattering.

It would seem that facilities for using XRMs are more remote from the general user than PMPs. But the situation is actually more promising than it appears. The only dedicated full-time XRM beamline now in operation is at the NSLS. In addition to time available through collaborations with NSLS staff members, 25% of the time is available to general users, upon application to the NSLS Scheduling Committee. There are several light sources now being built throughout the world, including advanced high energy storage rings at Argonne National Laboratory (APS), Grenoble, France (ESRF), and Japan (Spr-8). There is good reason to believe that general user access will follow similar ground rules to those of the NSLS.

Techniques for computer-assisted image processing and image enhancement have been developed for photographic images and for images obtained by different types of microprobe. The EPMA has been highly developed, and image processing has been applied most extensively within that area. For example, digital filters are routinely used for edge enhancement, smoothing, and sharpening. Since the images obtained with PIXE and SRIXE are generally noisier than EPMA images, image enhancement has seen less use with the former techniques. In the future, greater use of modern image processing will occur as the quality of the fluorescence images is improved.

## ACKNOWLEDGMENTS

This research was supported by the U.S. Department of Energy, Office of Basic Energy Sciences, Division of Chemical Sciences under contract DE-AC02-76CH00016.

We thank our colleagues M. L. Rivers, S. R. Sutton, and G. Schidlovsky for helpful discussions and the use of unpublished data as examples for this chapter.

## REFERENCES

1. Barkla, C. G., and Sadler, C. A. (1908). Homogeneous secondary Röntgen radiations, *Phil. Mag. 16*:550.
2. Moseley, H. G. J. (1913). The high-frequency spectra of the elements, *Phil. Mag. 26*:1024.

3. Bergström, and Nordling, C. (1965). The Auger effect, in *Alpha-, Beta- and Gamma-Ray Spectroscopy* (Siegbahn, K., Ed.), North-Holland Publishing, Amsterdam, p. 1533.

4. Nichols, M. C., and Boehme, D. R. (1988). X-ray microfluorescence: A new elemental imaging tool that complements the SEM, in *Microbeam Analysis—1988* (Newbury, D. E., Ed.), San Francisco Press, San Francisco, p. 389.

5. Golijanin, D. M., and Wittry, D. B. (1988). Microprobe x-ray fluorescence: New developments in an old technique, in *Microbeam Analysis—1988* (Newbury, D. E., Ed.), San Francisco Press, San Francisco, p. 391.

6. Newbury, D. E., Fiori, C. E., Marinenko, R. B., Myklebust, R. L., Swyt, C. R., and Bright, D. S. (1990). Compositional mapping with the electron probe microanalyzer: I and II, *Anal. Chem.*, *62*:1159A, 1245A.

7. Goldstein, J. I., Newbury, D. E., Echlin, P., Joy, D. C., Fiori, C. E., and Lifshin, E. (1981). *Scanning Electron Microscopy and X-ray Microanalysis*, Plenum Press, New York.

8. Newbury, D. E., Joy, D. C., Echlin, P., Fiori, C. E., and Goldstein, J. I. (1986). *Advanced Scanning Electron Microscopy & X-ray Microanalysis*, Plenum Press, New York.

9. Cookson, J. A., Ferguson, A. T. G., and Pilling, F. D. (1972). Proton microbeams, their production and use, *J. Radioanal. Chem. 12*:39.

10. Cahill, T. A. (1980). Proton microprobes and particle-induced x-ray analytical systems, in *Annual Review of Nuclear Particle Science* (Jackson, J. D., Gove, H. E., and Schwitters, R. F., Eds.), Annual Reviews, Palo Alto, CA, p. 211.

11. Kirkpatrick, P., and Baez, A. V. (1948). Formation of optical images by x-rays, *J. Opt. Soc. Am. 38*:766.

12. Wolter, H. (1952). Spiegelsysteme streifenden Einfalls als abbildende Optiken für Röntgenstrahlen, *Ann Phys. 10*:94.

13. Compton, A. H., and Allison, S. K. (1935). *X-Rays in Theory and Experiment*, Van Nostrand, Princeton, NJ, p. 274.

14. Johannson, T. B., Akselsson, R., and Johannsson, S. A. E. (1970). X-ray analysis: Elemental trace analysis at the $10^{-12}$ g level, *Nucl. Instrum. Methods*, *84*:141.

15. Cramer, S. P., Tench, O., Yocum, M., and George, G. N. (1988). A 13-element Ge detector for fluorescence EXAFS, *Nucl. Instrum. Methods Phys. Res. A266*:586.

16. Sparks, Jr., C. J. (1980). X-ray fluorescence microprobe for chemical analysis, in *Synchrotron Radiation Research* (Winick, H., and Doniach, S., Eds.), Plenum Press, New York, p. 462.

17. Pella, P. A., Newbury, D. E., Steel, E. B., and Blackburn, D. H. (1986). Development of National Bureau of Standards thin glass films for x-ray fluorescence spectrometry, *Anal. Chem. 58*:1133.

18. Gordon, B. M., Hanson, A. L., Jones, K. W., Pounds, J. G., Rivers, M. L., Schidlovsky, G., Spanne, P., and Sutton, S. R. (1989). The application of synchrotron radiation to microprobe trace-element analysis of biological samples, *Nucl. Instrum. Methods. Phys. Res. B45*:527.

19. Sparks, Jr., C. J. (1980). X-ray fluorescence microprobe for chemical analysis, in *Synchrotron Radiation Research* (Winick, H., and Doniach, S., Eds.), Plenum Press, New York, p. 466.

20.  Gordon, B. M., and Kraner, H. W. (1972). On the development of a system for trace element analysis in the environment by charged particle x-ray fluorescence, *J. Radioanal. Chem. 12*:181.

21.  McKee, J. S. C., Durocher, J. J. G., Gallop, D., Haldren, N. M., Mathur, M. S., Mirzai, A., Smith, G. R., and Yeo, Y. H. (1990). Micro- and milli-PIXE analysis at 40 MeV—Why, how and when? *Nucl. Instrum. Methods Phys. Res. B45*:513.

22.  Sparks, Jr., C. J. (1980). X-ray fluorescence microprobe for chemical analysis, in *Synchrotron Radiation Research* (Winick, H., and Doniach, S., Eds.), Plenum Press, New York, p. 508.

23.  Johannson, S. A. E., and Campbell, J. L. (1988). *PIXE*, John Wiley & Sons, Chichester, p. 274.

24.  Franks, A. (1977). X-ray optics, *Sci. Prog. Oxf. 64*:371.

25.  Rivers, M. L. (1988). Characteristics of the advanced photon source and comparison with existing synchrontron facilities, *Proceedings of a Workshop on Synchrotron X-Ray Sources and New Opportunities in the Earth Sciences* (Smith, J. V., and Manghnani, M. H., Eds.), Argonne National Laboratory, Argonne, IL, ANL/APS-TM-3. p. 5.

26.  Sparks, C. J., and Ice, G. E. (1990). X-ray microprobe microscopy, in *Proceedings of the 15th International Conference on X-Ray and Inner-Shell Processes*, Knoxville, TN.

27.  Rarback, H., Buckley, C., Goncz, K., Ade, H., Anderson, E., Attwood, D., Batson, P., Hellman, S., Jacobson, C., Kern, D., Kirz, J., Lindaas, S., McNulty, I., Oversluizen, M., Rivers, M., Rothman, S., Shu, D., and Tang, E. (1990). The scanning transmission microscope at the NSLS, *Nucl. Instrum. Methods Phys. Res. A291*:54.

28.  Saitoh, K., Inagawa, K., Kohra, K., and Hayashi, C. (1989). Characterization of sliced multilayer zone plates for hard x-rays, *Rev. Sci. Instrum. 60*:1519.

29.  Carpenter, D. A., Lawson, R. L., Taylor, M. A., Poirier, D. E., Morgan, K. Z., and Haney, G. W. (1988). A scanning x-ray microprobe with glass capillary collimation, in *Microbeam Analysis—1988* (Newbury, D. E., ed.), San Francisco Press, San Francisco, p. 391.

30.  Horowitz, P., and Howell, J. (1972). A scanning x-ray microscope using synchrotron radiation, *Science, 178*:608.

31.  Sparks, Jr., C. J., Raman, S., Ricci, E., Gentry, R. V., and Krause, M. O. (1978). Evidence against superheavy elements in giant-halo inclusions re-examined with synchrotron radiation, *Phys. Rev. Lett. 40*:507.

32.  Wu, Y., Thompson, A. C., Underwood, J. H., Giauque, R. D., Chapman, K., Rivers, M. L., and Jones, K. W. (1990). A tunable x-ray microprobe using synchrotron radiation, *Nucl. Instrum. Methods Phys. Res. A291*:146.

33.  Hayakawa, S., Iida, A., Aoki, S., and Gohshi, Y. (1989). Development of a scanning x-ray microprobe with synchrotron radiation, *Rev. Sci. Instrum. 60*:2452.

34.  Gordon, B. M., Hanson, A. L., Jones, K. W., Pounds, J. G., Rivers, M. L., Schidlovsky, G., Spanne, P., and Sutton, S. R. (1990). The application of synchrotron radiation to microprobe trace-element analysis of biological samples, *Nucl. Instrum. Methods Phys. Res. B45*:527.

35. Giauque, R. D., Thompson, A. C., Underwood, J. H., Wu, Y., Jones, J. W., and Rivers, M. L. (1988). Measurement of femtogram quantities of trace elements using an x-ray microprobe, *Anal. Chem. 60*:855.

36. Janssens, K. H., van Langeveld, F., Adams, F. C., Vis, R. D., Sutton, S. R., Rivers, M. L., and Jones, K. W. (1991). To be submitted.

37. Johansson, S. A. E., and Campbell, J. L. (1988). *PIXE*, John Wiley & Sons, Chichester, p. 52.

38. Gordon, B. M., and Jones, K. W. (1985). Design criteria and sensitivity calculations for multielemental trace analysis at the NSLS x-ray microprobe, *Nucl. Instrum. Methods Phys. Res. B10/11*:293.

39. Grodzins, L. (1982). Electron, proton, and photon-induced x-ray microprobes: Analytical sensitivity versus spatial resolution, *Neurotoxicology, 4*:23.

40. Ohashi K., Iida, A., and Gohshi, Y. (1991). Wavelength dispersive x-ray fluorescence analysis with a position sensitive proportional counter, in *Photon Factory Activity Report No. 8* (Amemiya, Y., Ed.), National Laboratory for High Energy Physics, Tsukuba, Japan, p. 55.

41. Cosslett, V. E., and Duncumb, P. (1957). A scanning microscope with either electron or x-ray recording, in *Electron Microscopy* (Sjöstrand, F. S., and Rhodin, J., Eds.), Academic Press, New York, p. 12.

42. O'Brien, P. M., and Legge, G. J. F. (1988). High speed acquisition and handling of scanning proton microprobe data, *Nucl. Instrum. Methods Phys. Res. B30*:312.

43. Newbury, D. E. (1988). Quantitative compositional mapping on a micrometer scale, *J. Res. Natl. Bur. Std. (U.S.) 93*:518.

44. Foster, M. C., and Saubermann, A. J. (1990). Personal computer-based system for electron beam x-ray microanalysis of biological samples. *J. Microsc. 161*:367.

45. LoPachin, R. M., Lowery, J., Eichberg, J., Kirkpatrick, J. B., Cartwright, Jr., J., and Saubermann, A. J. (1988). Distribution of elements in rat peripheral axons and nerve cell bodies determined by x-ray microprobe analysis. *J. Neurochem. 51*:764.

46. Watt, F., Grime, G. W., Brook, A. J., Gadd, G. M., Perry, C. C., Pearce, R. B., Turnau, K., and Watkinson, S. C. (1991). Nuclear microscopy of biological specimens. *Nucl. Instrum. Methods Phys. Res. B54*:123.

47. Jamieson, D. N. (1991). Nuclear microscopy of semiconductor heterostructures. *Mater. Forum, 15*:51.

48. Lu, F. -Q. (1991). The Bishop Tuff: Origins of the high-silica rhyolite and its thermal and compositional zonations, Ph.D. dissertation, Department of Geophysical Sciences, University of Chicago.

49. Jones, K. W., Schidlovsky, G., Bornschein, G., and Cheeks, C. M. (1991). To be published.

50. Sutton, S. R., Jones, K. W., Gordon, B. M., Rivers, M. L., and Smith, J. V. (1991). Reduced chromium in individual lunar olivine grains: X-ray absorption near edge structure (XANES). Submitted to *Geochim. Cosmochim. Acta.*

51. Nakai, I., and Iida, A. (1991). Application of SR-XRF imaging and micro-XANES to meteorites, archaeological objects and animal tissues, in *Advances in X-ray Analysis*, Vol. 35, Plenum Press, New York.

52. Bai, Y. Z., Fukushima, S., and Gohshi, Y. (1985). Coordination analysis by high resolution x-ray spectrometry, in *Advances in X-ray Analysis*, Vol. 28 (Barrett, C. S., Predocki, P. K., and Leyden, D. E., Eds.), Plenum Press, New York, p. 45.

53. Jones, K. W., and Gordon, B. M. (1989). Trace element determinations with synchrotron-induced x-ray emission, *Anal. Chem. 61*:341A.

54. Petersen, W., Ketelsen, P., Knöchel, A., and Pausch, R. (1986). New developments of x-ray fluorescence analysis with synchrotron radiation (SYXFA), *Nucl. Instrum. Methods Phys. Res. A246*:731.

55. Rivers, M. L., Sutton, S. R., and Jones, K. W. (1991). Synchrotron x-ray fluorescence microscopy, *Synchrotron Radiat. News, 4*:23.

56. Hayakawa, S., Gohshi, Y., Iida, A., Aoki, S., and Ishikawa, M. (1990). X-ray microanalysis with energy tunable synchrotron x-rays, *Nucl. Instrum. Methods Phys. Res. B49*:555.

57. Suzuki, Y., Uchida, F., and Hirai, Y. (1989). X-ray microprobe with a pair of elliptical mirrors, in *Photon Factory Activity Report No. 7* (Amemiya, Y., Ed.), National Laboratory for High Energy Physic, Tsukuba, Japan, p. 183.

58. Van Langevelde, F., Bowen, D. K., Tros, G. H. J., Vis, R. D., Huizing, A., and De Boer, D. K. G. (1990). Ellipsoid x-ray focussing for synchrotron radiation microprobe analysis at the SRS, Daresbury, UK. *Nucl. Instrum. Methods Phys. Res. A292*:719.

59. Baryshev, V. B., Gavirlov, N. G., Daryin, A. V., Zolotarev, K. V., Kulipanov, G. N., Mezentsev, N. A., and Terekhov, Ya. V. (1989). Status of x-ray fluorescence elemental analysis at VEPP-3, *Nucl. Instrum. Methods Phys. Res. A282*:570.

# 10

## Secondary Ion Mass Spectrometry Imaging

**Robert W. Odom**   *Charles Evans & Associates, Redwood City, California*

## I.  INTRODUCTION

Secondary ion mass spectrometry (SIMS) is a chemical analysis technique that employs mass spectrometry to analyze solid and low volatility liquid samples [1]. Although there are numerous configurations of SIMS instrumentation, the fundamental basis of SIMS analyses is the measurement of the mass and intensity of secondary ions produced in a vacuum by sputtering the surface of the sample with energetic ion or neutral beams. The sputtering beam is referred to as the primary beam and typically has a kinetic energy of several thousand electronvolts (keV). The primary beam removes atomic or molecular layers at a rate determined principally by the intensity, mass, and energy of the primary species and the chemical and physical characteristics of the sample [2]. Particle sputtering at the kiloelectronvolt level produces a variety of products including electrons, photons, atoms, atomic clusters, intact molecules, and distinctive molecular fragments. A small fraction of these sputter products are ionized, and these ions are the secondary ions in secondary ion mass spectrometry.

Two forms of SIMS have found extensive application in two distinctive analytical disciplines. The more commonly utilized form is referred to as fast atom bombardment (FAB-SIMS) [3] or liquid SIMS [4]. This SIMS technique provides unique molecular and elemental analysis of chemical compounds dissolved in low volatility solvents such as glycerol or thioglycerol. The FAB-SIMS technique utilizes keV neutrals as the primary species, while the liquid SIMS (LSIMS) technique employs keV ion beams to sputter the atoms and molecules of the analyte from the surface of the liquid matrix. FAB-SIMS and LSIMS are most commonly used to analyze thermally labile organic molecules.

The other common form of SIMS analysis performs chemical characterization of solid samples such as microelectronic materials and devices, and geological, metallurgical, biological, and organic solids and polymers [5]. Solids SIMS techniques can achieve high detection sensitivities within very small analytical volumes. For example, elemental SIMS analyses have demonstrated part- per-billion atomic (ppba) detection sensitivities in the analysis of $10^{-8}$ $cm^3$ of material. SIMS solids analysis also provides unique capabilities for localizing atomic and molecular distributions in two and three dimensions. Two-dimensional ion localization or ion image analysis is accomplished by registering the $X$ and $Y$ positions of mass selected secondary ions formed in the sputtering process [6]. Three-dimensional images are accumulated by imaging $X,Y$ ion intensity distributions from discrete depths as a function of primary ion sputtering depth into the sample [7]. State-of-the-art SIMS ion imaging instrumentation can resolve elemental and small fragment ions produced by ion beam sputtering at resolutions approaching several hundred angstroms [8], while molecular or structurally significant organic or inorganic ions are currently imaged at lateral resolutions approaching 2000 Å [9]. Although solid SIMS, FAB-SIMS, and LSIMS all produce secondary ions by keV particle beam sputtering, FAB and LSIMS cannot perform ion imaging because the analyte molecules are dispersed in and move throughout the liquid matrix.

The unique imaging capability of secondary ion mass spectrometry is the subject of this chapter. This chapter contains sections on the physical principles of SIMS and SIMS ion imaging, methods of quantifying the concentration of specific chemical species from SIMS ion images, and applications of SIMS ion imaging to materials analysis.

## II. SIMS PRINCIPLES

The formation of sputtered ions by ion beam bombardment of a solid surface is a relatively complex process dependent on a number of variables. The discussion that follows is intended to serve as an introduction to the subject, and the interested reader is referred to more detailed articles on these ion formation processes [10].

An energetic ion beam impacting a solid surface sets up a collision cascade in which chemical bonds are ruptured and a variety of chemical species are ejected. The primary ions are implanted into the solid, and the primary ion penetration depth is determined by the ion energy and mass. Figure 1 schematizes this process, in which the collision cascade ejects atoms and molecules or molecular fragments. Since the primary species are embedded into the solid, they become chemical components of the solid generally at relatively low concentration levels ($\leq$ 0.1%). SIMS analyses are performed over a range of primary ion current intensities (or current densities) and primary ion energies using a variety of primary ion species. The type of secondary ions formed and the nature of the anal-

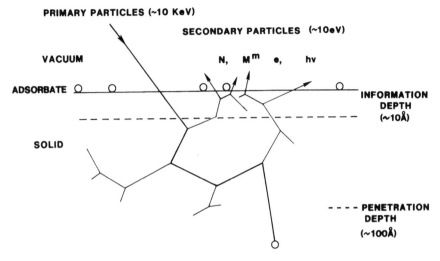

**Figure 1**   Schematic diagram of keV ion sputtering of a solid surface.

ysis depend strongly on these primary ion variables. Consequently, no treatment of SIMS phenomena is complete without a discussion of neutral sputter yields and ion formulation efficiencies. The instantaneous secondary ion current produced by primary beam bombardment of a solid is given by

$$I_A^\pm = \alpha^\pm \, \gamma^\pm \, C_A \, S \, I_p \tag{1}$$

where $I_A^\pm$ is the secondary ion intensity, $\alpha^\pm$ is the ion transmission efficiency for positive $(+)$ or negative $(-)$ ions, $\gamma^\pm$ is the ionization of efficiency of species A, $C_A$ is the concentration of species A in the solid, $S$ is the sputter yield of the solid (number of sputtered species per incident ion), and $I_p$ is the primary ion intensity [11].

The transmission efficiency $\alpha^\pm$ is determined principally by the physics of the sputtering process and the characteristics of the mass spectrometer. Transmission efficiencies typically range between 0.5 and $10^{-4}$. The ionization efficiency $\gamma^\pm$ is a complex function of the electronic and vibrational states of the sputtered neutrals and the solid surface. This ionization efficiency can vary by orders of magnitude (typical ranges are $0.1-10^{-5}$) for different chemical species, and this variability complicates quantification of secondary ion signals [12]. The sputter yield is determined primarily by the energy and mass of the primary ions, and the chemical bonding of the solid and typical sputter yields are 1–10 atoms per primary particle [13].

SIMS analyses are generally performed at either relatively high or very low primary ion intensities. High intensity primary beams are employed for elemental solids analysis calling for the highest detection limits. These *dynamic* SIMS conditions erode or sputter the solid surface at relatively high rates ($\geq$ 100Å/s), resulting in high volume consumption and high atomic sampling, hence high detection sensitivity [14]. Low intensity primary beams are utilized to sputter the constituents of the top surface of a solid. This *static* SIMS analysis produces molecular, organic, or chemical structure analysis of the top few monolayers of the sample, with detection limits ranging from 0.1% to a few parts per million [15].

Figure 2 illustrates the relationship between primary ion current, sample erosion or sputter rate, and detection sensitivity in elemental (dynamic) SIMS analyses using nominal values for the parameters in Eq. (1) [11]. These plots were generated for continuous primary ion bombardment of the sample surface (dc primary ion beam) and clearly illustrate that SIMS elemental detection sensitivities are proportional to the volume of material (i.e., number of atoms) sampled in an analysis. Primary ion intensities are often expressed in current densities, which are represented at the top of Fig. 2. The current density is simply the ion current per unit area and is given for a circular primary beam by

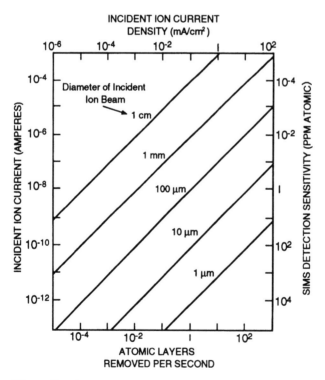

**Figure 2** Diagram of relationship between primary ion current, beam diameter, sputter rate, and detection sensitivity. (From Ref. 11.)

$$\phi_p = \frac{4I_p}{\pi d^2} \tag{2}$$

where $d$ is the beam diameter.

SIMS detection sensitivities ultimately depend on the number of atoms or molecules sampled or consumed during an analysis. Table 1 lists areal and volume densities for a typical solid in which the atomic densities and number of moles are given in centimeter and micrometer dimensions.

Dynamic SIMS elemental detection sensitivities range between 1 part per million atomic (ppma) and 1 ppba using current state-of-the-art instrumentation. These detection limits can be achieved on a microanalytical scale in which the sampling volume is quite small. For example, SIMS analysis for Na contamination in silicon wafers used in the microelectronics industry is on the order of 10 ppb in analysis volumes of $2 \times 10^{-8}$ cm$^3$ (20,000 $\mu$m$^3$), corresponding to an analysis area 150 $\mu$m in diameter and 1 $\mu$m deep [14]. This analysis consumes $2 \times 10^{14}$ atoms (0.3 nmol) of total sample, and the Na detection limit is approximately 0.1 femtogram (1 fg = $10^{-15}$ g).

Static SIMS analyses of molecular or structurally significant fragment ion generally consume less than 1 monolayer ($\sim 4$ Å) of the surface. Since ion detection sensitivities scale with the analytical volume, static SIMS detection limits for molecular and elemental constituents are typically 100–1000 times higher than elemental detection sensitivities observed in dynamic SIMS.

For example, a static SIMS analysis that consumes a complete monolayer over a 100 $\mu$m diameter area will sputter approximately $10^{11}$ atoms. Thus, if one were performing a sodium analysis in the top monolayer of the sample described above, the static SIMS detection limit would be 2000 higher (less sensitive) than a dynamic SIMS analysis.

## III. SECONDARY ION MASS SPECTROMETRY INSTRUMENTATION

The basic components of a SIMS instrument are illustrated schematically in Fig. 3. This instrument contains a primary ion source, sample holder and manipulator, secondary ion extraction optics, some form of mass spectrometer or mass filter, and secondary ion detector or detectors. The complete system is

**Table 1** Areal and Volumetric Solid Densities

|  | Area | Volume |
|---|---|---|
| Atoms | $10^{15}$ atoms/cm$^2$, $10^7$ atoms/$\mu$m$^2$ | $10^{22}$ atoms/cm$^3$, $10^{10}$ atoms/$\mu$m$^3$ |
| Moles | $1.6 \times 10^{-9}$cm$^2$, $1.6 \times 10^{-17}$/$\mu$m$^2$ | 0.016 mol/cm$^3$, $1.6 \times 10^{-14}$ mol/$\mu$m$^3$ |

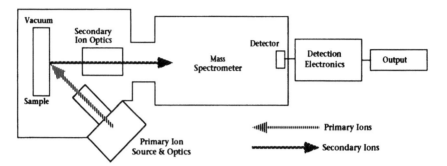

**Figure 3**   Schematic of basic SIMS instrumentation.

typically under a vacuum ranging from $10^{-6}$ to $10^{-11}$ torr. The operational vacuum depends ultimately on the type of SIMS analysis. For example, since a monolayer of residual gas can form on a surface in 1 second at $10^{-6}$ torr, static SIMS analyses are generally performed under ultrahigh vacuum (UHV) conditions ($< 10^{-8}$ torr) to minimize vacuum contamination of the surface. By contrast, since dynamic SIMS typically erodes 10–50 monolayers per second, the vacuum requirements for dynamic SIMS are less stringent, except for the analysis of ''atmospheric'' elements such as C, N, and O. Common primary ion species include inert gas ions such as $Ar^+$ or $Xe^+$, chemically reactive species such as $O_2^+$ or $Cs^+$, and liquid metals such as $Ga^+$ and $In^+$. Oxygen ions are typically employed for electropositive element analysis in dynamic SIMS applications. The oxygen implanted into the solid increases the work function of surface and the higher work function increases positive ion formation efficiencies by several orders of magnitude compared to rare gas ion bombardment [16]. Similarly, $Cs^+$ bombardment enhances the formation of negative ions from electronegative elements by lowering the work function of the surface, which increases negative ion formation efficiencies [17]. Typical primary ion energies range from 1 to 20 keV, and primary ion current densities range from 0.2 A/cm$^2$ for dynamic SIMS analyses to a few femtoamperes (1 fA $= 10^{-15}$ A) per square centimeter for static SIMS applications.

The most common mass spectrometer designs utilized for SIMS analyses include double-focusing sector instruments, quadrupole mass filters, and time-of-flight mass spectrometers [18]. Each of these instrumental configurations have their unique applications in SIMS analyses, and the choice of instrument type depends on such factors as mass resolution and mass range, detection sensitivity, sample throughput, image resolution, ease of operation, and instrument cost. The magnetic sector and quadrupole instruments are currently the most common configurations used for both dynamic and static SIMS analyses. The recent development of high mass range, high mass, and ion image resolution capabilities in time-of-flight secondary ion mass spectrometers (TOF-SIMS) has opened up unique applications for static SIMS analysis of the near-surface regions of a wide variety of materials [19]. Table 2 lists general features of these three mass spectrometer configurations for SIMS applications.

Secondary ion imaging is performed in two different modes: ion microscope and ion microprobe imaging. SIMS ion imaging can be performed under either dynamic or static analysis conditions. The ion microscope employs a large diameter (typically $> 200 \, \mu$m) primary ion beam to sputter the surface of the sample. Specialized secondary ion optics (stigmatic ion optics) transport the secondary ions from the sample surface through a double-focusing mass spectrometer to a two-dimensional secondary ion detector [20]. Figure 4 schematically illustrates ion transport and image formation in the ion microscope. The ion microprobe utilizes a finely focused primary ion beam, which is rastered about the sample surface [21]. Ion images are formed by synchronizing the mass selected ion detection with the primary ion beam position, similar to the synchronization employed in scanning electron microscopy (SEM) [22]. Figure 5 illustrates image formation on the ion microprobe. Each of these methods of forming secondary ion images is discussed in more detail below. This discussion describes dedicated ion microscope and ion microprobe instrumentation as well as SIMS instruments that combine the features of both types of ion imaging technique.

The ion microscope and microprobe techniques have different strengths and weaknesses in ion image analysis. The important factors to be considered with regard to technique performance are mass resolution, image resolution, analysis area or volume, and analysis time. In general, ion microprobe techniques achieve the highest spatial resolution, which currently approaches 200 Å; however, this high spatial resolution is obtained at some compromise in mass resolution and image area or analysis time. By contrast, the ion microscope technique generally performs rapid, high mass resolution analyses over relatively large areas, at image resolutions on the order of 1 $\mu$m. In addition, ion microprobe image resolution is not adversely effected by sample topographies that have reliefs on the scale of tens of micrometers. Ion microscope imaging, however, is very sensitive to sample smoothness, and high peaks and valleys in the sample surface having dimensions of a few micrometers will degrade the image resolution.

To understand these factors in more detail, consider an analytical volume (or voxel) $V$ comprised of an area $A$ and a depth $d$. Since the erosion or sputter rate of a keV primary ion beam is proportional to the ion beam current [Eq. (1)], the total number of atoms consumed in a time $t$ is given by

$$I_p \, S \, t = A \phi_p \, S \, t \tag{3}$$

where $A$ is the analysis area and $\phi_p$ is the primary current density. A major difference between ion microscope and ion microprobe analysis is the difference in

**Table 2** SIMS Commercial Instrumentation

| Mass spectrometer | Mass range (amu) | Mass resolution | Transmission efficiency (%) | Imaging mode | Image resolution (Å) | Manufacturers |
|---|---|---|---|---|---|---|
| Double-focusing | 0–500 | 10,000 (10% valley) | 0.1–0.5 | Ion microscope Ion microprobe | 5,000–10,000 5,000 | Cameca |
| Quadrupole | 0–1000 | Unit mass | 0.01–0.1 | Ion microprobe | 500 (LMIG) | Riber (Cameca) Physical Electronics Vacuum Generators Shimadzu (Kratos) Charles Evans & Associates |
| Time-of-flight | 0–10,000 | 10,000 (FWHM) at mass 30 | 0.5–1.0 | Ion microscope  Ion microprobe | 10,000  2,000 (LMIG) | Charles Evans & Associates Ion TOF (Cameca) Shimadzu (Kratos) Vacuum Generators Physical Electronics |

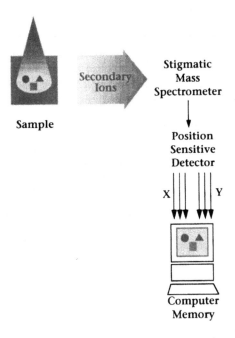

**Figure 4**   Schematic of ion microscope image formation.

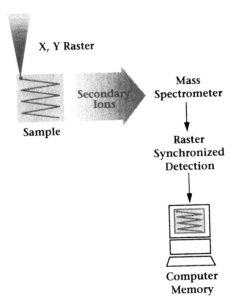

**Figure 5**   Schematic of ion microprobe image formation.

current intensity each technique can deliver to the sample. In general, $\phi_p$ is similar for both techniques in which the maximum current density is determined by space charge limitations in the primary ion optics [23]. However, since the ion microscope performs analyses over much larger areas, it can deliver much higher primary ion currents. Thus the sampling volume per unit time is much higher with this technique. For example, in a SIMS image analysis of a 100 $\mu$m$^3$ volume having a 100 $\mu$m diameter area and a depth of 1 $\mu$m, the analysis can be performed with either the ion microscope operating with a primary ion beam 100 $\mu$m or more in diameter or an ion microprobe using a 1.0 $\mu$m diameter or smaller beam. The ion microscope current is approximately 10,000 times that of the ion microprobe, and thus the ion microscope analysis would proceed at orders of magnitude faster rates. Alternatively, ion microscopes can achieve orders of magnitude higher sensitivities in similar analysis times. Thus, if the image analysis requires large sample areas or volumes, or relatively fast data acquisition times, the ion microscope is the preferred technique. However, if the analysis requires submicrometer lateral resolutions or the analytical areas are micrometers in diameter, the ion microprobe will generally provide the better analysis.

## A. SIMS Ion Microscopes

SIMS ion microscope optics maintain the lateral distribution of the secondary ions sputtered from the sample surface through the mass spectrometer and project a magnified image of the secondary ions onto a two-dimensional ion image detector. Ion microscopes are ion optical analogs of conventional optical microscopes and are equipped with both contrast diaphragms and field apertures for limiting the angular divergence of the imaged ions and the field of view of the analysis, respectively. Figure 6 is a schematic of the most popular of the ion microscopes, the Cameca IMS ion microanalyzer. This instrument was originally developed by Slodzian and Castaing [20] and is manufactured by Cameca Instruments (Courbevoie, France). Cameca ion microanalyzers are double-focusing mass spectrometers having a mass range of 1 to approximately 500 amu and a mass resolution as high as 10,000 (10% valley definition); they are normally equipped with $O_2$ and Cs primary ion sources. Although these instruments can be operated in both dynamic and static SIMS analysis modes, the vast majority of applications are in dynamic SIMS.

Secondary ion images are formed in the ion microscope by projecting a magnified image of the mass-resolved secondary ions onto some form of two-dimensional detector. Several discrete image fields are selected, using different field aperture sizes and projection lens magnifications. Typical image fields range between 25 and 400 $\mu$m in diameter. Image resolutions under optimal conditions are on the order of 1.0 $\mu$m. Ion image detectors are commonly comprised of a two-dimensional ion signal intensifier such as one- or two-microchannel

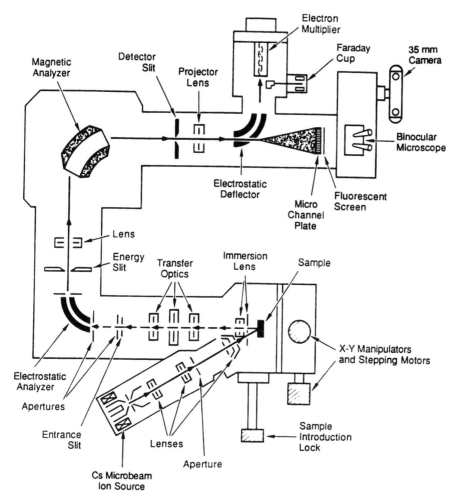

**Figure 6**  Schematic of the Cameca IMS-4f ion microanalyzer.

plate (MCP) ion-to-electron converters and some form of image sensor. The MCP intensifiers convert each spatially resolved positive or negative secondary ion signal into a spatially correlated, amplified electron pulse in which the ion-to-electron conversion gain is approximately 1000 for each MCP [24]. This localized electron pulse subsequently strikes an image sensor such as a phosphor screen, a charge-coupled device (CCD) array, or a resistive anode encoder (RAE). If a phosphor screen detector is utilized, the visible light output from the phosphor can be photographed or coupled into a fast video camera [25]. The camera output can be subsequently digitized, and the image information (position and intensity) can be stored in computer memory [26]. A CCD image sensor provides direct digitized output in which the position of the electron charge pulse exiting the MCP assembly is determined by the cell position in the CCD sensor [27]. The RAE image sensor is a two-dimensional pulse counting detector that computes the $X$ and $Y$ positions of the individual electron pulses arriving at a specially designed resistive film (the resistive anode) [28]. The centroid of the electron pulse striking the resistive anode is calculated by ratioing the charge collected at four corners of the anode. The RAE detector output includes the $X$ and $Y$ positional information (calculated to 8 or 10 bit accuracy) and a count pulse signifying the detection of an individual secondary ion. Figure 7 illustrates the general layout of the ion microscope image detector using the RAE as the image sensor.

All these detectors can detect single secondary ions; hence they have similar detection sensitivities. The main differences between these three types of image detector entail the following characteristics:

1. Quantification of the secondary ion intensity
2. Speed of operation
3. Dynamic range

The phosphor screen–camera system and the CCD array require rather detailed calibration of readout levels (light or charge) with the incident secondary ion intensity [27]. The RAE detector is a pulse counting, position-sensitive detector

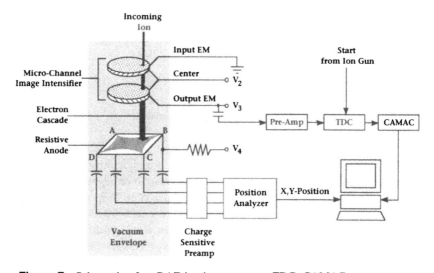

**Figure 7** Schematic of an RAE ion image sensor: TDC; CAMAC.

and as such provides direct quantification of secondary ion signals and calculated positional information [28]. However, the maximum count rate capacity of current RAE detectors is limited to about $10^5$ counts per second (cps), while the video camera and CCD sensors can operate at video frame rates ($\sim$8 Mhz or $8 \times 10^6$ cps). Phosphor screen detectors with MCP intensifiers can accommodate large secondary ion intensity ranges, since the gain in the ion-to-electron conversion in the intensifier can be adjusted by changing the voltage drop across the MCPs. Judicious control of this gain setting can accommodate signals over a dynamic range of approximately $10^9$ [25].

These three image sensors have nearly equivalent image resolutions. The ultimate image resolution in the SIMS ion microscope is not limited by the resolution of the image sensors but rather by aberrations in the ion optics and the transverse kinetic energy of the secondary ions. This latter effect generally dominates the secondary ion image resolution. The sputtering process produces secondary particles (ions and neutrals), which have a discrete kinetic energy distributions. For example, in the keV primary ion sputtering of crystalline silicon, the Si ions typically have a peak kinetic energy of 10 eV and Gaussian kinetic energy distribution [29]. Cluster or molecular ions typically have lower average kinetic energies than atomic ions [30]. The ion microscope employs an immersion lens ion optic to extract the secondary ions from the sample surface, and this immersion lens is also an integral component of the ion image formation optics [20]. The ultimate image resolution achievable using the ion microscope is given by

$$R = \frac{ke_T}{|E|} \tag{3}$$

where $ke_T$ is the transverse kinetic of the secondary ions and $|E|$ is the electric field strength in the extraction region. Typical extraction field strengths in the ion microscope are 1 V/$\mu$m; hence the ultimate resolution of a secondary ion having only 1 eV of kinetic energy is 1.0 $\mu$m. Since this is a very low kinetic energy value for sputtered secondary ions, the image resolutions for atomic ions would generally be much greater than 1.0 $\mu$m if this resolution depended only on the electric field strength in the ion extraction region. However, the ion microscope also utilizes contrast diaphragms of various sizes located at an image focal point, to limit the angular divergence of the secondary ions. By limiting this angular divergence, the contrast diaphragms improve the image resolutions of ions having relatively high transverse kinetic energies. This improved resolution is achieved, of course, at the expense of detection sensitivity, since the contrast diaphragms limit the number of secondary ions reaching the image detector (i.e., the instrument transmission efficiency is reduced at smaller contrast diaphragm diameters).

Figure 8 illustrates SIMS ion microscope imaging using two different contrast diaphragm sizes. These images were produced from a semiconductor test pattern in which a <100> silicon (Si) wafer was implanted with a relatively high dose of $^{11}$B. The intentional doping of Si is a very common method of producing semiconductor materials with specific electrical properties [31]. Ion implantation doping is one method of introducing a known concentration and in-depth distribution of a specific dopant into Si and other semiconductor materials. Ion implantation is performed by forming an ion beam of the intended dopant and implanting or driving this dopant into the Si material by accelerating the dopant ions to a specific kinetic energy. The dopant kinetic energy determines the depth and breadth of the implant, and this energy is typically greater than 50 keV [32]. The test device whose results are illustrated in Fig. 8 was produced by patterning the $^{11}$B$^+$ distribution into strips approximately 20 $\mu$m wide. The $^{11}$B$^+$ dose was approximately $10^{15}$ ions/cm$^2$ and the kinetic energy was 50 keV. The peak in-depth position of this implant is approximately 5000 Å below the Si surface, and the peak concentration is approximately $10^{19}$ atoms/cm$^3$ or 0.1% of the Si concentration.

The ion images in Fig. 8 were acquired on an RAE image detector by summing the $^{11}$B$^+$ ion intensity as a function of depth for two different contrast diaphragm sizes. The field of view in both images was 150 $\mu$m in diameter. The ion intensity in these images is represented using a gray scale (visible at the side of each image). Improved visualization of these ion images can be achieved using false- or pseudocolor scales such as the thermal image scale discussed by Newbury [33]. The upper image was acquired using a 150 $\mu$m diameter contrast diaphragm. Although the $^{11}$B$^+$ intensity is relatively high, the image is somewhat blurred because this contrast diaphragm setting does not significantly limit the angular divergence of the boron secondary ions. The image in Fig. 8b was produced from another region of this test device using a 30 $\mu$m diameter contrast diaphragm. This image shows significantly higher image resolution, approaching the ion microscope limit of 1.0 $\mu$m; however, the ion intensity in this image is lower than that in the upper image by approximately a factor of 5.

The ion microscope can be operated in a nonimaging mode in which no projection of the laterally resolved secondary ion signal is performed. Depending on the signal intensity, these secondary ions are detected using a Faraday cup or conventional electron multiplier detector.

## B. SIMS Ion Microprobe

One of the first commercially available SIMS instruments was the ion microprobe at the Applied Research Laboratory (Waltham, MA), which coupled a microbeam primary ion source with a double-focusing mass spectrometer [34]. This instrument utilized a primary beam probe size ranging from 300 to 2 $\mu$m in

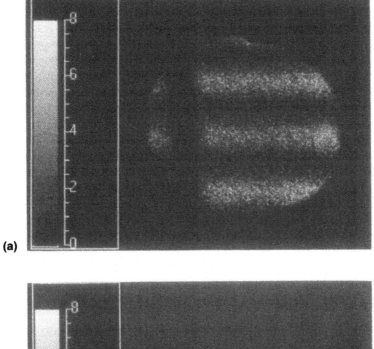

**(a)**

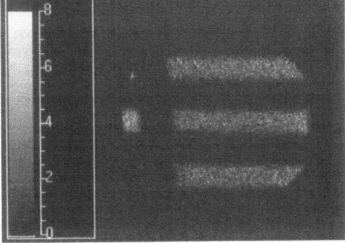

**(b)**

**Figure 8** Ion microscope images of $^{11}$B ions produced from Si test pattern: (a) image acquired with 150 $\mu$m diameter contrast diaphragm (CD) and (b) image acquired with 30 $\mu$m CD.

diameter. The secondary ions were mass analyzed using a double-focusing mass spectrometer, and mass resolutions were on the order of 10,000. The ARL instrument is no longer commercially available, and the highest spatial resolution SIMS ion microprobe instruments currently in operation are the quadrupole system developed by Levi-Setti and coworkers [8] and the magnetic sector instrument developed by Slozdian et al. [35]. Levi-Setti's SIMS microprobe is referred to as the UC-HRL SIMS instrument in acknowledgement of the contributions made to its development by the Levi-Setti group at the University of Chicago as well as researchers at Hughes Research Laboratories in Malibu, California. A schematic of this ion microprobe appears in Fig. 9. The UC-HRL microprobe is comprised of a microfocused $Ga^+$ liquid metal ion (LMIG) source coupled to a quadrupole mass spectrometer. This instrument forms SIMS ion images by rastering the $Ga^+$ over the surface of the sample and synchronizing the detection of mass-selected secondary ions with the position of the primary beam. This ion microprobe typically employs a 40–60 keV primary ion beam energy, and the $Ga^+$ beam can be focused to a spot size ranging from 200 to 5000 Å in diameter. Typical image areas are 20 $\mu$m × 20 $\mu$m or smaller, and the mass resolution of the quadrupole mass filter is approximately unit mass, which means that at mass $M$ the mass resolution is ±1 amu. Thus, this mass spectrometer is not capable of resolving isobaric interferences such as $C_2H_2^-$ and $CN^-$ or the silicon dimer ion ($^{28}Si_2^+$ from $^{56}Fe^+$.

Slodzian's ion microprobe utilizes a $Cs^+$ ion beam focused to 1000 Å spot size and a double-focusing mass spectrometer capable of mass resolutions better than 10,000. This instrument is currently configured to analyze only negative secondary ions.

High secondary ion image resolution with an ion microprobe requires forming tightly focused, high current density primary ion beams. The UC-HRL microprobe achieves 200 Å primary beam spot sizes using a combination of small diameter LMIG source, three sets of electrostatic octupole deflecting lenses, a 5 $\mu$m diameter beam defining aperture, and an Einzel focusing lens. Ion images can be acquired either on photographic film using a high resolution CRT display or acquired into computer memory. Typical raster sizes are 512 or 1024 picture elements (pixels) across, corresponding to image fields of 10–20 $\mu$m on a side.

Although the $Ga^+$ primary ion current is typically very low (~1.0 pA), the current density is high (> 0.1 A/cm$^2$), and this instrument generally performs dynamic SIMS analyses. Levi-Setti and coworkers have investigated a wide range of applications with this ion microprobe and several of these applications are discussed in Section IV. Examples of microbeam ion images produced with UC-HRL instrument are given in Fig. 10. The image in Fig. 10a is the mass 40 positive ion ($^{40}Ca^+$ or $C_3H_4^+$) produced in the analysis of linen (flax) fibers [8]. The scale bar in the image is 10 $\mu$m, and the image resolution is approximately 0.5 $\mu$m. The image illustrates a nearly uniform distribution of the mass 40 ion

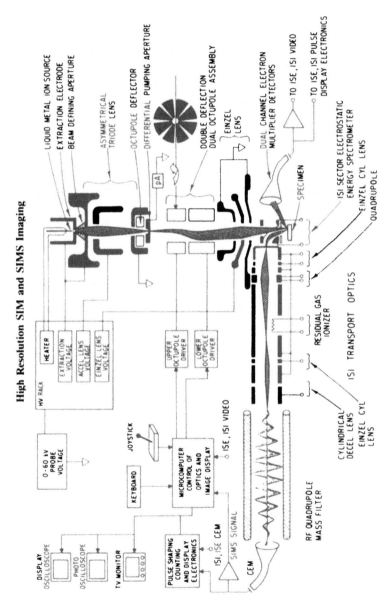

**Figure 9** Schematic of the UC-HRL ion microprobe: (From Ref. 8.)

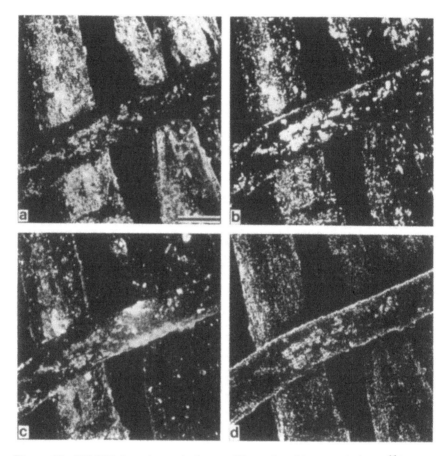

**Figure 10** UC-HRL ion microprobe image of linen (flax) fibers: (a) Ca$^+$, (b) $^{16}$O$^-$, (c) K$^+$, and (d) $^{35}$Cl$^-$. Bar = 10 $\mu$m. (From Ref. 8.)

along the surface of the two lower fibers and nonuniform distribution along the top fiber. As will be discussed in Section III on quantification, this nonuniform ion intensity distribution can reflect both differences in the concentration of the neutral species producing the ions and topographic differences in secondary ion formation probability.

Another version of the SIMS ion microprobe is the time-of-flight secondary ion mass spectrometer (TOF-SIMS) ion microprobe [36]. Figure 11 offers a generalized schematic of this instrument. TOF-SIMS ion microprobes also utilize an LMIG (Ga$^+$) primary ion beam to produce secondary ions. The TOF-SIMS ion microprobe generates mass spectra by pulsing the microfocused primary ion beam and measuring the flight time required for the different mass ions to

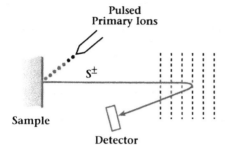

**Figure 11**   Schematic of the TOF-SIMS ion microprobe equipped with an ion reflector.

traverse the distance between the sample surface and the secondary ion detector [5]. Secondary ion images are produced by rastering the primary ion beam over the selected analysis area and synchronizing the location of the detected secondary ions with the position of the primary ion beam. Ion images are stored in computer memory in an array of $X$, $Y$ position and secondary ion intensity values. Time-of-flight mass and image analysis requires pulsing the primary ion beam at relatively low duty cycles $(T \sim 10^{-4})$, and since the direct current (dc) levels in the microfocused $Ga^+$ beam are in the picoampere range, TOF-SIMS analyses are generally performed in the static SIMS mode of operation. Typical primary ion doses are less than $10^{12}$ primary ions/cm$^2$, which corresponds to instantaneous primary ion currents of 0.5 nA/cm$^2$. The rapid pulsing of the LMIG $Ga^+$ beam (typical pulse parameters are 1–50 ns pulse widths at $10^4$ Hz repetition rates) causes the beam to broaden compared to the dc beam utilized in the UC-HRL ion microprobe. As a consequence, the typical $Ga^+$ spot sizes in a TOF-SIMS ion microprobe are 1500 Å or larger [36]. Time-of-flight mass spectra are acquired in TOF-SIMS analysis by measuring ion flight times using either fast time-to-digital converters (TDCs) or transient recorders [37]. The mass analysis range is variable over masses from 1 to 1000 amu or higher.

The mass resolution required for specific TOF-SIMS applications is intimately related to ion image resolution in the ion microprobe technique. In general, mass resolution in time-of-flight mass analysis is given by

$$\frac{M}{\Delta M} = \frac{t}{2\Delta t} = \frac{\ell}{v} = \frac{\ell(m/2qV)^{1/2}}{2\Delta t} \tag{4}$$

where $t$ is the total ion flight time, which is given by the total flight path $\ell$ divided by the ion velocity $v$, and $\Delta t$ is the time width of given mass peak. The ion velocity for singly charged ions is $(2qV/m)^{1/2}$, where $q$ is the electronic charge of the ion, $V$ is the acceleration voltage applied to the ions, and $m$ is the mass of the ion. The time width $\Delta t$ is determined principally by the sum of the ion formation time (primary ion pulse width) and the kinetic energy spread of the secondary ions. Thus, shorter ion formation pulses produce higher mass resolution in TOF-SIMS instruments for constant values of the drift distance and ion acceleration voltage. These two variables are normally held constant for a particular instrument design. Pulsing the microfocused beam on a TOF-SIMS ion microprobe at shorter pulse durations tends to broaden the size of the beam. This broadening decreases the spatial resolution. Thus, one often is faced with a mass resolution/image resolution compromise in TOF-SIMS microprobe analyses. For example, 2000 Å image resolutions can be achieved at full width at half-maximum (FWHM) resolutions between 1000 and 1500 at mass 41. Mass resolutions in excess of 10,000 FWHM can be achieved at mass 41 approximately 5 $\mu$m image resolution.

It is also worth noting that since the mass resolution in time-of-flight mass analyzers is proportional $m^{1/2}$, mass resolutions are usually given at a specific mass.

Commercial TOF-SIMS instrumentation such as the equipment illustrated in Fig. 11 employ some form of ion kinetic energy compensation to minimize the effects of secondary ion kinetic energies on mass resolution. For example, the TOF-SIMS instruments manufactured by Kratos, VG, and Cameca all utilize an ion reflector at one end of the flight path to minimize ion kinetic energy broadening of the detected ion time width [38]. The ion reflector concept was developed by Mamyrin and coworkers in the early 1970s, and optimization of this technique during the 1980s revolutionized the performance of time-of-flight mass spectrometers. For example, before the advent of the ion reflector, mass resolutions in time-of-flight analysis were on the order of several hundred at mass 100. Resolutions with ion reflectors have recently been reported in excess of 10,000 at mass 100 [39].

An example of a TOF-SIMS ion microprobe analysis appears in Fig. 12, which illustrates the distribution of $Na^+$, $Li^+$, and total secondary ions along the surface of a single optical fiber 50 $\mu$m in diameter. The analysis was performed at a mass resolution of approximately 500. The image resolution is less than 0.5 $\mu$m, and the images illustrate that sodium is localized on the surface of the fiber cross section while lithium is distributed along the fiber shaft.

## C. Combined Ion Microscope–Ion Microprobe Instrumentation

There are two instrumental SIMS designs that combine ion microscope and ion microprobe image analysis capabilities. The more recent versions of the Cameca ion microanalyzers (the IMS-4f and IMS-5f models) are equipped with the stigmatic ion optics of the ion microscope as well as a microfocused $Cs^+$ primary ion source. The microbeam source can achieve image resolutions on the order of 0.5 $\mu$m under high mass resolution analysis conditions.

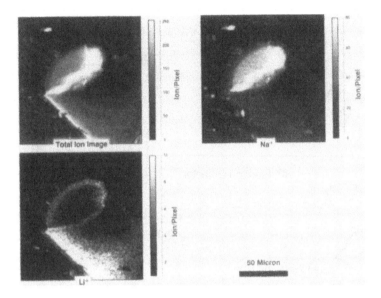

**Figure 12**   TOF-SIMS ion microprobe images of free-standing optical fiber.

The latest development in combined SIMS ion imaging instruments is the ion microscope/microprobe TOF-SIMS system manufactured by Charles Evans & Associates [40]. Figure 13 is a schematic of this instrument. The system is equipped with a large-diameter $Cs^+$ ion source for ion microscope analysis and an LMIG ($Ga^+$) microbeam source for ion microprobe analysis. Ion images are formed in the ion microscope mode using stigmatic optics very similar to those utilized on the Cameca ion microscope, and ion images are acquired on an RAE imaging detector. Energy compensation for secondary ion kinetic energy distributions is achieved by using a triple electrostatic energy analyzer (ESA), system which accommodates a several hundred eV kinetic energy spread while maintaining the lateral distribution of secondary ions in the ion microscope. Ion microprobe analysis employs synchronization of the beam raster position with secondary ion detection.

## IV.   QUANTIFICATION

Although quantitative SIMS analyses can be performed using the fundamentals of ion sputtering and ion yields, realistic SIMS quantification having analytical accuracies better than ±20% requires the analysis of standard samples of the analyte element or molecule in the chemical matrix of interest. These standards are necessary because secondary ion yields depend on both the analyte species and its chemical environment. The following is a brief explanation of these effects.

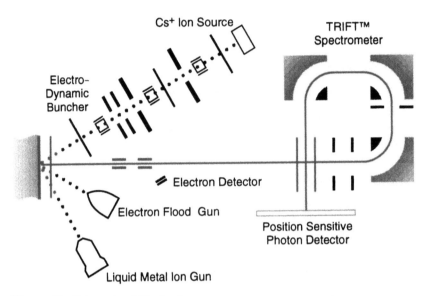

**Figure 13** Schematic of Charles Evans & Associates combined ion microscope and ion microprobe TOF-SIMS.

Different chemical matrices have different sputter yields and thus, the ion yields for a given element or molecule may vary significantly depending on the chemical matrix. This effect is referred to as the substrate matrix effect [41]. This effect has been rationalized for chemically reactive primary ions such as $O_2^+$ or $Cs^+$ based on differences in the steady state concentration of the primary ion species in the sample matrix. In addition, secondary ion yields [$\gamma\pm$ in Eq. (1)] for different elements or molecules contained within the same chemical matrix vary widely. This ion yield variation, known as the SIMS matrix effect, is a function of the electronic (and vibrational) states of both the sputtered species and the surface as well as the chemical bonding of the analyte to the surface. Positive elemental SIMS ion yields have been shown to be strongly dependent on the ionization potential (IP) of the sputtered element, while negative ion formation probabilities are strongly dependent on the electron affinity (EA) of the element. The SIMS matrix effect can be described approximately by the Saha–Eggert equations, which are given by [42]

$$I_{M^+} = \frac{M^+}{M_0} = A^+ \exp\left(-\frac{IP_M}{B^+}\right) \tag{5}$$

and

$$I_{N^-} = \frac{N^-}{N_0} = A^- \exp\left(\frac{EA_N}{B^-}\right) \tag{6}$$

where $M_o$ and $N_o$ are the neutral concentrations of M and N, $A^\pm$ and $B^\pm$ are constants, and $IP_M$ and $EA_N$ are the ionization potential and electron affinity of M and N, respectively. Ion yields in SIMS can vary by several orders of magnitude, depending on the exact values for the IP or EA of the analyte species and the chemical identity of the primary ion species. For example, the most sensitive elemental SIMS analyses are achieved with the alkali and alkaline earth elements (those with relatively low values of IP), while the most sensitive negative elemental analyses are achieved for the halides (high electron affinity values).

Although these effects can be reasonably well parametrized for any type of analysis, they effectively prevent standardless SIMS quantitative analysis.

SIMS quantitative analyses are performed by generating relative sensitivity factors (RSFs) for elements of interest in substrates of interest.

The RSF is defined as

$$RSF\ (X/Y) = \frac{I_X}{I_Y}\bigg/\frac{f_Y C_Y}{f_X C_X} \tag{7}$$

where $I_X$ and $I_Y$ are the ion currents of elements $X$ and $Y$, $f_i$ are the isotopic abundances (where appropriate) for these ions, and $C_X$ and $C_Y$ are the elemental concentrations.

A suite of SIMS elemental standards have been developed over the last two decades, and these standards permit tabulation of RSFs for a large collection of sample types [43]. These RSF values are, of course, specific to specific analyte and chemical matrices. To date, no satisfactory molecular standards have been developed for static SIMS analyses. However, several research groups are actively pursuing the development of such standards [44].

A third sputter yield effect is operative in the quantification of SIMS ion images. This effect is produced by the dependence of sputter yield on the angle of incidence of the primary ion beam with the sample surface. In its simplest form, the sputter yield is proportional to the cosine of the angle between the primary ion beam and the sample surface measured from the normal to the surface [45]. Since most SIMS instruments have a fixed, nonnormal orientation of the primary ion beam with respect to the plane of the sample surface, different areas on structured samples having three-dimensional relief will present differing angles of incidence to the primary beam. Thus, ion yields can vary across a structured sample solely because of this sputter yield effect. The variation in the mass 40 ion intensity across the flax fibers shown in Fig. 10 is most certainly due to both concentration differences and topographic effects.

In principle, this topographic ion yield effect can be determined from standard samples, and quantitative SIMS ion images can be produced [46]. Image quantification requires applying RSF values to an ion image on a pixel-by-pixel or voxel-by-voxel basis [47]. In practice, however, it is often difficult to prepare good structured standards for most analytically useful applications. Several research groups are developing imaging standards, and data produced from these groups are discussed in the next section [48]. In the absence of such standards and standardization procedure, SIMS ion images must be considered to be a qualitative measurement of the relative concentration of various constituents.

## V. SIMS IMAGING APPLICATIONS

The power of SIMS imaging is its ability to localize elemental and molecular or structurally significant secondary ions produced from a wide range of sample materials. This section provides an overview of SIMS imaging applications in biological, microelectronic, geological, polymeric, and organic materials. Biological ion imaging is probably the most mature of these applications, and results obtained in this field are a good measure of both the capabilities and limitations of SIMS ion microscopy. A number of biological applications are presented because these applications center about improving our understanding of the chemistry of living systems and because these analyses often generate spectacular ion images. Solid state physicists and engineers have employed SIMS imaging to elucidate the details of microelectronic fabrication chemistry and to study device failures resulting from improper chemical controls or unwanted contaminants. Researchers in geology and metallurgy have also employed SIMS imaging to determine the presence of minor constituents in ores and metals and to determine the composition and distribution of grain boundaries. Polymer and organic SIMS imaging, a relatively new field of application, has enjoyed vigorous growth since the development of TOF-SIMS imaging techniques and instrumentation.

## A. Biology

SIMS ion imaging has probably found its greatest application in the analysis of biomedical or biological materials [49]. The high elemental detection sensitivity and isotope selectivity of SIMS ion imaging make this technique a natural extension of other biomaterials microanalysis techniques such as scanning electron microscopy (SEM), transmission electron microscopy (TEM), and electron probe microanalysis (EPMA) [50]. In addition, static SIMS ion image analyses can, in principle at least, achieve molecular or chemical structure imaging at micrometer and submicrometer lateral resolutions [51]. Although SIMS elemental ion imaging of biological materials has been actively pursued by a number of research groups over the past 15–20 years, quantitative image analysis in which elemental concentrations have been measured to better than ±20% accuracies have been achieved for only a few biological systems [48]. Thus, SIMS ion imaging of biological materials provides by and large qualitative chemical mapping of the major, minor, and trace components in soft biological tissue, bone, and teeth.

Even as a qualitative imaging technique, SIMS imaging of biological materials requires specialized sample preparation to provide at cellular and subcellular dimensions elemental and molecular distributions that accurately reflect in vivo distributions. There are, of course, a variety of biological tissue preparation techniques for various types of microscopy. Figure 14 illustrates several of these methods of preparing soft biological tissue for light microscopy, SIMS, TEM, EPMA analyses [52]. Different sample preparation protocols maintain the cellular structure at differing levels of resolution and chemical integrity. The objective of all these sample preparation procedures is to remove or immobilize

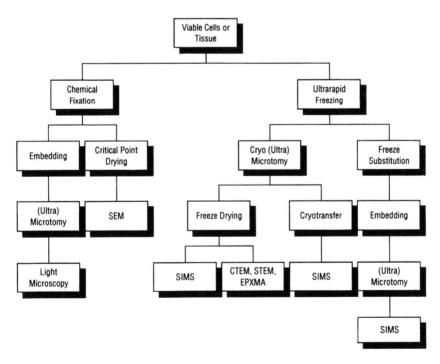

**Figure 14**   Schematic of biological tissue preparation methods

(freeze) tissue water so that it does not expand in the instrument vacuum system, destroying cell walls and membranes and transporting chemical species from their indigenous locations.

Chemical fixation of soft biological tissue generally introduces chemical contaminants into the tissue and may transport mobile chemical species throughout the tissue. Thus, chemical fixation is generally not a good technique for SIMS tissue analyses, although in certain specific applications, chemical fixation does not seriously compromise sample integrity. For example, Kupke and colleagues have demonstrated the utility of SIMS analysis of fixed tissue specimens for the identification of $BaSO_4$ inclusions in perirectal abscesses [53]. These SIMS imaging results were correlated with light optical and SEM imaging, providing a detailed microscopic characterization of these inclusions. Larras-Regard and coworkers have also prepared useful biological tissue samples for the analysis of Fe and other relatively immobile elements using a combination of chemical fixation, cryoembedding, and microtomy [54].

The optimum soft tissue preparation maintains the ultrastructural (i.e., subcellular) composition and morphology of the tissue. This type of sample preparation requires rapid (shock) freezing of the specimen, cryomicrotomy of selected regions of the frozen material to expose regions of interest, and controlled freeze-drying of the cut, frozen section. Alternatively, the cryomicrotomed section can be analyzed frozen if the instrument is equipped with a cryotransfer system and cryostage for sample mounting in the instrument [55]. If the specimens are rapidly frozen (freezing rates $> 10^4$ K/s) and the frozen material is kept below 130 K, ice crystal formation is minimized because ice can be maintained in a vitreous (noncrystalline) state below this temperature. Ice crystals disrupt cellular membranes, and their formation is one of the primary causes of chemical translocation in frozen biological specimens [56].

Cryogenic sample preparation techniques require specialized equipment such as shock freezing systems (e.g., liquid nitrogen or liquid helium slammers, cryopliers), cryomicrotomes or ultracryomicrotomes, and low temperature freeze-dryers or cold stages. Most laboratories are not equipped with all this apparatus; hence, many of the applications discussed below have employed higher temperature freezing and freeze-drying procedures or some form of chemical fixation.

Chandra and colleagues have developed a novel method of preparing freeze-dried cell culture samples for SIMS analysis [57]. This technique involves growing cells on the native oxide ($SiO_2$) surfaces of pieces of high purity silicon. Another piece of silicon is placed over the culture cells, forming a sandwich, which is plunged into a liquid nitrogen slush and frozen. The frozen silicon pieces are cleaved, producing frozen regions of the intact culture cells with the cell membrane removed. This sample is freeze-dried and analyzed.

The general capabilities of SIMS ion imaging in biological materials analysis are illustrated in Figs. 15–20. Figure 15 is an ion microscope image of the $^{40}Ca^+$

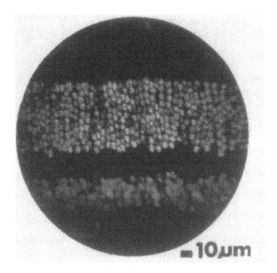

**Figure 15**   Ion microscope image of $^{40}Ca^+$ produced from cat retina; 150 $\mu$m image field. (From Ref. 58.)

ion produced from a dehydrated and epoxy-embedded section of cat retina [58]. The image field is 150 $\mu$m in diameter, and the image was recorded with photographic film. The bilobed region of the sample contains intense Ca signal intensities where the separation between the lobes is approximately 5000 Å. Burns and coworkers have investigated the distribution of a variety of alkali and alkaline earth elements in retinal tissue using SIMS ion imaging. This research is directed at understanding the chemistry of light adaption in photoreceptors [59].

Galle and coworkers have performed extensive ion image analysis of marine organisms and have identified in some cases distributions of trace elements or unusual elemental localization in these organisms [60]. For example, Fig. 16 shows positive secondary ion images over a 60 $\mu$m diameter field of view produced from a radiolaria skeleton 200 million years old [61]. The most intense ion emission is produced by Si, Al, Ca, Na and K, and Mn, with distinct, low intensity signals for V distributed throughout the skeletal region. These images were also recorded on film, and the numbers along the bottom refer to the film exposure time at constant $f$ stop.

The ion microscope is capable of high mass resolution imaging, as illustrated in Fig. 17, which presents positive ion images of $^{56}Fe^+$, $^{40}CaO^+$, and $^{40}Ca^+$ produced from fixed, embedded, and microtomed sections of mouse thyroid cells. This analysis attempted to localize iron enrichment in thyroid cells after slight manganese overloading [54]. The mass resolution was approximately 2000, and the images demonstrate that Fe is localized only in cytoplasmic regions (arrows

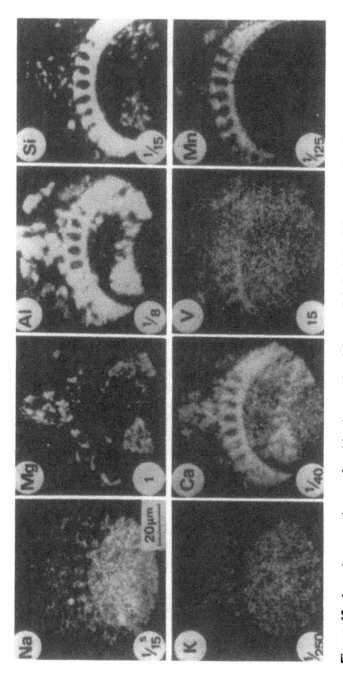

**Figure 16** Ion microscope image of positive ions produced from radiolaria 200 million years old; 60 μm image field. (From Ref. 61.)

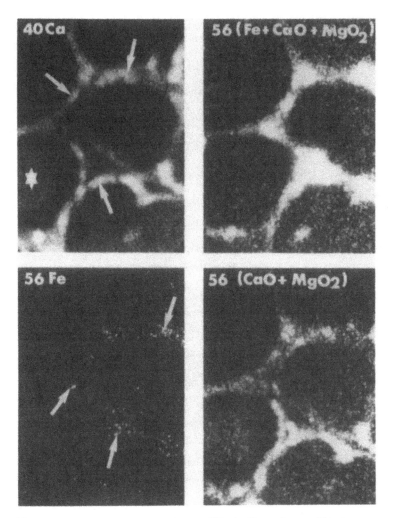

**Figure 17** Ion microscope images of $^{56}Fe^+$, $^{40}CaO^+$, and $^{40}Ca^+$ of mouse thyroid tissue; image field 110 × 60 μm. (From Ref. 54.)

in the images), while the Ca and CaO are formed from both ep ithelial and luminal regions. The image field is approximately 60 μm × 110 μm, and images were recorded on photographic film.

Ion microprobe images of biological materials are shown in Figs. 18 and 19 [62]. Figure 18 contains four microprobe images produced from healthy human teeth samples. This work was performed on the UC-HRL ion microprobe in collaboration with Professor A. Lodding of Chalmers University. Lodding and coworkers have developed a number of SIMS quantitative analytical methods for

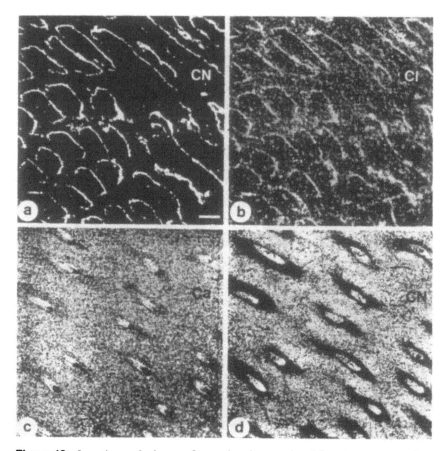

**Figure 18**  Ion microprobe image of secondary ions produced from transverse sections of healthy human teeth: (a and b) $CN^-$ and $^{35}Cl^-$ in enamel; (c and d) $Ca^+$ and $CN^-$ in dentine; bar = 4 $\mu$m. (From Ref. 62.)

the characterization of various minor and trace elements in hard tissues such as teeth and bone [63]. The images in Fig. 18a and d illustrate the distribution of organic constituents (as monitored by the $CN^-$ intensity) of healthy human enamel and dentine, respectively. Figure 18b maps the $Cl^-$ intensity in enamel regions, while Fig. 18c illustrates the distribution of Ca in dentine. The $Ca^+$ and $CN^-$ images of dentine show teardrop-shaped tubules in which the cone region is devoid of Ca and contains mostly organic constituents.

Figure 19 illustrates a very intriguing application of SIMS ion imaging. The samples in this study were chromosomes incubated in U $^{14}$C-thymidine containing 10 atoms of $^{14}$C per molecule [64]. These chromosomes originated from

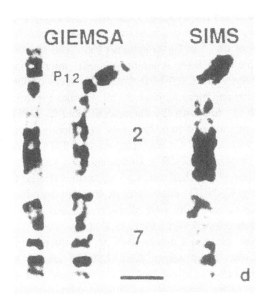

**Figure 19** Ion microprobe images of $^{14}C^{14}N^-$ of chromosomes 2 and 7 from human peripheral blood lymphocytes; bar $= 2\ \mu$m. (From Ref. 64.)

human peripheral blood lymphocytes. The samples were chemically fixed in methanol/acetic acid (3:1), and suspensions were deposited on gold substrates and air-dried. Mitotic cells were localized using total ion images produced from the irradiation of the samples with the microfocused (600 Å in diameter) $Ga^+$ beam on the UC-HRL ion microprobe. Dynamic SIMS sputtering of the $^{14}C$-labeled base produces a unique negative ion at mass 28 which is $^{14}C^{14}N^-$. The CN species efficiently forms negative ions, since it has one of the highest measured electron affinities [65]. Thus, $^{14}C^{14}N^-$ is a very sensitive marker for the labeled base. The $^{14}C^{14}N^-$ ion images at the right in Fig. 19 are tentatively identified as chromosomes 2 and 7, and these ion images are compared to the optical images produced by Giemsa staining of corresponding chromosomes (left, Fig. 19). The band structures in each form of chromosome image are quite similar, with the possible exception of band p12 (not visible in the SIMS image) and the weak or absent SIMS centromere band in chromosome 7. What makes this SIMS imaging technique most interesting, however, is the ability of SIMS imaging to detect band patterns in the distribution of specific nucleotides. Since the association of banding patterns produced by staining methods with the distribution of specific nucleotides is often not well understood, SIMS ion imaging may provide a better understanding of the preferential affinity of certain stains for specific base pairs.

Another exciting application of SIMS ion imaging to biomaterials analysis is illustrated in Fig. 20. These images were produced in a study correlating cellular morphology with the distribution of $Ca^{+2}$ in Golgi apparati [66]. The microscopies employed in this study included laser scanning confocal microscopy (LSCFM) and SIMS ion microscopy. The LSCFM technique provides optical identification of the Golgi apparatus and measurement of cell volumes for three types of cultured cell line stained in vitro with the fluorescent label $C_6$-NBD-ceramide. Each of the cell lines was divided into three sets: cells treated only with stain, stained cells treated with a mild Ca chelating agent (EGTA), and stained cells treated with a Ca ionophore (A23187). Unstained cells were also prepared and analyzed. These unstained cells had total $Ca^+$ intensities similar to the stained cells, indicating that the $C_6$-NBD-ceramide stain does not introduce spurious Ca contamination. These culture cells were shock frozen and freeze-dried using the sandwich method developed by Chandra et al. [57].

Ion microscopy was performed using a Cameca IMS-3f ion microscope equipped with a CCD array as the image sensor [27]. LSCFM images of LLC-PK$_1$ porcine kidney epithelial cells are illustrated on the left-hand side in Fig. 20, and the $^{40}Ca^+$ ion images of the same cells are illustrated on the right-hand side. The field of view in these images is approximately 150 $\mu$m in diameter. The images in the top row were produced from cells incubated with stain only; those in the middle row were produced from cells exposed to the EGTA chelating agent, and those in the bottom row were produced from cells exposed to the Ca ionophore in EGTA. High fluorescence intensities are observed for the perinuclear regions of the Golgi apparatus in all images. Exposure of the cells to EGTA enhances the localization of Ca in the perinuclear regions of the Golgi (middle SIMS images), which indicates that the Golgi apparatus is resistant to Ca depletion. Exposure of cells to the calcium ionophore releases Ca from the Golgi, as illustrated in the bottom SIMS image.

In addition to providing $Ca^+$ intensity distributions in these culture cells, Ausserer and coworkers quantitatively measured the Ca concentration in various cellular regions using RSF values developed from tissues of these types [48]. Quantitative SIMS imaging of biological tissues requires standard samples. Of the various methods proposed for preparing SIMS imaging standards of biomaterials, the most successful generate RSFs for specific elements or molecules from homogenates of the cellular material of interest. Ausserer and coworkers demonstrated good elemental accuracy and precision in the analysis of freeze-dried culture cells using such an RSF approach with $^{12}C^+$ as the matrix ion [48]. Standards for these imaging analyses were homogenates of the same cellular material. Absolute concentrations of the major physiological cations (Na, Mg, K, and Ca) in these homogenates were determined using inductively coupled plasma–atomic emission spectroscopy (ICP-AES). Accuracies were on the order of a 10%, while precisions ranged between 1 and 5%. The RSFs developed

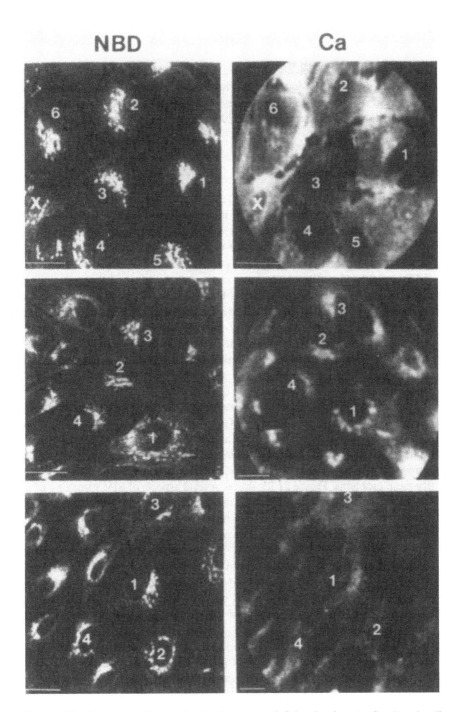

**Figure 20** Laser scanning confocal microscopy (left-hand column) of cultured cell Golgi apparatus and ion microscope images of $^{40}Ca^+$ of same regions. (From Ref. 66.)

from these standards yielded SIMS imaging concentration values for Na, Mg, and K of *individual* cells equivalent to those determined by EPMA. Since the Ca localization studies discussed above used culture cells of the same types, the RSFs developed by Ausserer et al. could be directly applied to quantification of Ca concentrations.

## B. Microelectronics

SIMS elemental ion imaging has found a variety of applications in semiconductor or microelectronics manufacturing including identification and localization of chemical contaminants in or on device circuits, localization of chemical "defects," and determination of the three-dimensional uniformity of ion implants [67]. Organic ion imaging applications performed on TOF-SIMS ion microscope and ion microprobe instrumentation include determination of the uniformity of lubricant layers on hard disk surfaces, identification of residual process chemicals such as photoresist, etch residues, and cleaning solvents, and more complete and detailed characterization of particulates introduced during various phases of processing [68]. Since device feature sizes are currently at submicrometer dimensions, there is continuing need for SIMS imaging techniques offering higher spatial resolution. Two applications of SIMS ion imaging in microelectronics material processing are discussed next.

Ion implication is a very common method to modify the electrical properties of high purity crystalline silicon and GaAs. Problems associated with ion implantation include poor control of the implanter conditions, introduction of unwanted impurities, and thermal diffusion of implant profiles during subsequent thermal processing of the material. SIMS elemental ion imaging can often identify these problems. Figure 21 illustrates several aspects of SIMS imaging of ion implants. The sample in this application is a <100> silicon wafer that contains a low level of bulk-doped boron. This wafer was patterned into two distinct regions: an inverted C and two small square areas. The inverted C was implanted with a high dose ($10^{15}$ atoms/cm$^2$) of $^{11}$B and the squares were implanted with an order of magnitude lower dose. The peak implant depth was approximately 1.0 $\mu$m. An ion microscope (Cameca IMS-4f) image depth profile was acquired from this sample in which ion images of $^{28}$Si$^+$ and $^{11}$B$^+$ were acquired on an RAE imaging detector as a function of the sputtering time (sampling depth) of the primary ion beam.

Figure 21a shows the $^{11}$B$^+$ distribution across the surface of this sample at a depth of approximately 0.5 $\mu$m. This image is a "frame" in the total image depth profile; the field of view for this analysis was 150 $\mu$m, and the image resolution is between 1 and 2 $\mu$m. The $^{11}$B lateral distribution is essentially identical to the pattern design for this test structure, indicating that the device masking is good. The $^{11}$B$^+$ intensity images in Fig. 21b and c were formed after the depth profile by summing intensities along two discrete lines drawn through the two-dimensional image in Fig. 21a. One cross-sectional image was formed from a line drawn along the long part of the inverted C. The other cross section was formed from a line that intersects the two square implant boxes. These images are essentially SIMS image cross sections in which the $^{11}B^+$ intensities are displayed in the $X$ or $Y$ surface dimension and the $Z$ depth dimension. These cross-sectional images illustrate the various structures in the mask pattern and exhibit no anomalous diffusion or migration of the boron implant. Three-dimensional images could be constructed from this image depth profile using three-dimensional reconstruction software [69].

A second example of SIMS imaging in microelectronics is illustrated in Fig. 22, which is a TOF-SIMS ion microprobe image of a particle contaminant on a semiconductor device. This image shows the $^{28}Si^+$ intensity distribution over an area approximately 50 $\mu$m square. The patterning in the device is obvious, and the dark bands are the Al electrical interconnects. The dark, irregularly shaped object is a particle on the device, probably introduced in the final stages of processing. Selected area mass spectra can be generated from this sample using this Si image as a template. Figure 23 illustrates mass spectra produced from the suspected particle (top spectrum). Al line (middle spectrum), and Si regions (bottom spectrum). The peaks in these spectra are quite distinctive, with the Al and Si spectra comprised primarily of cations of these elements along with low mass (adventitious) hydrocarbon peaks. The particle spectrum exhibits much more organic character and is comprised of a series of higher mass $C_xH_y^+$ peaks. These peaks match TOF-SIMS positive ion spectra of polyethylene, suggesting that the foreign particle was introduced during storage of the device before final packaging.

## C.  Geology and Metallurgy

There has been a vast amount of SIMS analysis performed on a variety of geological and metallurgical materials [70]. SIMS imaging has played a large role in many of these applications, since it is capable of localizing inclusions, voids, grain boundaries, and trace elements in samples of these types [71].

SIMS imaging can determine the presence and spatial distribution of trace elements in geological materials. The localization of gold in pyrite grains is a case in point. This gold has been dubbed "invisible" gold, since its concentration (1–100 ppma) is below the detection limit of most microanalytical techniques. Figure 24 shows SIMS ion microprobe images of $^{34}S^-$ and $Au^-$ produced in the analysis of several pyrite grains from the Carlin district of Nevada. These images were acquired on a Cameca IMS-4f ion microanalyzer using a microfocused $Cs^+$ primary ion beam and raster image acquisition. The $^{34}S^-$ image delineates the shape of the pyrite grains, as illustrated in the top left-hand

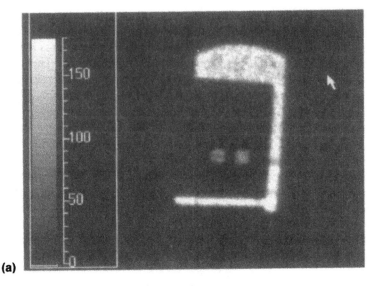

**(a)**

**Figure 21**  Ion microscope images of $^{11}B^+$ implanted into silicon test pattern. (a) Lateral $^{11}B^+$ distribution at peak of implant. (b) $^{11}B^+$ cross-sectional image along vertical line spanning long implant line at right of image in (a). (c) $^{11}B^+$ cross-sectional image along vertical line drawn through one small square in center of image in (a). Image field 150 $\mu$m diameter. (Images from author's laboratory.)

image. The $Au^-$ image intensity is localized on the edges of these grains, as is apparent in the top right-hand image. Both these images are approximately 400 $\mu$m square. This localization is even more apparent in the two lower images, which are 20 $\mu$m across. The $Au^-$ intensity is localized within 1–2 $\mu$m of the pyrite surface. The observation that the gold is located along the outer surface of the pyrite suggests that it can be liberated by oxidizing only a small fraction of the total pyrite.

Another unique application of SIMS imaging is the identification and localization of flotation agent *molecules* on the surfaces of mineral particles [72]. This application was performed with a TOF-SIMS ion microprobe using a microfocused $Ga^+$ primary ion beam and raster ion imaging. The flotation agent investigated was diisobutyl dithiophosphinate (Aero 3418A, manufactured by American Cyanamid Company), which is used as a flotation agent in ore processing. The formula of this agent is $[(CH_3)_2(CH)CH_2]_2P(S)S^-Na^+$ and its molecular weight is 232 amu. The mineral was sized galena particles to which the surfactant was added in a 5 $\mu$M solution of the Na salt. The analyzed particles were the flotation fraction produced by passing $N_2$ gas through a Hallimond fil-

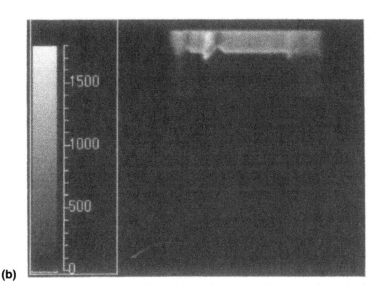

(b)

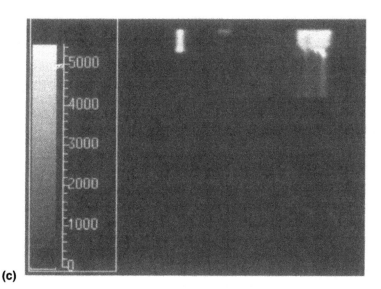

(c)

ter containing the particle–surfactant solution. This flotation fraction was rinsed in deionized water, air-dried, and pressed into indium foil. The negative ion TOF-SIMS images produced from several of these particles (Fig. 25) were formed from the $(M-Na)^-$ ion of the flotation agent ($m/z$ 209), is $PS_2^-$ ($m/z$ 95) also produced from the agent, and the $O^-$ ($m/z$ 16) and $O_2^-$ ($m/z$ 32). The oxygen

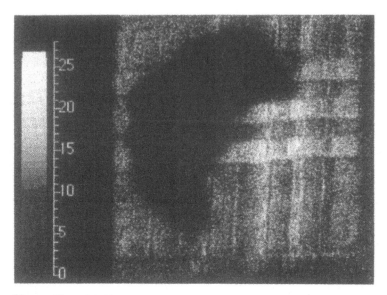

**Figure 22** TOF-SIMS ion microprobe image of total positive secondary ions produced from particle on silicon device; image field 50 $\mu$m. (Images from author's laboratory.)

ions were produced from uncoated galena surfaces. Since the TOF-SIMS analysis was performed under static SIMS conditions, less than a monolayer of the surface was removed. The anion of diisobutyl dithiophosphinate is observed to localize on distinct faces of the galena crystals, which are spatially correlated with the oxygen ion intensities.

## D.  Polymers and Organic Materials

The localization of specific chemical compounds or structurally significant fragment ions of such compounds is tantamount to the development of a number of very important industrial products and processes. For example, the bonding, distribution, and uniformity of organic coatings (e.g., paints and lubricants) on metals and polymers often determine the ultimate performance of these materials in diverse environments and applications [73]. The localization of specific organic compounds or their fragments on the surfaces of physically or chemically treated polymers is also essential to more complete understanding of the bonding of polymers to organic (polymeric) and inorganic surfaces. This type of analysis is also important for the processing of polymers for printing applications, as well as for identification of surface bonding for chemical modification of polymer surfaces and for biopolymer biocompatibility. In addition, molecular ion imaging holds the promise of providing a viable method of analyzing organic

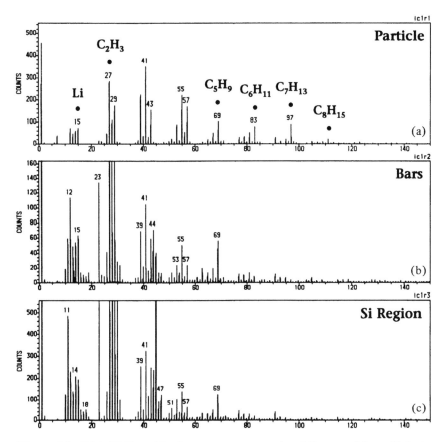

**Figure 23** TOF-SIMS positive ion mass spectra produced from particle on Si device: (a) particle only, (b) Al run, and (c) field oxide (SiO$_2$) regions. (Data from author's laboratory.)

and inorganic species at the cellular and subcellular dimensions [51]. Molecular ion imaging could also provide valuable, detailed insight into the chemistry of airborne particulates [74]. SIMS ion imaging is one potential method for performing this type of chemical structure localization, as the following examples illustrate.

Paper products are often coated with a variety of chemicals to enhance their physical or chemical properties [75]. The distribution of these coatings as a function of time, processing, and handling conditions can determine the ultimate integrity and uniformity of the coated fibers. One method of reducing potential static buildup on paper products is to coat the surface with a thin layer

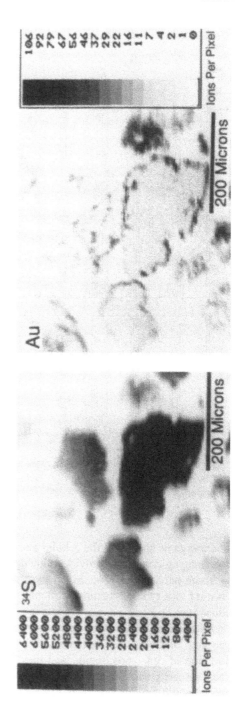

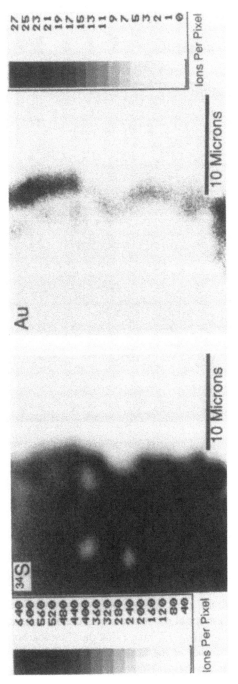

**Figure 24** Ion microscope images of ³⁴S and Au negative ions produced from pyrite grains. (Images from author's laboratory.)

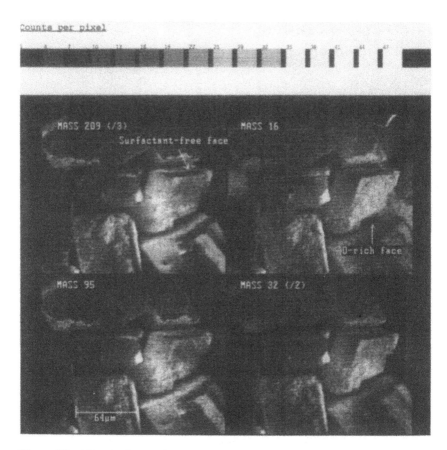

**Figure 25** TOF-SIMS ion microprobe images of negative ions produced from flotation agent on galena particles. (From Ref. 72.)

(one to several monolayers) of fluorocarbon. Fluorocarbons are known to be very mobile, and thus one might expect fluorocarbon migration and possible segregation on coated paper fibers. As a test of this hypothesis, fluorocarbon-coated paper fibers having an average cord diameter of 25 $\mu$m were analyzed using TOF-SIMS ion microprobe imaging. Figure 26 shows the total positive ion image and the $CF_3^+$ ion images of this paper sample. The image field is approximately 100 $\mu$m square. The $CF_3^+$ cation is a good measure of the presence of the fluorocarbon, since it is one of the end groups of the particular formulation used to coat the paper fibers. It is apparent from the $CF_3^+$ image that the fluorocarbon is not uniformly distributed across the surface of the fibers. The fibers have discrete topography, which will also affect ion yields. However, the $CF_3^+$ intensity

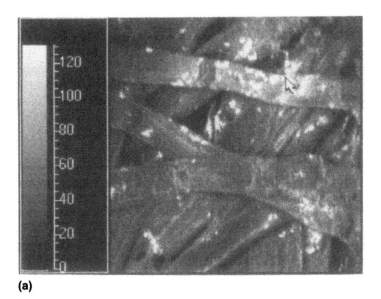

(a)

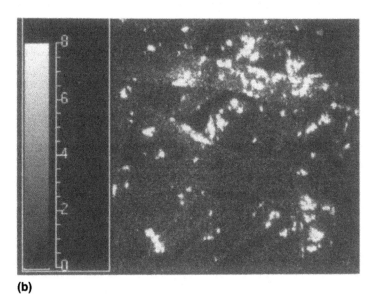

(b)

**Figure 26** TOF-SIMS ion microprobe images of (a) total positive ions and (b) $CF_3^+$ produced from fluorocarbon coating on paper fibers. (Images from author's laboratory.)

is much too high to be explained solely on the basis of topographic enhancement of ion emission. For example, dividing the $CF_3^+$ image by the total ion image produces an image that exhibits hot or intense spots attributable to the $CF_3^+$ species. Furthermore, if the $CF_3^+$ intensity is subtracted from the total ion image, the hot spots disappear. These observations indicate that the intense regions of $CF_3^+$ emission correspond to higher concentrations of fluorocarbon in these microregions.

Airborne particulates produced from coal-fired power plants can contain toxic inorganic and organic chemicals [73]. Environmental researchers have determined the bulk elemental and organic compositions of a variety of these particulates [76]. Elemental microanalysis has also determined the elemental composition of individual particles as well as the compositional variation of discrete particles. Researchers are currently attempting to determine the organic surface composition of different types of coal fly ash and Fig. 27 gives TOF-SIMS ion microprobe images produced from NIST standard coal fly-ash treated with a thin coating of 2-aminoanthroquinone. Figure 27a illustrates the $Al^+$ ion distribution on the surface of individual particles. Figure 27b illustrates the sum of higher mass species (masses 200–1000 amu) produced from the same particles. These higher mass ions are composed of C, H, N, and O and indicate the presence of higher molecular weight organics on the surfaces of the particles. None of these organic signals is sufficiently intense to form a useful ion image; however, it is apparent that the higher mass ion signals are produced from the same regions on the particle surface as the $Al^+$ ions.

The ultimate utility of TOF-SIMS organic imaging in this type of application will require methods of producing higher intensity signals. Since the ion yields for organic species are typically less than $10^{-3}$, any method that manages to increase the organic ion yield will improve detection sensitivity and image contrast. One possible method of achieving higher organic ion yields is to ionize the neutral organic fragments sputtered from the sample surface with a high intensity laser pulse [77]. This process is referred to as laser postionization, since the ionization occurs after the neutrals have been sputtered from the sample surface. Probably the most attractive laser postionization scheme for organic neutrals is the single-photon, laser vacuum ultraviolet postionization technique developed by Pallix and coworkers [78]. This technique utilizes a nantupled Nd:YAG output (118 nm), which is pulsed on after the primary ion pulse in a TOF-SIMS instrument. Laser postionization may well provide the necessary sensitivity to localize single monolayers of organic molecules at micrometer and submicrometer dimensions.

## VI. CONCLUSIONS

SIMS imaging is a versatile technique that localizes elemental, fragment ions and in some cases molecular ions produced in ion beam sputtering of a solid

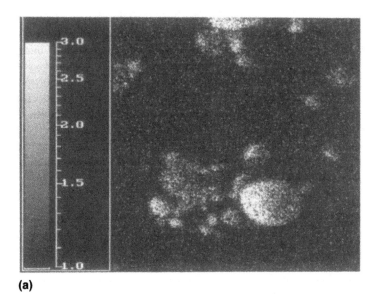

**(a)**

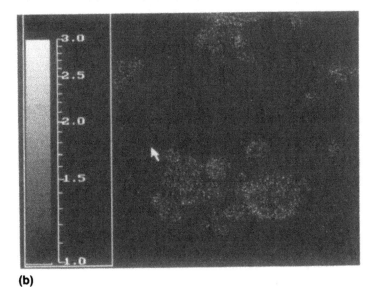

**(b)**

**Figure 27**    TOF-SIMS ion microprobe images of (a) Al$^+$ and (b) high mass ($m/z$ 200–1000 amu) produced from coal fly ash particles coated with 2-aminoanthroquinone. (Images from author's laboratory.)

surface. Elemental ion signals can be resolved down to dimensions of several hundred angstroms. New developments in ionizations methods could well provide increased molecular detection sensitivity, leading to submicrometer imaging of molecular species in films only a monolayer thick. This capability will lead to the development of even more sophisticated analytical methods of molecular microanalysis. New applications could include localization of specific molecules or metabolites at cellular or subcellular levels and detailed analysis of membranes. New developments in sample handling and preparation methods could also lead to SIMS imaging analytical methods that characterize the occurrence and distribution of individual layers of contaminants on semiconductor devices and other ultrahigh purity materials.

## ACKNOWLEDGMENTS

The author acknowledges the valuable contributions made to this chapter by his colleagues Drs. John A. Chakel, Constance M. John, Patricia M. Lindley, D. Fraser Reich, and B. Schueler. The assistance of Ms. Leticia G. Rivas in the preparation of the images and Ms. Nancy B. Church and Jill Vosburg for the preparation of the manuscript is also greatly appreciated. The support of grant GM39005-031A from the National Institute of Health is also gratefully acknowledged.

## REFERENCES

1. Evans, C. A., Jr., and Blattner, R. J. (1978). Modern experimental methods for surface and thin-film chemical analysis, *Ann. Rev. Mater. Sci. 8*:181.
2. Williams P. (1979). The sputtering process and sputtered ion emission, *Surf. Sci. 90*:588.
3. Barber, M., Bordoli, R. S., Elliott, G. J., Sedgwick, R. D., and Tyler, A. N. (1982). Fast atom bombardment mass spectrometry, *Anal. Chem. 54*:645.
4. Aberth, W., Straub, K. M., and Burlingame, A. L. (1982). Secondary ion mass spectrometry with cesium ion primary beam and liquid target matrix for analysis of bioorganic compounds, *Anal. Chem. 54*:2029.
5. Benninghoven, A., Rüdenaur, F. G., and Werner, H. W., Eds. (1987). *Secondary Ion Mass Spectrometry: Basic Concepts, Instrumental Aspects, Applications and Trends*, John Wiley & Sons, New York.
6. Fleming, R. H. (1991). Three-dimensional SIMS in materials analysis, in *Microbeam Analysis—1991* (Howitt, D. G., Ed.), San Francisco Press, San Francisco, p. 344.
7. Linton, R. W., Ro, C. U., Wilson, D. C., Hunter, J. L., and Corcoran, S. F. (1989). Molecular microanalysis by application of pattern recognition techniques to SIMS images, in *Secondary Ion Mass Spectrometry: SIMS V* (Benninghoven, A., Colton, R. J., Simons, D. S., and Werner, H. W., Eds.), Springer-Verlag, Berlin.

8. Levi-Stetti, R., Crow, G., and Wang, Y. L. (1985). Progress in high resolution scanning ion microscopy and secondary ion mass spectrometry imaging microanalysis, *Scanning Electron Microsc. II*:535.

9. Schueler, B., Sander, P., and Reed, D. A. (1990). A time-of-flight secondary ion microscope, *Vacuum, 41*:1661.

10. Williams, P. (1982). On mechanisms of sputtered ion emission, *Appl. Surf. Sci. 13*:241.

11. McHugh, J. A. (1975). Secondary ion mass spectrometry, in *Methods of Surface Analysis* (Czanderna, A. W., Ed.), Elsevier, Amsterdam, p. 223.

12. Wittmack, K. (1981). Implications in the use of reactive ion bombardment for secondary ion yield enhancement, *Appl. Surf. Sci. 9*:315.

13. Sigmund, P. (1981). In *Sputtering by Particle Bombardment, Vol. I* (Behrisch, R., Ed.), Springer-Verlag, Berlin, Sect. 47, p. 9.

14. Blattner, R. J., and Evans, C. A., Jr. (1980). High-performance secondary ion mass spectrometry, in *Scanning Electron Microscopy* (Jahari, O., Ed.), SEM, Chicago.

15. Briggs, D. (1989). Characterization of surfaces, in *Comprehensive Polymer Science*, Vol. 1 (Booth, C., and Price, C., Eds.), Pergamon Press, Oxford.

16. Morgan, A. E., de Grefte, H. A. M., and Tolle, H. J. (1981). Effects of oxygen implantation upon secondary ion yields, *J. Vac. Sci. Technol. 18*:164.

17. Storms, H. A., Brown, K. F., and Stein, J. D. (1977). Evaluation of a cesium positive ion source for secondary ion mass spectrometry, *Anal. Chem. 49*:2023.

18. Vickerman, J. C., Brown, A., and Reed, N. M., Eds. (1989). *Secondary Ion Mass Spectrometry: Principles and Applications*, Clarendon Press, Oxford.

19. Niehuis, E., Heller, T., Feld, H., and Benninghoven, A. (1985). High-resolution TOF secondary ion mass spectrometer, in *Ion Formation from Organic Solids (IFOS III): Mass Spectrometry of Involatile Materials* (Benninghoven, A., Ed.), Springer-Verlag, Berlin, p. 198.

20. Soldzian, G. (1975). Looking at the collection efficiency problem through the ion microscope optics, NBS Special Publication 427, U.S. National Bureau of Standards, Gaithersburg, Md.

21. Liebl, H. (1975). The ion microprobe—instrumentation and techniques, in *NBS Special Publication 427*, U.S. National Bureau of Standards, Gaithersburg, Md.

22. Wells, O. C. (1974). *Scanning Electron Microscopy*, McGraw-Hill, New York.

23. Septier, A., Ed. (1980). *Applied Charged Particle Optics*, Academic Press, New York.

24. Wiza, J. L. (1979). Microchannel plate detectors, *Nuc. Instrum. Methods, 162*:587.

25. Odom, R. W., Wayne, D. H., and Evans, C. A., Jr. (1984). A comparison of camera-based and quantized detectors for image processing on an ion microscope, in *Secondary Ion Mass Spectrometry: SIMS IV* (Benninghoven, A., Okano, J., Shimizu, R., and Werner, H. W., Eds.), Springer-Verlag, Berlin, p. 186.

26. Furman, B. K., and Morrison, G. H. (1980). Direct digitization system for quantification in ion microscopy, *Anal. Chem. 52*:2305.

27. Mantus, D. S., and Morrison, G. H. (1990). Ion image detection with a microchannel plate evaluated by using a charge coupled device camera, *Anal. Chem. 62*:1148.

28. Odom, R. W., Furman, B. K., Evans, C. A., Jr., Bryson, C. E., Petersen, W. A., Kelly, M. A., and Wayne, D. H. (1983). Quantitative image acquisition system for ion microscopy based on the resistive anode encoder, *Anal. Chem. 55*:574.
29. Wittmack, K. (1975). Energy dependence of the secondary ion yield of metals and semiconductors, *Surf. Sci. 53*:626.
30. Blaise, A., and Slodzian, G. (1973). *Rev. Phys. Appl. 8*:105.
31. Wilson, R. G., and Brewer, G. R. (1973). *Ion Beams with Applications to Ion Implantation*, John Wiley & Sons, New York.
32. Mayer, J. W., Eriksson, L., and Davies, J. A. (1970). *Ion Implantation in Semiconductors: Silicon and Germanium*, Academic Press, New York.
33. Newbury, D. (1986). Strategy for Interpretation of Contrast Mechanisms in Scanning Electron Microscopy, in *Microbeam Analysis—1986* (Romig, A. D., and Chambers, W. F., Eds.), San Francisco Press, San Francisco, p. 1.
34. Liebel, H. (1971). Design of a combined ion and electron microprobe apparatus, *Int. J. Mass Spectrom. Ion Phys. 6*:401.
35. Soldzian, G., Diagne, B., Girard, F., and Boust, F. (1987). High sensitivity and high resolution ion probe instrument, in *Secondary Ion Mass Spectrometry: SIMS VI* (Benninghoven, A., Huber, A. M., and Werner, H., Eds.), John Wiley & Sons, New York, p. 189.
36. Mullock, S. J., Reich, D. F., and Dingle, T. (1989). Ultra low dose static SIMS with TOF analysis and a pulsed source, in *Secondary Ion Mass Spectrometry: SIMS VII* (Benninghoven, A., Evans, C. A., McKeegan, K. D., Storms, H. A., and Werner, H. W., Eds.), John Wiley & Sons, New York, p. 847.
37. Schueler, B., Sander, P., and Reed, D. A. (1989). A new time-of-flight secondary ion microscopy, in *Secondary Ion Mass Spectrometry: SIMS VII* (Benninghoven, A., Evans, C. A., McKeegan, K. D., Storms, H. A., and Werner, H. W., Eds.), John Wiley & Sons, New York, p. 851.
38. Karataev, V. I., Mamyrin, B. A., and Shmikk, D. V. (1972). A new method for focusing ion bunches in time-of-flight mass spectrometers, *Sov. Phys.-Tech. Phys. 16*:1177.
39. Bletsos, I. V., Hercules, D. M., van Leyen, D., Niehuis, E., and Benninghoven, A. (1985). TOF-SIMS of polymers in the high mass range, in *Ion Formation from Organic Solids (IFOS III): Mass Spectrometry of Involatile Material* (Benninghoven, A., Ed.), Springer-Verlag, Berlin, p. 41.
40. Schueler, B. (1989). TOF-SIMS with a stigmatic ion microscopy, in *Secondary Ion Mass Spectrometry: SIMS VII* (Benninghoven, A., Evans, C. A., McKeegan, K. D., Storms, H. A., and Werner, H. W., Eds.), John Wiley & Sons, New York, p. 311.
41. Deline, V. R., and Evans, C. A. (1978). A unified explanation for secondary ion yields, *Appl. Phys. Lett. 33*:578.
42. Anderson, C. A., and Hinthorn, J. R. Thermodynamic approach to the quantitative interpretation of sputtered ion mass spectra, *Anal. Chem. 45*:1421.
43. Wilson, R. G., and Novak, S. W. (1987). Systematics of SIMS relative sensitivity factors, in *Secondary Ion Mass Spectrometry: SIMS VI* (Benninghoven, A., Huber, A. M., and Werner, H. W., Eds.), John Wiley & Sons, New York, p. 57.
44. Gillen, G., Simons, D. S., and Williams, P. (1990). Molecular ion imaging and dynamic secondary ion mass spectrometry, *Anal. Chem. 62*:2122.

45. Sigmund, P. (1969). Theory of sputtering. I: Sputtering yield of amorphous and polycrystalline targets, *Phys. Rev. 184*:383.
46. Newbury, D. E., and Bright, D. S. (1989). Concentration histogram images: A digital imaging method for analysis of SIMS compositional maps, in *Secondary Ion Mass Spectrometry: SIMS VII* (Benninghoven, A., Evans, C. A., McKeegan, K. D., Storms, H. A., and Werner, H. W., Eds.), John Wiley & Sons, New York.
47. Rüdenaur, F. G. (1984). Spatially multidimensional SIMS analysis, *Surf. Inter. Anal. 6*(3):132.
48. Ausserer, W. A., Ling, Y.-C., Chandra, S., and Morrison, G. H. (1989). Quantitative imaging of boron, calcium, magnesium, potassium, and sodium distributions in cultured cells with ion microscopy, *Anal. Chem. 61*:2690.
49. A good overview of SIMS imaging of biological is in *Biology of the Cell, 74* (1992).
50. Lechene, C. P., and Warner, R. R., Eds. (1979). *Microbeam Analysis in Biology*, Academic Press, New York.
51. John, C. M., Chakel, J. A., and Odom, R. W. (1992). Time-of-flight secondary ion mass spectrometry analysis of biological materials, in *Secondary Ion Mass Spectrometry: SIMS VIII* (Benninghoven, A., et al., Eds.) John Wiley & Sons, New York.
52. Robards, A. W., and Sleytr, U. B. (1985). In *Low Temperature Methods in Biological Electron Microscopy* (Glauert, A. M., Ed.), Elsevier, Oxford.
53. Kupke, K. G., Pickett, J. P., Ingram, P., Griffis, D. P., Linton, R. W., Burger, P. C., and Shelbourne, J. D. (1984). Preparation of biological tissue sections for correlative ion, electron, and light microscopy, *J. Electron. Microsc. Tech. 1*:299.
54. Aioun, J., and Larras-Regard, E. (1990). Effect of slight manganese overload on thyroid function in mice, *J. Trace Elem. Biol. 3*:91.
55. Bernius, M. T., Chandra, S., and Morrison, G. H. (1985). Cryogenic sample stage for the Cameca IMS-3f ion microscope, *Rev. Sci. Instrum. 56*:1347.
56. Dubochet, J., and McDowall, A. W. (1984). Cryoultramicrotomy: Study of ice crystals and freezing damage, *Proc. 8th Eur. Reg. Conf. Electron Microsc.*, Budapest, *2*:1407.
57. Chandra, S., and Morrison, G. H. (1925). Imaging elemental distributions and ion transport in cultured cells with ion microscopy, *Science, 228*:1543.
58. Burns, M. S. (1982). Applications of secondary ion mass spectrometry (SIMS) in biological research: A review, *J. Microsc. 127*:237.
59. Burns, M. S., File, D. M., Brown, K. T., and Flaming, D. G. (1981). Localization of calcium and barium in toad retina by secondary ion mass spectrometry, *Brain Res. 220*:173.
60. Galle, P. (1985). La microscopie ionique analytique des tissues biologiques, *Ann. Phys. Fr. 10*:287.
61. Galle, P., Berry, J. P., and Escaig, F. (1979). Biomedical applications of secondary ion emission microanalysis, secondary ion mass spectrometry, SIMS II, (Benninghoven, A., et al., Eds.)Springer-Verlag, N.Y., p. 238.
62. Levi-Setti, R., Chabala, J. M., Wang, Y. L., and Hallégot, P. (1988). High resolution ion probe imaging and analysis, *Microbeam-Analysis-1988* (Newbury, D. E., Ed.), San Francisco, CA, p. 93.

63. Lodding, A. (1983). Quantitative ion probe microanalysis of biological mineralized tissues, *Scanning Electron Microprobe Microscopy*, AMF O'Hare, Chicago, p. 1229.

64. Levi-Setti, R., and LeBeau, M. (1992). Cytogenic applications of high resolution secondary ion imaging microanalysis: Detection and mapping of tracer isotopes in human chromosomes, *Biol. Cells*, *74*:51.

65. Janousek, B. K., and Brauman, J. I. (1979). Electron affinities, in *Gas Phase Ion Chemistry*, Vol. 2 (Bowers, M. T., Ed.), Academic Press, New York.

66. Chandra, S., Kable, E. P. W., Morrison, G. H., and Webb, W. W. (1991). Calcium sequestration in the Golgi apparatus of cultured mammalian cells revealed by laser scanning confocal microscopy and ion microscopy, *J. Cell. Sci. 100*: 747.

67. Lee, J. J., Linton, R. W., Lin, W. J., Hunter, J. L., Jr., and Griffis, D. P. (1990). Three-dimensional display of secondary ion images, in *Microbeam Analysis—1990* (Armstrong, J. T., Ed.), San Francisco Press, San Francisco, p. 101.

68. Schueler, B. W., Odom, R. W., and Chakel, J. A. (1992). Surface contamination analysis of semiconductor by time-of-flight SIMS, in *Secondary Ion Mass Spectrometry: SIMS VIII* (Benninghoven, A., et al., Eds.), John Wiley & Sons, New York.

69. Castleman, K. R. (1979). *Digital Image Processing*, Prentice-Hall, Englewood Cliffs, NJ.

70. Lovering, J. F. (1975). Applications of SIMS microanalysis techniques to trace element and isotopic studies in geochemistry and cosmochemistry, in *NBS Special Publication 427*, U.S. National Bureau of Standards, Gaithersburg, Md.

71. Fleming, R. H., Meeker, G. P., and Blattner, R. J. (1987). Three-dimensional secondary ion mass spectrometry of surface-modified materials and the application to $^{13}C$-implanted films, *Thin Solid Films*, *153*:197.

72. Brinen, J. S., and Reich, F. (1992). Static SIMS imaging of the adsorption of diisobutyl dithiophosphinate on galena surfaces, *Surf. Inter. Anal. 18*: 448.

73. Briggs, D., Hearn, M. J., Fletcher, I. W., Waugh, A. R., and McIntosh, B. J. (1990). Charge compensation and high-resolution TOFSIMS imaging of insulating materials, *Surf. Inter. Anal. 15*:62.

74. Cabaniss, G. E., and Linton, R. W. (1984). Correlation surface analysis studies of environmental particles, *Environ. Sci. Technol. 18*(5):319.

75. Brinen, J. S., and Greenhouse, S. (1990). Applications of a gallium liquid metal ion gun in surface analysis, *Vacuum*, *42*:205.

76. Keyser, T. R., Natusch, D. F. S., Evans, C. A., Jr., and Linton, R. W. (1978). Characterizing the surfaces of environmental particles, *Environ. Sci. Technol. 12*(7):768.

77. Lin, S. H., Fujimura, Y., Neusser, H. J., and Schlag, E. W. (1984). *Multiphoton Spectroscopy of Molecules*, Academic Press, Orlando, FL.

78. Pallix, J. B., Schuhle, U., Becker, C. H., and Huestis, D. L. (1989). Advantages of single-photon ionization over multiphoton ionization for mass spectrometric surface analysis of bulk organic polymers, *Anal. Chem. 61*:805–811.

# 11

## Electron Paramagnetic Resonance Imaging

**Gareth R. Eaton and Sandra S. Eaton**   *University of Denver,*
*Denver, Colorado*

## I.  INTRODUCTION

This chapter is an introduction to electron paramagnetic resonance (EPR) imaging with an emphasis on the approaches that have been used and the types of sample to which the technique has been applied. Comprehensive treatments and complete sets of literature citations can be found in Refs. 1–7. For the benefit of readers who are not familiar with EPR, a brief introduction to the principles of EPR is given in Section II. The emphasis is on the aspects of EPR that impact the way that imaging experiments are performed.

## II.  PRINCIPLES OF ELECTRON PARAMAGNETIC RESONANCE (EPR)

### A.  Resonance Condition

EPR [also known as electron spin resonance (ESR)] is a technique for studying unpaired electrons by measuring the absorption of microwave energy by unpaired electrons. An electron has quantized angular momentum with both spin ($S$) and orbital ($L$) contributions. For the purposes of this discussion we consider only the spin contribution and only a single unpaired electron. More comprehensive introductions, including orbital angular momentum and multielectron systems, are available in standard texts [8,9].

When an electron is placed in a magnetic field, the projections of the magnetic moment on the axis defined by the external magnetic field take on two discrete values $+ \frac{1}{2}$ and $- \frac{1}{2}$. The separation between these two energy levels is proportional to the magnetic field strength $B$. When an electromagnetic field irradiates the sample with energy equal to the separation between the spin energy levels, transitions between the spin states occur. This resonance condition is defined by $h\nu = g\beta B$, where $h$ is Planck's constant, $\nu$ is the microwave frequency, $\beta$ is the Bohr magneton, $B$ is the magnetic field strength, and $g$ is a characteristic value for a particular paramagnetic species. Many EPR experiments are performed at microwave frequencies of 9.0–9.5 GHz (X-band). For organic radicals $g$ values typically are close to 2.0, so resonance at X-band occurs at magnetic fields of about 3200–3400 G [1 gauss (G) = 0.1 millitesla (mT)]. The "spin-only" Lande $g$-factor is 2.0023. Contributions of orbital angular momentum in heavy atoms shift the $g$-factor away from 2, as do the angular momentum contributions of multiple unpaired electrons. In a continuous wave (CW) experiment the microwave frequency and power are held constant and the magnetic field is swept through resonance to record the spectrum. For instrumental reasons [10], the first derivative of the absorption spectrum is recorded.

## B. Hyperfine Splitting

Very few unpaired electrons are so isolated physically that one could observe the simple situation described in the preceding section. Nuclear spins in the vicinity of an unpaired electron contribute to the net magnetic field experienced by the electron spin. The energy required for the electron spin transition depends on the quantized spin states of neighboring nuclei, which result in splitting of the EPR signal into multiple lines. This is called hyperfine splitting. The number of lines in the EPR signal is equal to $2nI + 1$ where $n$ is the number of equivalent nuclei and $I$ is the nuclear spin. Thus interaction with 1 nitrogen ($I = 1$) causes splitting into 3 lines, and interaction with 3 equivalent protons ($I = \frac{1}{2}$) causes splitting into 4 lines. The splitting between adjacent lines is called the hyperfine splitting constant. The magnitude of the splitting constant in fluid solution (a system tumbling rapidly enough to average electron–nuclear dipolar interactions to zero) depends on the extent to which the unpaired electron spin is delocalized onto that nucleus. Hyperfine splitting constants of 10–20 G are common for organic radicals, and splitting constants of 100 G or more are common for transition metal ions. Thus a typical EPR spectrum of an organic radical may extend over tens of gauss. Since the hyperfine splitting constants are independent of magnetic field, the hyperfine splittings become increasingly large fractions of the resonant field as the microwave frequency (and corresponding resonant magnetic field) is decreased.

## C. Rigid Lattice Spectra

When a paramagnetic center is rapidly tumbling in solution, the $g$ anisotropy is averaged away and a single $g$ value is observed. For most systems the $g$ values are anisotropic, which means that the $g$ value is different for different orientations of the molecule with respect to the magnetic field. Thus, when the sample is immobilized in a rigid lattice (a solid or a frozen solution), the EPR spectrum is more complicated and extends over a wider range of magnetic fields than when the same material is in solution. The anisotropic dipolar contributions to hyperfine splitting, which were averaged away in fluid solution, also contribute to spectral complexity in the solid state.

## D. Contributions to Line Widths

As discussed below, for most approaches to EPR imaging, the spatial resolution of the image is inversely proportional to the line width of the EPR signal. Thus it is important to have an appreciation of the more common factors that contribute to the line width. EPR line widths for organic radicals typically are of the order of a few gauss (1 G at $g = 2$ is 2.8 MHz). In some cases the line widths are determined by relaxation time. Frequently, unresolved nuclear hyperfine splitting and a distribution of chemical environments make a significant contribution to line widths.

In fluid solution, collisions between paramagnetic species (including oxygen) can cause EPR spectral line broadening, so solutions with radical concentrations higher than a few millimolar typically have line widths greater than those observed at lower concentrations. At concentrations higher than about 10 mM, collision broadening usually is severe and can cause loss of hyperfine structure. At very high concentrations, exchange narrowing can cause collapse of the spectrum to a single line. Solvent viscosity is a factor for samples with significant $g$ anisotropy because incomplete motional averaging of the anisotropy results in broadening of the EPR signal.

In rigid lattice samples, unresolved hyperfine splitting and distributions in $g$ values and nuclear hyperfine splittings are major contributors to the line widths. As the concentration of the paramagnetic species increases, there is increasing dipolar interaction between neighboring paramagnetic centers, which causes broadening of the lines.

Thus for both fluid solution and rigid lattice spectra, concentrations of paramagnetic centers greater than a few millimolar causes significant broadening of the spectra and results in a loss of spatial resolution in EPR images. Since, ultimately, the resolution achievable in EPR imaging is limited by signal-to-noise ratio $(S/N)$, the broadening that occurs at high concentration (high $S/N$) places physical limits on achievable resolution [11].

## III. CONTRASTS WITH NMR IMAGING

Since NMR and EPR are both magnetic resonance techniques, one might assume that NMR and EPR imaging would be performed similarly. Although it is true that both techniques examine samples in the presence of a magnetic field, there are many differences in the practical aspects of the two techniques.

The most striking contrast between the typical NMR imaging experiments and most EPR imaging experiments is that the NMR experiments are pulsed whereas the EPR experiments are CW. The signal–noise advantages of pulsed experiments in NMR are well known. Several factors make pulsed techniques less widely applicable in EPR than in NMR.

1.   Electron spin relaxation times typically are much shorter than nuclear spin relaxation times. At room temperature, electron spin–spin (transverse) relaxation times $(T_2)$ typically are of the order of microseconds for organic radicals with much shorter values for transition metal ions, whereas nuclear spin–spin relaxation times frequently are of the order of milliseconds or longer. Thus pulsed EPR experiments must be conducted on time scales about 1000 times faster than NMR experiments (microseconds instead of milliseconds). It is difficult to pulse magnetic field gradients of the magnitude required for EPR imaging (see below) at this speed.

2.   As noted in the preceding section, EPR spectra frequently extend over tens of megahertz (10 G = 28 MHz), whereas fluid solution NMR spectra for many nuclei extend over only tens of kilohertz. To excite such a wide range of frequencies requires short high power pulses. Pulsed EPR spectrometers typically require kilowatt microwave amplifiers.

3.   The signal-to-noise ratio that can be obtained from an EPR spectrometer is inversely proportional to the bandwidth of the resonator in which the sample is studied. The resonators used in CW X-band spectrometers have a bandwidth (full width at half-height) of about 1 G at $g = 2$. To observe spectral widths of tens of gauss requires much broader banded resonators, with concomitant loss of signal-to-noise ratio. The current pulsed EPR spectrometers can observe spectra with widths up to about 70 G by using large bandwidth resonators [12].

Recently two-dimensional pulsed EPR imaging has been reported at X-band [13,14]. Both the microwaves and the magnetic field gradients were pulsed. A bridged loop-gap resonator with a much wider bandwidth than traditional cavities was used for these experiments. Spectra–spatial images were obtained for a pair of tubes—one contained [$^{15}$N]nitroxyl (2-line EPR spectrum) and the second contained [$^{14}$N]nitroxyl (3-line EPR spectrum) in fluid solution at room temperature [13]. The tubes had 1.0 mm internal diameter and were separated by 1.5 mm. The magnetic field gradient had rise and fall times of 45 ns. Data were obtained by phase encoding (gradient pulse between the microwave pulse and collection of the free induction decay) and by frequency encoding (gradient pulse applied during the collection of the free induction decay). The gradient range was ±80 G and the spectral window was 120 MHz. The phase-encoded results were comparable in quality to results obtained in a CW experiment on the same sample [13]. For the same sample, two-dimensional electron–electron double resonance (ELDOR) was obtained by phase-encoded Fourier transform (FT) imaging [14]. Two-dimensional spectral–spatial and spatial–spatial images based on spin echoes have also been obtained for a γ-irradiated quartz tube [14]. These experiments are likely to be the beginning of a new generation of EPR imaging experiments.

Most EPR spectra consist of multiple lines due to nuclear hyperfine splitting. The hyperfine splitting can be used to characterize the species that are present, but the spectral dimension must be included as an additional dimension in the image. Several of the approaches discussed below were developed to distinguish between multiple lines due to hyperfine splitting and multiple lines due to spatially separated signal sources.

Although NMR imaging is readily performed on relatively large samples, most current techniques for EPR imaging are limited to smaller samples for several reasons. In conventional EPR spectrometers, the sample is positioned in a cavity or other structure that is resonant at the microwave frequency of the instrument. The dimensions of the resonator are determined by the wavelength of the microwaves. The standard cavity that is used for EPR at X-band accepts cylindrical tubes with diameters up to about 11 mm and permits observation of about a 2.0 cm length sample. At lower frequencies, the wavelength is longer, so larger samples can be accommodated. For example, at 250 MHz a resonator has been designed that is large enough to hold a mouse [15]. Objects too large to fit inside a resonator have been examined by some of the localized detection techniques that are described in Section IV.B.

The tendency for nonresonant absorption of microwaves is denoted as the dielectric lossiness of a sample. Dielectric loss is much greater at the microwave frequencies used in EPR than at the radio frequencies used in NMR, which limits the amount of a sample can be put in a resonator. For example in a standard X-band resonator, a lossy aqueous sample is limited to a cylindrical tube with internal diameter of about 0.5 mm or a flat cell with a maximum thickness of about 0.3 mm. Larger quantities of aqueous or other lossy samples can be used at lower microwave frequencies.

The depth penetration into a sample is smaller for microwaves than for radiofrequencies [16]. It has been estimated that penetration into aqueous samples decreases from tens of centimeters at about 10 MHz to less than a millimeter at 10 GHz. This decrease occurs for samples inside a resonator or monitored with localized detection devices.

## IV. APPROACHES TO IMAGING

## A. Magnetic Field Gradients

The majority of EPR imaging experiments have been performed with magnetic field gradients. The general principle in one dimension is illustrated in Fig. 1 for

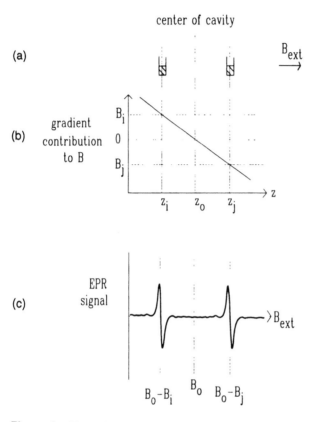

**Figure 1** Illustration of the basic features of EPR imaging with a magnetic field gradient. (a) Two point samples with a single-line EPR spectrum are positioned at $z_i$ and $z_j$. The two samples have the same $g$ value. The center of the cavity, which is also the nodal plane for the gradient, is at $z_0$. (b) The polarity of the gradient is defined such that the gradient adds to the magnetic field at $z_i$ and subtracts at $z_j$. The resonance condition for the sample is satisfied when the magnitude of the magnetic field is $B_0$. (c) In the presence of the gradient, as the external magnetic field $B_{ext}$ is swept from low field to high field, resonance is observed for the sample at $z_i$ and then for the sample at $z_j$. Since $B_j$ is negative, $B_0 - B_j$ is greater than $B_0$.

a pair of point samples with a single-line EPR spectrum. The direction of the main (external) magnetic field is defined as the $z$ axis. A pair of coils [17] is used to generate a magnetic field gradient along the $z$ axis. The direction of the magnetic field remains the same, but the magnitude of the magnetic field is dependent on position along the $z$ axis. For the polarity shown in Fig. 1 ($\partial B_z/\partial z$ is negative), the contribution to the magnetic field due to the gradient coils adds to the main magnetic field at the left-hand side of the cavity and subtracts at the right of the cavity. For example, the contribution to the magnetic field at $z_i = \partial B_z/\partial z \times (z_i - z_0)$, where $z_0$ is the center of the cavity, which is the nodal plane for the gradient generated by the coils. Suppose that the resonant condition is satisfied at $B_0$ (i.e., in the absence of a gradient, a single line is observed when $B_{ext} = B_0$). In the presence of the gradient, resonance will occur when the net magnetic field at $z_i = B_0$, which is achieved when $B_{ext} = B_0 - B_i$. Similarly resonance at $z_j$ will occur when $B_{ext} = B_0 - B_j$. Thus when $B_{ext}$ is scanned from low field to high field, the sample at $z_i$ achieves resonance at smaller $B_{ext}$ than the sample at $z_j$, and the magnetic field scan becomes a scan of the spatial location of sample along the $z$ axis. The scan in Fig. 1c is a one-dimensional image of the spatial distribution of unpaired electrons. The gradient is the conversion factor between magnetic field and spatial units.

Figure 1 illustrates the principle for imaging with a gradient in one dimension. Imaging in a second or third dimension is achieved with additional sets of coils to create $B_z$ gradients along the $x$ and $y$ directions [18,19]. Gradients have also been generated with ferromagnetic wedges instead of gradient coils [20,21].

Two important features of EPR imaging with gradients can be understood by further consideration of Fig. 1c.

1. The horizontal axis in Fig. 1c is generated by scanning the external magnetic field, so the line width of the signal in the image depends on the inherent line width in the absence of the gradient, the magnitude of the gradient, and the spatial extent of the sample along the direction of the gradient. The separation between the signals from the two samples in Fig. 1 is determined by the gradient multiplied by the spatial separation. If the inherent line width of the sample is twice as large, a magnetic field gradient twice as large is required to achieve the same spatial resolution. As noted above, EPR line widths are orders of magnitude greater than NMR line widths, so the gradients required for EPR imaging are orders of magnitude greater than those required for NMR imaging for the same spatial resolution.

For most EPR imaging experiments that have been performed to date, the gradient coils were located outside the resonator with distances of several centimeters between the two coils. Gradients typically are 20–400 G/cm (0.2–4.0 T/m). Higher gradients for the same current can be generated if the distance between the coils is smaller. A pair of miniature coils with a coil diameter of 1 mm and a separation of 0.5 mm gave a gradient of about 2 kG/cm (20 T/m) for a current of 5 A [22]. Miniature coils of superconducting Nb(Ti) wire gave a gradient of 6 kG/cm (60 T/m) [23]. The linearity of the gradient is strongly dependent on the uniformity of the winding of the coils and the positioning of the coils [22].

2. If the sample used to obtain the image in Fig. 1c had given an EPR signal with nuclear hyperfine splitting, that hyperfine splitting would have been present in the one-dimensional image. In a single one-dimensional image it is not possible, in general, to distinguish between multiple peaks due to hyperfine

splitting and ones due to samples with different spatial locations. Since most EPR samples exhibit multiline spectra, the development of EPR imaging has required procedures that take account of hyperfine splitting.

## 1. Constant Gradients

One-dimensional images with a magnetic field gradient in one or more directions can be used to obtain images for samples in which there is no hyperfine splitting or one hyperfine line is well separated from other lines (an isolated line). In the latter case, the magnitude of the gradient and the range over which the external magnetic field is scanned are restricted such that only one line of the EPR spectrum contributes to the image. For example, if the EPR spectrum consists of two lines with 1 G line widths separated by 25 G and the maximum dimension of the sample is 0.5 cm, then a gradient of 40 G/cm will cause the signal to spread over 20 G. A carefully selected 25 G scan centered at the nongradient resonant position of one line of the EPR spectrum would permit imaging without interference from the second line. Note however that higher gradients could not be used to improve resolution because higher gradients would cause overlap of the contributions from the two lines. By rotating the sample relative to the direction of the gradient [24,25] or using additional sets of gradient coils [26,27], a multidimensional spatial image can be obtained by mathematical image reconstruction techniques.

## 2. Constant Gradients with Deconvolution of Constant Line Width or Constant Hyperfine Splitting

If the line shape of the EPR signal is constant, the resolution of the image can be improved by deconvolution of the spectrum in the presence of the gradient by the nongradient spectrum [28]. Similarly if the nuclear hyperfine splitting is constant, it can be removed by deconvolution. In this case it is important to select a filter that avoids division by zero or values close to zero [28–30]. If the line shape and/or hyperfine splitting is not constant for a sample and it is deconvoluted from imaging data, distortion of the spin density distribution is observed [31,32].

## 3. Modulated Gradients

Modulated gradients have been used to obtain images of samples that contain more than one paramagnetic species and samples with hyperfine splittings [33–36]. To avoid distortion of the spatial distribution, however, the hyperfine splittings and line widths for any individual component must be uniform throughout the sample. Modulation of the gradient is used to broaden the EPR signals from portions of the sample that are outside the plane of interest (the ''sensitive'' plane). By controlling the current in the gradient coils while the external magnetic field is held constant at the resonance for the signal of interest, the plane can be swept across a dimension of the sample to determine the spatial distribution of a particular species, even if there are other EPR signals in the sample. If the sensitive plane is held constant and the external magnetic field is swept, the EPR signal from a particular slice of the sample is recorded. Since it is not possible to achieve infinitely large gradients, there is some interference from neighboring spins, for which partial correction can be performed mathematically [37]. A disadvantage of this technique and other techniques for localized detection described below is that these techniques typically have poorer sensitivity than observation of the complete sample in the presence of a gradient with subsequent image reconstruction.

## 4. Multiple Stepped Gradients, Spectral–Spatial Imaging

The approaches discussed above work well for particular cases. They share the feature that hyperfine structure and variations in line width are viewed as a problem to work around. However, the hyperfine structure and varied line widths provide useful information concerning the nature of the paramagnetic center and its chemical environment. For certain types of sample it is useful to obtain an image with an explicit spectral dimension. With these ''spectral–spatial'' images, it is possible to determine what species are present in various regions of the sample and how the concentration of each species varies spatially.

If it is known that there are $n$ species in the sample and that the line shape for each species is invariant through the sample, EPR spectra at $n$ different gradients can be treated as a system of equations in $n$ unknowns to determine the spatial distribution of the $n$ species [38].

For the more general case in which the number of species is unknown and the line shape may vary continuously through the sample, a spectral–spatial image can be generated that explicitly includes the spectral information as a dimension [39–41]. Field-swept EPR spectra are recorded as a function of magnetic field gradient. The particular values of the gradient selected are those that permit reconstruction of the spectral–spatial image with standard image reconstruction techniques. Two-dimensional images with one spectral dimension and one spatial dimension [39–41] and three-dimensional images with one spectral dimension and two spatial dimensions [42–44] have been reported. Provided sufficient projections are used to prevent the number of projections from limiting resolution, the maximum gradient determines the spatial resolution of the image. There are experimental realities, including heat removal and space in the magnet gap, that limit the magnitude of the gradient that can readily be achieved. For a given maximum gradient, the spatial resolution can be improved by about a factor of 5 by using an incomplete set of projections with 6% missing information [45].

## 5. Field-Swept Electron Spin Echo Detected Imaging

The preceding subsections have emphasized spatial and spectral dimensions in an image. Electron spin relaxation times also can be important dimensions that can be exploited through Fourier transform EPR imaging [14] or by field-swept electron spin echo imaging [46]. Species with short relaxation times that may dominate the CW image can be largely removed from the pulsed EPR image by selection of a long time between the pulses in the echo-forming sequence [47]. Species with long relaxation times can be dominant in the echo-detected image but have negligible intensity in the CW image.

## B. Localized Detection

An alternative approach to imaging is to use localized detection to sequentially monitor the EPR signal from particular regions of the sample.

### 1. Surface Coils

Imaging with surface coils has been demonstrated at L-band (1–2 GHz) [48–50]. A flat loop with a diameter of 7 mm gave a depth penetration of about 3 mm for an aqueous sample at 1.83 GHz with 100 mW of incident power [48]. For a toroidal surface coil with dimensions of 10 mm × 40 mm the depth penetration at 1.35 GHz was up to 10 mm [49]. A double split-ring surface coil with a resonant frequency of 1.1 GHz gave a maximum penetration depth of about 8 mm with an effective radius of about 5 mm [50]. For each of these devices, the sensitivity decreases with increasing distance from the surface in a complex manner that depends on the sample geometry and microwave absorption [50]. The spatial resolution obtained with these surface coils was relatively low, but it could be improved by the use of gradient coils to encode spatial information while observing the signal with the surface coil.

### 2. Localized Magnetic Field Modulation

Signal detection in most CW EPR spectrometers uses magnetic field modulation and phase-sensitive detection at the modulation frequency [10]. The usual modulation coils are located outside the resonator and produce a modulation field that encompasses the entire sample. Miniature modulation coils with a radius of 120 $\mu$m have been used for localized detection of the EPR signal (Fig. 2) [51,52]. The magnetic field is held constant at the resonant condition for the signal of interest. One- and two-dimensional spatial images were generated by moving the sample relative to the modulation coil. The resolution of the image is determined by the coil geometry and the accuracy of positioning the sample. This technique can be used to study the spatial distribution of the EPR signal for one component of a multicomponent spectrum.

### 3. Pinhole in Resonator (EPR Scanning Microscopy)

A 1 mm pinhole in a rectangular resonator permits the microwaves to exit the resonator and enter a sample positioned above the pinhole [53]. A modulation coil near the sample provides for the usual phase-sensitive detection (Fig. 3). A magnetic field sweep centered at the resonance condition for a particular line in the EPR spectrum gives the spectrum for that region of the sample [53,54]. The

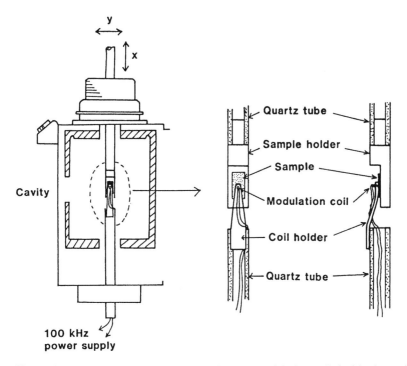

**Figure 2** Arrangement of sample and miniature modulation coils inside the cavity for localized detection of the EPR signal. The sample is moved relative to the modulation coils. (Reproduced with permission from Ref. 51.)

sample is moved on an *x-y* stage relative to the pinhole, and a magnetic field sweep is recorded for each position of the sample. The spatial variation of the intensity for a particular line in the spectrum is plotted to generate an image. The microwave magnetic field is not uniform across the pinhole [54]. The resolution of the image can be improved by measuring the microwave magnetic field distribution (the impulse response function) and deconvoluting it from the experimental data [54,55]. The minimum number of detectable spins was estimated to be $6 \times 10^{16}$ spins/cm$^3$ [56]. Without deconvolution, the resolution is determined by the size of the pinhole [54,56]. With deconvolution, it was estimated that the resolution was limited by the sensitivity of the spectrometer and that the minimum detectable number of spins was about $10^{10}$ spins with a line width of a 1.0 G [23]. A resolution of 0.3 mm was estimated for a test sample of a stable organic radical [55]. Only a small depth near the surface of the sample is imaged in this technique.

A slit in one wall of a rectangular resonator also has been used for localized detection of radicals produced by irradiation of teeth [57].

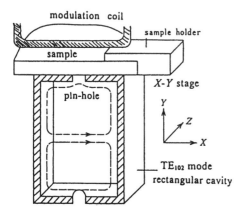

**Figure 3** Sample arrangement for EPR imaging with a pinhole in a resonant cavity. The direction of the external magnetic field parallels the $z$ axis. The dotted lines with arrows denote the microwave magnetic field in the cavity ($TE_{102}$ mode). The sample is attached to the computer-controlled mechanical stage and placed in contact with the cavity. The sample is scanned two-dimensionally in the $x$-$z$ plane. The EPR signal is detected only for the portion of the sample adjacent to the pinhole. (Reproduced with permission from Ref. 55.)

### 4. Localized Contribution to Magnetic Field

An array of closely spaced wires (diameter $= 50\ \mu$m) supported on a quartz holder was inserted into an X-band resonator, adjacent to a sample [58,59]. Current passing through $i$th and $(i + 1)$th wires makes a contribution to the magnetic field in the immediate vicinity of the wires. With the external magnetic field held constant, the current through the wires was adjusted to selectively satisfy the resonance condition for various regions of the sample with a one-line EPR spectrum. The magnetic field contribution falls off rapidly with increasing distance from the wire, so only the EPR signals in the surface of the sample near the wires were detected [58,59].

## V. EXAMPLES OF APPLICATIONS

The sections that follow present examples of the types of system that have been examined by EPR imaging. More extensive discussions of applications can be found in Ref. 1. The focus here is on the imaging of small samples and/or high spatial resolution.

### A. Radiation-Induced Defects

A sample of NaCl was irradiated with soft unfiltered x-rays through a pinhole in a lead shield [22]. The spatial distribution of defect centers in a 1.0 mm thick sample was determined by one-dimensional EPR imaging with miniature gradient coils inside the cavity. Fourier deconvolution of the nongradient spectrum gave the spatial distribution of defects. The attenuation coefficient obtained from the image agreed within a factor of 2 with the expected mass attenuation coefficient for photons at 50 keV [22].

Slices of human tooth were irradiated with x-rays or gamma rays [60]. The two-dimensional spatial distribution of carbonate radical was determined by moving the slice relative to a pinhole in an X-band cavity in 0.25 mm steps. The defects produced by the gamma rays were observed in the enamel on the front and back of the slice, but the defects produced by the x-rays (30kV) were observed only on the front surface.

An alanine dosimeter was irradiated with 40 kV x-rays through a lead shield with three pinholes separated by 600 and 700 $\mu$m, respectively [61]. The spatial distribution of alanine radicals in the dosimeter was imaged by moving the sample relative to a miniature magnetic field modulation coil. The three regions of radiation damage were well resolved in the one-dimensional image.

The depth profile of radiation defects in alanine dosimeters and in polypropylene-containing hindered amine light stabilizers (HALS) has been determined by spectral–spatial imaging [62]. The sample dimensions were 3–5 mm, the maximum gradient was 4 T/m, and 36 projections were used to reconstruct the image.

The dominant signal in irradiated silicon dioxide is the $E'$ signal, although other broader signals also are present. Spectral–spatial imaging permits determination of the spatial distribution of the $E'$ signal without interference from the broader signals [63]. The spectral–spatial imaging also provides signal averaging, which facilitates imaging of samples with low signal-to-noise ratio. An example of the spatial distribution of $E'$ centers obtained as a slice from the spectral–spatial image is shown in Fig. 4. A sample (ca. $10 \times 3.7 \times 1$ mm$^3$ plate of vitreous silicon dioxide) was irradiated with x-rays from a copper anode with an exposure of 1.7 J/cm$^2$. It was oriented with the magnetic field gradient along the direction of irradiation (the ca. 1 mm dimension of the sample). The apparent thickness of the plate in the image is in good agreement with the physical dimensions of the sample. A major source of uncertainty in defining the spatial distribution of the defects is the difficulty in aligning the sample such that the direction of the gradient is precisely parallel to the direction of irradiation. The combination of the imperfection in the sample alignment and the broadening due to the finite number of projections (64 for this sample) can be viewed as the experimental response function. The experimental spatial distribution is a convolution of the response function with the defect spin distribution. For these optically flat plates, at the spatial resolution of this image, the leading edge of the theoretical spin distribution is a delta function. Thus the shape at the edge can be used to estimate the experimental response function, which is shown as trace b in Fig. 4. Since the x-rays were not fully monochromatic, the

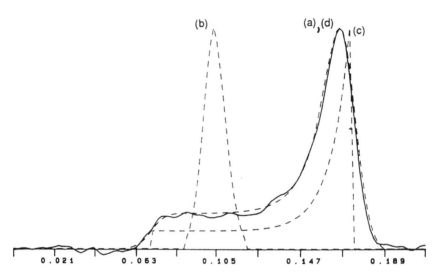

**Figure 4** (a) Spatial slice through a spectral–spatial image of a sample of irradiated silicon dioxide at the magnetic field that corresponds to the peak of the signal for the $E'$ center (solid curve). The sample was oriented such that the irradiated face is toward the right side of the figure. The maximum gradient was 300 G/cm. The image was reconstructed on a 128 × 128 grid with 5 iterations of an iterative back-projection algorithm. (b) Experimental response function estimated from the shape of the spatial distribution at the irradiated face of the sample. (c) Calculated spatial distribution of spins. (d) Convolution (dashed curve) of spatial distribution and experimental response function. The $x$ axis is in centimeters. The $y$ axes are in arbitrary units selected to give the same $y$-axis amplitudes for the four curves. (Reproduced with permission from Ref. 63.)

spatial distribution of defects was assumed to be the sum of a component that decayed exponentially with distance into the sample and a component that was uniform through the sample. The calculated distribution and the fit to the experimental data are shown as traces c and d in Fig. 4. These data suggest that when the experimental response function can be estimated, the definition of the spatial distribution can be improved by deconvolution of the response function from the spatial distribution [63].

## B. Diffusion

Many of the initial applications of EPR imaging have used stable nitroxide radicals to measure diffusion [4,6,64]. The experiments typically were designed such that the nitroxyl line shape was invariant through the sample and the diffusion coefficient could be determined from one-dimensional images as a function of time. Magnetic field gradients have been used for most of these experiments. Diffusion coefficients for the nitroxide radical 4-oxo-2,2,6,6-tetramethylpiperidine-1-oxyl (tempone) in the nematic phase of the liquid crystal *p*-pentylbenzylidine-*p*-butylaniline were anisotropic with a larger value of $D$ for motion perpendicular to the director axis ($9.0 \times 10^7 \text{ cm}^2/\text{s}$) than parallel to the director axis ($6.4 \times 10^7 \text{ cm}^2/\text{s}$) at 27°C [30]. Values of $D$ were calculated from one-dimensional images obtained with a magnetic field gradient. A rapid method for measuring translational diffusion constants by analysis of the Fourier transforms of the concentration profiles has been developed [64] and applied to cholestane spin labels in liquid crystal solvents between 11 and 60°C [21]. The activation energies for two systems were 6.3 and 8.63 kcal/mol [21]. Values of $D$ between $\sim 10^{-7}$ and $10^{-9} \text{ cm}^2/\text{s}$ can be measured in an hour by this method [65]. The effects of cholesterol on the dynamics and structural properties of two spin probes in multilayer model membranes between 15 and 60°C was examined [66,67]. There is a strong interrelationship between the dynamic and thermodynamic properties of these systems [67]. The experiments assumed that the nitroxyl line shape was invariant through the sample. Line shape variation can be included with spectral–spatial imaging. Macroscopic and microscopic diffusion were measured for the same sample of nitroxide in a model membrane by obtaining spectral–spatial images as a function of time [68]. The microscopic diffusion was found to be faster than the macroscopic diffusion.

## C. Geological and Mineral Samples

One of the first EPR imaging experiments used a magnetic field gradient in one dimension (50 G/cm) and rotation of the sample to determine the two-dimensional spatial distribution of paramagnetic nitrogen defects in two diamonds with diameters of about 2.5 mm [24]. In a subsequent experiment the nitrogen defect concentration was shown to be higher in the "coating" of a natural diamond than in the interior [69].

The spatial distribution of substitutional nitrogen (100 ppm) and nickel (10 ppm) in a single crystal of synthetic diamond ($\sim 1$ cm diameter, 0.2 mm thick slice) were measured by EPR imaging at liquid nitrogen temperatures. The sample was contained in a Dewar flask that was moved relative to a pinhole in an X-band rectangular cavity [70]. The resolution of the image was about 0.5 mm. The concentrations were different for different facets of the crystal. It was proposed that the variations were due to differences in the growth conditions. The line width of the nickel signal varied with the local nickel concentration.

A carbonate fossil of a crinoid (15 mm diameter and 1 mm thickness) was imaged by moving the sample relative to a pinhole in an X-band cavity [53]. The spatial distribution of the Mn(II) signal showed that the Mn(II) concentration was lower in the inner quinqueradiate part of the fossil than in the outer region (Fig. 5). The spatial distribution of Mn(II) in an ammonite fossil has also been determined by this technique [52,53]. The Mn(II) signal consists of six lines that extend over about 600 G. An organic radical signal is also present in varying concentrations. The presence of two species would invalidate imaging methods that employ deconvolution of the hyperfine split spectrum. The large spectral width would require the use of large gradients to obtain a spectral–spatial image with high spatial resolution. The individual lines are about 30 G wide, which

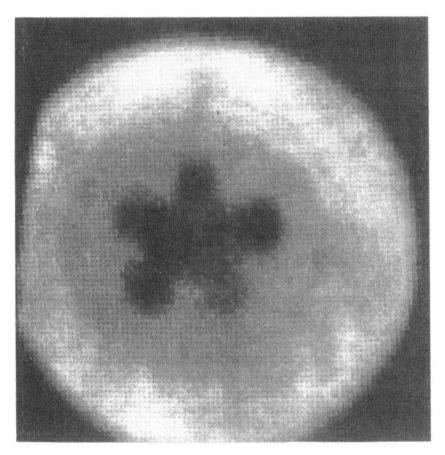

**Figure 5** EPR image of the spatial distribution of the Mn(II) signal in a slice of a stem of a carbonate crinoid. The image was obtained by moving the fossil relative to a pinhole (1 mm diameter) in an X-band cavity in 0.1 mm steps. The signal intensity at 101 × 101 points was converted to a 16-level gray scale. (Reproduced with permission from Ref. 53.)

would make imaging with gradients of isolated lines difficult. Thus localized detection is the method of choice for this type of sample.

The two-dimensional distribution of carbonate radicals in aragonite samples was measured by moving the sample relative to a pinhole in an X-band cavity [56]. Images of 1 cm crystals had a resolution of 0.5 mm. For one sample the region of the sample with an intense EPR signal was red, but for other samples there was no correlation between the EPR signal intensity and the color intensity. Two EPR signals in barite were observed to have different spatial distributions [56].

Hole centers in zircon are thought to have been produced by alpha decay [71]. Two-dimensional images of the hole centers and $Gd^{3+}$ centers in a 1 mm thick slice of a zircon crystal were obtained by moving the sample relative to a pinhole in an rectangular X-band resonator (Fig. 6) [55,56,71]. The spatial distribution of the hole centers correlated more closely with that observed for fission track density than did the spatial distribution of the $Gd^{3+}$ centers. This observation supports the proposal that the production of hole centers is closely related to alpha decay of radionuclides [71].

An aqueous solution of a stable perdeuterated [$^{15}$N] nitroxide radical was diffused into the pores of a pumice sample [72]. The $^{15}$N isotope was chosen because it has two hyperfine lines instead of the three lines observed for the naturally occurring isotope $^{14}$N. Deuteration reduces the line width of the EPR

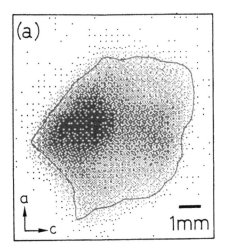
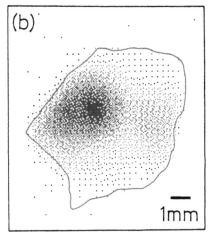

**Figure 6** EPR images of a zircon single crystal obtained by moving the crystal relative to a 1 mm diameter pinhole in an X-band cavity in 0.25 mm steps: (a) Gd(III) centers and (b) hole centers. The density of dots in each image is proportional to the intensity of the EPR signal. The closed curve in each image indicates the border of the zircon sample. (Reproduced with permission from Ref. 71.)

signal, which increases signal amplitude for the same concentration of radical and improves the resolution of the image. Magnetic field gradients were used to obtain two-dimensional and three-dimensional images of the radical distribution by imaging one of the lines of the EPR spectrum (an isolated line). The resolution estimated from the ratio of the gradient to the EPR line width was 69 $\mu$m for the two-dimensional images and 46 $\mu$m for the three-dimensional images. This resolution was not achieved for the three-dimensional image because of the limited number of projections used (22 projections in 22 planes). The image showed the open and interconnected pore structure of the pumice.

## D. Catalysts

Molybdenum ($H_3PMo_{12}O_{40}$) was dispersed in silica and pressed into a 20 mm pellet. Images of the $Mo^{5+}$ signal were obtained by moving the sample relative to a pinhole in an X-band resonator [73]. Under reducing conditions, the $Mo^{5+}$ concentration was found to be higher at the edges of the pellet than in the center and higher on the lower surface that was exposed to the reducing gas than on the upper surface [73]. Imaging of $V^{4+}$ in zeolite by the same technique was complicated by the small $g$-value difference between the $V^{4+}$ and organic radical signals [73].

The diffusion of a stable nitroxide radical into a zeolite catalyst was studied by EPR imaging with modulated gradients [74]. The nonuniform distribution was monitored as a function of time for four different zeolites. The spatial distribution of coke formed by cracking n-hexane at 773 K or cracking mesitylene at 803 K on a zeolite catalyst containing aluminum oxide has also been examined [74]. The coke formed from n-hexane was uniformly distributed through the sample, but the coke from mesitylene was irregularly distributed. It was proposed that because of its smaller size, the hexane could diffuse into channels that the larger mesitylene could not enter.

## E. Polymers

EPR imaging was used to examine defects in polypropylene due to thermal oxidation under stretching load [32]. Samples were oxidized under constant loading until fracture occurred. The broken samples were immersed in a solution of a stable nitroxide radical for 2–3 hours. Excess solution was removed from the surface, and the radical distribution was imaged with magnetic field gradients in two dimensions. The nongradient spectrum was deconvoluted. Defects were detected as higher than average nitroxide concentrations. Two regions of defects were observed to grow together near the location of the break. The results were compared with a model of catastrophic failure. The distribution of defects was in good agreement with infrared studies of the location of carbonyl groups produced by the oxidation process.

## F. Electrochemically Generated Radicals

Two-dimensional spectral–spatial imaging has been used to examine several systems of electrochemically generated radicals [75–77]. In these images one dimension is spectral and the second dimension is spatial—the image displays the EPR spectrum as a function of position in the sample. When *p*-benzosemiquinone was produced by reduction of *p*-benzoquinone in dimethylformamide (DMF), the characteristic five-line EPR spectrum was observed in the vicinity of the working electrode (75,76). The line width of the EPR signal on the side of the electrode toward bulk solution was broader than for the EPR signal in a restricted volume at the end of the electrochemical cell [76]. The line broadening was attributed to electron transfer between *p*-benzoquinone and the *p*-benzosemiquinone radical. Electrolysis of a mixture of diphenylanthracene and *p*-benzohydroquinone in DMF with the working electrode set at 1.54 V resulted in production of the broad signal for the diphenylanthracene cation at the working electrode and the characteristic five-line spectrum for *p*-benzosemiquinone at the auxiliary electrode (Fig. 7) due to the potential drop between the two electrodes [75]. The simultaneous generation of two radicals at different electrodes

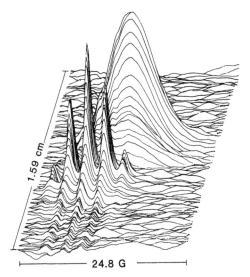

1.59 cm

24.8 G

**Figure 7** Spectral–spatial EPR image of *p*-benzosemiquinone radical and 9,10-diphenylanthracene cation radical produced electrochemically in DMF. A horizontal slice through the image is the EPR spectrum at a particular location in the sample. The maximum gradient was 213 G/cm. The image was reconstructed from 62 experimental and 2 estimated projections with an iterative back-projection algorithm. (Reproduced with permission from Ref. 75.)

would have been difficult to analyze without EPR imaging. Production of $p$-benzosemiquinone at the auxiliary electrode was shown to occur by oxidation of $p$-benzohydroquinone at the working electrode, diffusion to the auxiliary electrode, and subsequent reduction to the semiquinone [76]. The two-step process, which had been proposed on the basis of kinetic evidence, was clearly demonstrated by the imaging experiment.

In any imaging experiment there are a finite number of projections and points per projection, and these may restrict the resolution of the resulting image. For spectral–spatial EPR imaging it is important to determine how much spectral information from an image is retained in the spectral slice. The $p$-phenylenediamine cation radical gives a multiline EPR spectrum [77]. In a $256 \times 256$ pixel spectral–spatial image obtained with 120 experimental projections and 8 projections estimated iteratively, the spectral features were in good agreement with the nongradient spectrum [77]. This accuracy of the spectral dimension is important for accurate identification of the radical species that contribute to the signals in an image.

## VI. PROGNOSIS FOR THE FUTURE OF EPR IMAGING

The utility of EPR imaging with magnetic field gradients has been demonstrated with spatial resolution of the order of $10–100$ $\mu$m for samples with narrow EPR lines. At X-band, almost any sample with single-electron paramagnetic centers that is small enough to fit in the resonator can be studied by EPR imaging. As shown by the work of Ikeya, even some samples that are too large to fit in the resonator can be examined. For biological and other lossy samples, lower frequencies will be used, with attendant loss of $S/N$ (hence loss of resolution), but with greater experimental flexibility.

A major challenge is the development of higher magnetic field gradients to permit imaging of samples with broader lines or to achieve improved spatial resolution. This is primarily a heat removal problem, which is increased by the need for gradient linearity. Faster data acquisition will also be needed to facilitate multidimensional imaging.

FT-EPR and pulsed magnetic field gradient methods will be increasingly important for samples with sufficiently long relaxation times and narrow spectral widths. The faster data acquisition that is possible with pulsed methods will facilitate multidimensional imaging.

Spectral–spatial imaging permits the analysis of samples with multiple species and variable line shapes. It expands the range of samples that can be imaged and provides insight into the nature of the species present. The technique will be enhanced by the development of improved image reconstruction algorithms for incomplete data sets.

Scanning EPR microscopy provides access to samples that are too large to fit in standard resonators and to samples with large spectral widths. The challenge here is to improve the resolution and sensitivity that can be achieved with this technique.

Each of the EPR imaging methodologies will be enhanced by constructing resonators [78] and gradient coil systems optimized for the sample of interest rather than using standard cavities (e.g., $TE_{102}$ rectangular resonators at X-band) and magnet-mounted gradient coil systems.

## ACKNOWLEDGMENT

The work in the authors' laboratory on EPR imaging is supported by National Science Foundation grant CHE9103262.

## REFERENCES

1. Eaton, G. R., Eaton, S. S., and Ohno, K., Eds. (1991). *EPR Imaging and in Vivo EPR*, CRC Press, Boca Raton, FL.
2. Ikeya, M. (1991). Electron spin resonance (ESR) microscopy in materials science, *Annu. Rev. Mater. Sci. 21*:45.
3. Eaton, S. S., and Eaton, G. R. (1991). EPR imaging, *Electron Spin Reson. Spec. Period. Rep. 12b*:176.
4. Eaton, S. S., and Eaton, G. R. (1990). Electron spin resonance imaging, in *Modern Pulsed and Continuous-Wave Electron Spin Resonance* (Kevan, L., and Bowman, M. K., Eds.), John Wiley & Sons, New York, p. 405.
5. Ohno K. (1987). ESR imaging, *Magn. Reson. Rev. 11*:275.
6. Ohno, K. (1986). ESR imaging and its application, *Appl. Spectrosc. Rev. 22*:1.
7. Sotgiu, A., Gualtieri, C., Momo, F., and Indovina, P. L. (1988). ESR imaging: An overview, *Phys. Med. 149*.
8. Wertz, J. E., and Bolton, J. R. (1972). *Electron Spin Resonance: Elementary Theory and Practical Applications*, McGraw-Hill, New York.
9. Pilbrow, J. R. (1991). *EPR of Transition Metal Ions*, Oxford University Press, London.
10. Eaton, G. R., and Eaton, S. S. (1990). Electron paramagnetic resonance, in *Analytical Instrumentation Handbook* (Ewing, G., Ed.), Marcel Dekker, New York, p. 467.
11. Ewert, U. (1991). Sensitivity in EPR imaging, in *EPR Imaging and in Vivo EPR* (Eaton, G. R., Eaton, S. S., and Ohno, K., Eds.), CRC Press, Boca Raton, FL, p. 161.
12. Gorchester, J., and Freed, J. H. (1988). Two-dimensional Fourier transform ESR correlation spectroscopy, *J. Chem. Phys. 88*:4678.
13. Ewert, U., Crepeau, R. H., Dunnam, C. R., Xu, X., Lee, S., and Freed, J. H. (1991). Fourier transform electron spin resonance imaging, *Chem. Phys. Lett. 184*:25.

14. Ewert, U., Crepeau, R. H., Lee, S., Dunnam, C. R., Xu, D., and Freed, J. H. (1991). Spatially-resolved two-dimensional Fourier transform electron spin resonance, *Chem. Phys. Lett. 184*:34.

15. Halpern, H. J., Spencer, D. P., van Polen, J., Bowman, M. K., Nelson, A. C., Dowey, E. M., and Teicher, B. A. (1989). Imaging radiofrequency electron-spin-resonance spectrometer with high resolution and sensitivity for in vivo measurements, *Rev. Sci. Instrum. 60*:1040.

16. Halpern, H. J., and Bowman, M. K. (1991). Low frequency EPR spectrometers: MHz range, in *EPR Imaging and in Vivo EPR* (Eaton, G. R., Eaton, S. S., and Ohno, K., Eds.), CRC Press, Boca Raton, FL, p. 45.

17. Quine, R. W., Eaton, G. R., Ohno, K., and Eaton, S. S. (1991). Gradient coils, in *EPR Imaging and in Vivo EPR* (Eaton, G. R., Eaton, S. S., and Ohno, K., Eds.), CRC Press, Boca Raton, FL, p. 15.

18. Colacicchi, S., Indovina, P. L., Momo, F., and Sotgiu, A. (1988). Low-frequency three-dimensional ESR imaging of large samples, *J. Phys. E: Sci. Instrum. 21*:910.

19. Woods, R. K., Bacic, G. C., Lauterbur, P. C., and Swartz, H. M. (1989). Three-dimensional electron spin resonance imaging, *J. Magn. Reson. 84*:247.

20. Galtseva, E. V., Yakimchenko, O. Ye., and Lebedev, Ya. S. (1983). Diffusion of free radicals as studied by tomography, *Chem. Phys. Lett. 99*:301.

21. Cleary, D. A., Shin, Y.-K., Schneider, D. J., and Freed, J. H. (1988). Rapid determination of translational diffusion coefficients using ESR imaging, *J. Magn. Reson. 79*:474.

22. Ikeya, M., and Miki, T. (1987). ESR microscopic imaging with microfabricated field gradient coils, *Jpn. J. Appl. Phys. 26*:L929.

23. Ikeya, M. (1989). A portable spectrometer for ESR microscopy, dosimetry, and dating, *Appl. Radiat. Isot. 40*:845.

24. Hoch, M. J. R., and Day, A. R. (1979). Imaging of paramagnetic centres in diamond, *Solid State Commun. 30*:211.

25. Hoch, M. J. R. (1981). Electron spin resonance imaging of paramagnetic centres in solids, *J. Phys. C: Solid State 14*:5659.

26. Ohno, K. (1985). Two-dimensional ESR imaging for paramagnetic species with anisotropic parameters, *J. Magn. Reson. 64*:109.

27. Ohno, K. (1986). An ESR image of nitrodioxide molecules adsorbed on a copper rod at 10 K, *Chem. Lett. 17*.

28. Ewert, U., and Thiessenhusen, K.-U. (1991). Deconvolution for the stationary gradient method, in *EPR Imaging and in vivo EPR* (Eaton, G. R., Eaton, S. S., and Ohno, K., Eds.), CRC Press, Boca Raton, FL, p. 119.

29. Ohno, K. (1982). ESR imaging: A deconvolution method for hyperfine patterns, *J. Magn. Reson. 50*:145.

30. Hornak, J. P., Moscicki, J. K., Schneider, D. J., and Freed, J. H. (1986). Diffusion coefficients in anisotropic fluids by ESR imaging of concentration profiles, *J. Chem. Phys. 84*:3387.

31. Sotgiu, A., Gazzillo, D., and Momo, F. (1987). ESR imaging: Spatial deconvolution in the presence of an asymmetric hyperfine structure, *J. Phys. C.: Solid State Phys. 20*:6297.

32. Yakimchenko, O. E., Smirnov, A. I., and Lebedev, Ya. S. (1990). The spin probe technique in the EPR-imaging of structurally heterogeneous media, *Appl. Magn. Reson. 1*:1.

33. Herrling, T. (1991). Modulated field gradients: Instrumentation, in *EPR Imaging and in Vivo EPR* (Eaton, G. R., Eaton, S. S., and Ohno, K., Eds.), CRC Press, Boca Raton, FL, p. 35.

34. Ewert, U. (1991). Modulated gradients: Software, in *EPR Imaging and in Vivo EPR* (Eaton, G. R., Eaton, S. S., and Ohno, K., Eds.), CRC Press, Boca Raton, FL, p. 127.

35. Herrling, T., Klimes, N., Karthe, W., Ewert, U., and Ebert, B. (1982). EPR zeugmatography with modulated magnetic field gradient, *J. Magn. Reson. 49*:203.

36. Ebert, B., Hanke, T., Klimer, N. (1984). Application of ESR zeugmatography, *Stud. Biophys. 103*:161.

37. Ewert, U., and Herrling, T. (1985). Numerical analysis in EPR zeugmatography with modulated gradient, *J. Magn. Reson. 61*:11.

38. Gorelkinski, Yu. V., and Kim, A. A. (1985). Measurement of spatial distributions of different types of paramagnetic centers by EPR tomography. *Akad. Nauk Kaz. SSR. Ser. Fiz-Mat.:24*; *Chem. Abstr. 103*:114780e.

39. Maltempo, M. M., Eaton, S. S., and Eaton, G. R. (1987). Spectral-spatial imaging, in *EPR Imaging and in Vivo EPR* (Eaton, G. R., Eaton, S. S., and Ohno, K., Eds.), CRC Press, Boca Raton, FL, p. 135.

40. Maltempo, M. M., Eaton, S. S., and Eaton, G. R. (1991). Spectral-spatial two-dimensional EPR imaging, *J. Magn. Reson. 72*:449.

41. Ewert, U., and Herrling, T. (1986). Spectrally resolved EPR tomography with stationary gradient. *Chem. Phys. Lett. 129*:516.

42. Eaton, S. S., Maltempo, M. M., Stemp, E. D. A., and Eaton, G. R. (1987). Three-dimensional EPR imaging with one spectral and two spatial dimensions, *Chem. Phys. Lett. 142*:567.

43. Sueki, M., Quine, R. W., Eaton, S. S., Eaton, G. R., and Maltempo, M. M. (1989). Three-dimensional EPR imaging with two spatial dimensions and a spectral dimension, obtained with unequal maximum gradients in two sets of coils. *Phys. Med. 77*.

44. Sueki, M., Eaton, G. R., and Eaton, S. S. (1990). Multi-dimensional EPR imaging of nitroxides, *Pure Appl. Chem. 62*:229.

45. Maltempo, M. M., Eaton, S. S., and Eaton, G. R. (1991). Algorithms for spectral–spatial imaging with a "missing angle," in *EPR Imaging and in Vivo EPR* (Eaton, G. R., Eaton, S. S., and Ohno, K., Eds.), CRC Press, Boca Raton, FL, p. 145.

46. Eaton, S. S., and Eaton, G. R. (1991). Pulsed EPR imaging, in *EPR Imaging and in Vivo EPR* (Eaton, G. R., Eaton, S. S., and Ohno, K., Eds.), CRC Press, Boca Raton, FL, p. 73.

47. Sueki, M., Eaton, G. R., and Eaton, S. S. (1990). Electron spin echo and CW perspective in 3D EPR imaging, *Appl. Magn. Reson. 1*:20.

48. Nishikawa, H., Fujii, H., and Berliner, L. J. (1985). Helices and surface coils for low-field in vivo ESR and EPR imaging applications, *J. Magn. Reson. 62*:79.

49. Sotgiu, A., Fujii, H., and Gualtieri, G. (1987). Toroidal surface coil for topical ESR spectroscopy, *J. Phys. E: Sci. Instrum. 20*:1428.

50. Bacic, G., Nilges, M. J., Magin, R. L., Walczak, T., and Swartz, H. M. (1989). In vivo localized ESR spectroscopy reflecting metabolism, *Magn. Reson. Med. 10*:266.

51. Miki, T., and Ikeya, M. (1987). Electron spin resonance microscopy by localized magnetic field modulation, *Jpn. J. Appl. Phys. 26*:L1495.

52. Miki, T., and Ikeya, M. (1988). A method for EPR imaging for differential detection of paramagnetic species distribution utilizing localized field modulation, *J. Magn. Reson. 80*:502.

53. Furusawa, M., and Ikeya, M. (1988). Electron spin resonance microscopic imaging of fossil crinoid utilizing localized microwave field, *Anal. Sci. 4*:649.

54. Furusawa, M., and Ikeya, M. (1990). Electron spin resonance imaging utilizing localized microwave magnetic field, *Jpn. J. Appl. Phys. 29*:270.

55. Ikeya, M., Furusawa, M., and Kasuya, M. (1990). Near-field scanning electron spin resonance microscopy, *Scanning Microsc. 4*:245.

56. Furusawa, M., Kasuya, M., Ikeda, S., and Ikeya, M. (1991). ESR imaging of minerals and its application to dating, *Nucl. Tracks Radiat. Meas. 18*:185.

57. Ishii, H., and Ikeya, M. (1990). An electron spin resonance system for in-vivo human tooth dosimetry, *Jpn. J. Appl. Phys. 29*:871.

58. Ikeya, M., Furusawa, M., Ishii, H., and Miki, T. (1990). ESR microscopy, *Appl. Magn. Reson. 1*:70.

59. Ikeya, M., and Ishii, H. (1990). A scanning ESR microscopy using microwire arrays on a quartz sample holder, *J. Magn. Reson. 88*:130.

60. Ikeya, M., and Furusawa, M. (1988). Microdosimetric imaging of a tooth irradiated by x- and gamma-rays with ESR microwave scanning microscope, *Oral. Radiol. 4*:133.

61. Miki, T. (1989). ESR spatial dosimetry using localized magnetic field modulation, *Appl. Radiat. Isot. 40*:1243.

62. Morita, Y., Ohno, K., Ohashi, K., and Sohma, J. (1989). ESR imaging investigation of depth profiles of radicals in organic solid dosimetry, *Appl. Radiat. Isot. 40*:1237.

63. Sueki, M., Eaton, S. S., and Eaton, G. R. (1993). Spectral–spatial EPR imaging of irradiated silicon dioxide. *Appl. Radiat. Isot. 44*: 377.

64. Moscicki, J. K., Shin, Y. K., and Freed, J. H. (1991). The method of dynamic imaging of diffusion by EPR, in *EPR Imaging and in Vivo EPR* (Eaton, G. R., Eaton, S. S., and Ohno, K., Eds.), CRC Press, Boca Raton, FL, p. 189.

65. Moscicki, J. K., Shin, Y.-K., and Freed, J. H. (1989). Dynamic imaging of diffusion by ESR, *J. Magn. Reson. 84*:554.

66. Shin, Y.-K., and Freed, J. H. (1989). Dynamic imaging of lateral diffusion of electron spin resonance and study of rotational dynamics in model membranes: Effect of cholesterol, *Biophys. J. 55*:537.

67. Shin, Y.-K., Moscicki, J. K., and Freed, J. H. (1990). Dynamics of phosphatidyl-choline–cholesterol mixed model membranes in the liquid crystalline state, *Biophys. J. 57*:445.

68. Shin, Y.-K., Ewert, U., Budil, D. E., and Freed, J. H. (1991). Microscopic vs. macroscopic diffusion in model membranes by electron spin resonance spectral–spatial imaging, *Biophys. J. 59*:950.

69. Zommerfelds, W., and Hoch, M. J. R. (1986). Imaging of paramagnetic defect centers in solids, *J. Magn. Reson. 67*:177.

70. Furusawa, M., and Ikeya, M. (1990). Distribution of nitrogen and nickel in synthetic diamond crystal observed with scanning ESR imaging, *J. Phys. Soc. Jpn. 59*:2340.

71. Kasuya, M., Furusawa, M., and Ikeya, M. (1990). Distributions of paramagnetic centers and alpha-emitters in a zircon single crystal, *Nucl. Tracks Radiat. Meas. 17*:563.

72. Kordas, G., and Kang, Y.-H. (1991). Three-dimensional electron paramagnetic resonance imaging technique for mapping porosity in ceramics, *J. Am. Ceram. Soc. 74*:709.

73. Xu, Y., Furusawa, M., Ikeya, M., Kera, Y., and Kuwata, K. (1991). ESR microscopic imaging on the spatial distribution of paramagnetic reactive centers in catalysts, *Chem. Lett. 296.*

74. Ulbricht, K., Ewert, U., Herrling, T., Thiessenhusen, K. U., Aebli, G., Volter, J., and Schneider, W. (1991). EPR imaging on zeolites and zeolite catalysts, in *EPR Imaging and in Vivo EPR* (Eaton, G. R., Eaton, S. S., and Ohno, K., Eds.), CRC Press, Boca Raton, FL, p. 241.

75. Sueki, M., Quine, R. W., Eaton, S. S., and Eaton, G. R. (1990). Spectral–spatial electron paramagnetic resonance imaging of electrochemically generated radicals, *J. Chem. Soc., Faraday Trans. 86*:3181.

76. Sueki, M., Eaton, S. S., and Eaton, G. R. (1991). Electron paramagnetic resonance imaging of electrochemically generated *p*-benzosemiquinone radicals, *Anal. Chem. 63*:883.

77. Sueki, M., Eaton, S. S., and Eaton, G. R. (1991). Electron paramagnetic resonance imaging of electrochemically generated *p*-phenylenediamine radicals, *Anal. Sci., 7*: 571.

78. Hyde, J. S., and Froncisz, W. (1986). Loop gap resonators, in *Electron Spin Resonance*, Vol. 10, *Specialist Periodical Reports* (Symons, M., Ed.) The Royal Society of Chemistry, London, p. 175.

# 12

# Chemical Aspects of NMR Imaging

**Marjorie S. Went and Lynn W. Jelinski**   *Cornell University,*
*Ithaca, New York*

## I.  SCOPE

This chapter is intended for researchers in chemistry and related fields who wish
to augment their knowledge of the possible application of microscopic nuclear
magnetic resonance (NMR) imaging to important problems in materials and life
sciences. We provide a cursory background on magnetic resonance, with a
somewhat more detailed description of specific imaging experiments most commonly used. The bulk of this chapter is devoted to selected applications of NMR
imaging that demonstrate its versatility and unique capability of obtaining chemical and functional information in a noninvasive way. Rather than an exhaustive
review of the literature, this chapter is intended as a starting point from which
an interested reader may further explore the realm of possible applications. As
opposed to the other molecular level techniques described here, the NMR imaging experiment yields information on a length scale of tens of micrometers.
We demonstrate that one can, within the constraints of this length scale, simultaneously elucidate chemical, spatial, and dynamic properties of both biological
and nonbiological materials.

## II.  INTRODUCTION

## A.  Magnetic Resonance

For a thorough treatment of the principles of magnetic resonance, several excellent texts are available [1–5]. In this section we give a brief, phenomenolog-
ical description of an NMR experiment. A sample containing a sufficient

number of NMR-active nuclei (Table 1) is placed in a strong magnetic field [1–10 tesla (T)]. The field lifts the degeneracy of the nuclear spin energy levels, as shown in Fig. 1, and the spins precess about the field at the Larmor frequency $\omega$ given by:

$$\omega = \gamma \, \mathbf{B}_0 \tag{1}$$

where $\mathbf{B}_0$ is the strength of the applied field, which is usually taken to be along the z axis, and $\gamma$ is the gyromagnetic ratio. In *pulse* NMR experiments, a *resonant* radiofrequency (rf) field is applied in a transverse direction. The sample lies in a coil, which is the inductor of the resonant circuit. The effect is to torque the net magnetization about the axis of the rf field. The duration and strength of the pulse controls the angle by which the magnetization vector is tipped. A 90° pulse tips the magnetization completely into the *xy* plane. The spins continue to precess about the z axis at their Larmor frequencies. The NMR signal is observed as an induction in the rf coil.

In a macroscopic sample, unlike spins will precess at different frequencies determined by the *effective* fields at the distinct nuclei. The effective field at a nucleus is determined by its local chemical environment (i.e., electronic orbitals), which may either shield or deshield the applied field. The modified Larmor equation is:

$$\omega = \gamma \, \mathbf{B}_0(1 - \sigma) \tag{2}$$

where $\sigma$ is the chemical shift, relative to some reference. The fact that different chemical species have distinct chemical shifts is the basis for chemical identi-

**Table 1**  Properties of Nuclei Relevant to NMR Imaging

| Nucleus | Spin | Natural abundance (%) | Relative sensitivity[a] | Gyromagnetic ratio[b] |
|---------|------|------------------------|--------------------------|------------------------|
| $^1$H    | 1/2  | 99.98                 | 1.00                     | 2.6752                 |
| $^{19}$F | 1/2  | 100                   | 0.83                     | 2.5167                 |
| $^7$Li   | 3/2  | 92.58                 | 0.29                     | 1.0397                 |
| $^{23}$Na| 3/2  | 100                   | $9.25 \times 10^{-2}$    | 0.7076                 |
| $^{31}$P | 1/2  | 100                   | $6.63 \times 10^{-2}$    | 1.0829                 |
| $^{13}$C | 1/2  | 1.108                 | $1.59 \times 10^{-2}$    | 0.6726                 |
| $^2$H    | 1    | $1.5 \times 10^{-2}$  | $9.65 \times 10^{-3}$    | 0.4107                 |
| $^{15}$N | 1/2  | 0.37                  | $1.04 \times 10^{-3}$    | 0.2711                 |
| $^{41}$K | 3/2  | 6.88                  | $8.40 \times 10^{-5}$    | 0.0685                 |

[a]At constant field for equal numbers of nuclei.
[b]In SI units, $\times 10^{-8}$.
*Source*: Refs. 2 and 4.

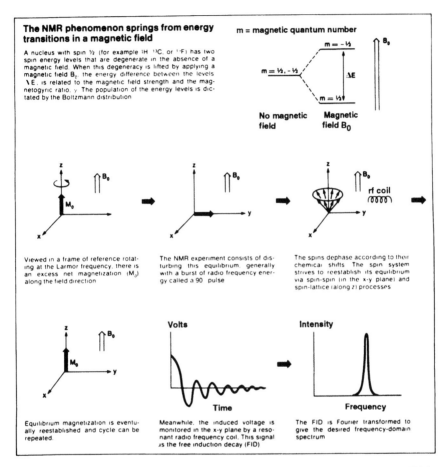

**Figure 1** Diagram of the NMR experiment. [Reprinted from Jelinski, L. W. (1984). Modern NMR spectroscopy, *Chem. Eng. News.* 62:26.]

fication by NMR spectroscopy. As the spins precess at different frequencies, *dephasing* will occur. Additional dephasing interactions may be present which govern the local fields at nuclei in a macroscopic sample. These include $B_0$ field inhomogeneities, rf field inhomogeneities, heteronuclear magnetic dipole interactions, and magnetic susceptibility effects, all of which behave analogously to chemical shift effects. In addition, *nonlinear* effects may be present. These include the homonuclear magnetic dipole interaction, where the local fields of neighboring like nuclei influence each other, and the nuclear quadrupolar interaction, when a nucleus has spin greater than $\frac{1}{2}$. All these interactions cause the induced signal to lose intensity when monitored over the time period in which the dephasing takes place, as shown in Fig. 1. This process is referred to as $T_2$ (or spin–spin) relaxation. The time constant ranges from tens of microseconds to seconds. The signal, called a free induction decay (FID), can be expressed as follows:

$$S(t) = \int \rho(\mathbf{r},t) \exp[-i\omega t]d\mathbf{r} \tag{3}$$

where, in this case, $\omega$ accounts for chemical shift effects only and $\rho(\mathbf{r},t)$ is the spin density at position $\mathbf{r}$ and time $t$. As normally viewed, in a reference frame rotating at the Larmor frequency, $\omega$ becomes simply $\gamma \mathbf{B_0}\sigma$. This signal can be Fourier-transformed from the time domain to the frequency domain to yield the NMR spectrum $s(\omega)$.

The spin system eventually will regain equilibrium and realign with the $\mathbf{B_0}$ field along the $z$ axis in a process called $T_1$ (or spin-lattice) relaxation. To signal-average, one must wait several times this relaxation time before repeating the experiment. Typical values for $T_1$ are 0.5–5 seconds for biological specimens, and from minutes to even hours for some solid examples.

In a liquid, rapid isotropic motion spatially averages the *orientation-dependent* interactions such as the chemical shift, the dipole–dipole interaction, and any electric quadrupole interaction. The chemical shift is averaged to an isotropic value, whereas the other interactions are averaged to zero. Thus their effect on the frequency spectrum is negligible. In a solid sample, however, the effect of these interactions is to broaden the spectrum by several orders of magnitude. The ramifications of this for performing imaging experiments are discussed further in this chapter.

## B. Resonance in a Gradient

The concepts introduced above can be applied in an *imaging* experiment to obtain a *spatial* map of the distribution of spin density, of other NMR parameters (such as $T_1$, $T_2$, and chemical shift), or functional parameters (such as flow rate and diffusivity) across a heterogeneous sample. The subject of NMR imaging has been described and reviewed extensively in the past few years [6–12]. Protons, being the most sensitive as well as the most ubiquitous, are by far the most imaged nuclei. The idea is to impose a known magnetic field *gradient* on the sample so that different regions in space are exposed to slightly different magnetic fields. The applied field $\mathbf{B_0}$ is still along the $z$ axis, but its magnitude is linearly varied in the $x$, $y$, or $z$ direction. Then different regions of the sample resonate at different frequencies, as shown in Fig. 2 for two capillaries filled with water. The Larmor equation can be expressed as follows:

$$\omega = \gamma \, (\mathbf{B_0} + \mathbf{G} \cdot \mathbf{r})(1 - \sigma) \tag{4}$$

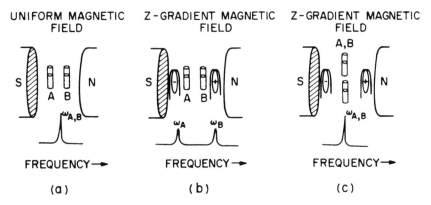

**Figure 2** Two capillaries of water side by side in a homogeneous magnetic field produce a single NMR line. (b) In the presence of a magnetic field gradient, these two capillaries produce separate lines. (c) However, if the capillaries experience the same field gradient, they give a single line. [Reprinted from Jelinski, L. W., Behling, R. W., Tubbs, H. K., and Cockman, M. D. (1989). *Am. Biotech. Lab. 34* (April).] (From Ref. 9.)

where **G** and **r** are the gradient and position vectors, respectively. It is usually desirable to experimentally ensure that the gradient term is much larger than the chemical shift term. Then the signal acquired in time–space, viewed in a reference frame rotating at the Larmor frequency, can generically be expressed as follows:

$$S(t) = \int \rho(\mathbf{r},t) \exp[-i\phi]d\mathbf{r} \tag{5}$$

where the phase shift $\phi$ is

$$\phi = \gamma \int_0^t \mathbf{G} \cdot \mathbf{r}\, d\tau \tag{6}$$

Consider a simple experiment consisting of a 90° pulse, as described above, with the FID now acquired in the presence of a gradient in the $x$ direction. The Fourier-transformed spectrum will be analogous to the chemical shift spectrum, but the frequency axis will correspond to position ($x$) rather than to chemical shift. The intensity will represent a one-dimensional *projection* of the three-dimensional spin density distribution onto the $x$ axis. To obtain spatial information in *two* or *three* dimensions, or to obtain functional information, this basic experiment must be modified with appropriate sequential application of rf pulses and magnetic field gradients.

To understand how imaging experiments work, it is helpful to begin with the concepts that are involved in putting these pieces together in a way that will yield the desired information.

## C. Building Blocks of Imaging Experiments

### 1. Dephasing and Refocusing

The concept of dephasing is directly applicable to the imaging experiment. Following a 90° pulse, the spins at different positions precess at different frequencies and will over time lose phase coherence. Viewed in a reference frame rotating at the Larmor frequency $\omega$ (i.e., corresponding to the center of the gradient, or the frequency of the applied rf), the magnetization vectors "fan out" in the $xy$ plane as some rotate faster and others more slowly than the applied frequency. The signal induced in the rf coil, viewed along one axis in the plane, decreases in intensity. The signal can be made to regain intensity if the magnetization vectors can be *refocused*, and the phase coherence regained. This can be accomplished via either spin echoes or gradient echoes, discussed below. A delay placed between the dephasing and refocusing periods, during which flow, diffusion, dephasing under other gradients, or other interactions are allowed to occur, modifies the echo signal, thereby allowing encoding of a new dimension of information.

### 2. Spin Echoes and Gradient Echoes

A spin echo refocuses magnetization with a 180° rf pulse applied at a time $\tau$ following the 90° pulse. The concept of an echo was conceived by Hahn in 1950 [13], while this particular type of echo was first demonstrated by Carr and Purcell in 1954 [14]. Figure 3 demonstrates the echo effect. The 180° pulse reverses the precessing magnetization vectors in the $xy$ plane such that those that were rotating clockwise rotate counterclockwise, and vice versa. The vectors that precessed the fastest in one direction will continue to do so in the reverse direction. At a time $2\tau$, the spins will once again be in phase and will refocus, and the signal will regain its maximum intensity, unless attenuated by any relaxation effects. The gradient that caused the dephasing following the 90° pulse remains on for the same period of time during the buildup of the echo to its maximum intensity. If the first (90°) pulse is phase-shifted by $\pi/2$ relative to the second pulse, the echo sequence will be more tolerant of errors in the 180° pulse setting. The second way to achieve an echo is simply to reverse the sign of the gradient. If spins at position $+x$ were precessing clockwise relative to the applied field under the influence of a positive gradient, they will precess counterclockwise during an equal negative gradient.

The spin echo technique has the advantage that it refocuses dephasing due to chemical shift and local inhomogeneities that otherwise would attenuate and broaden the signal. It is preferable for images of samples with long $T_2$'s. It also

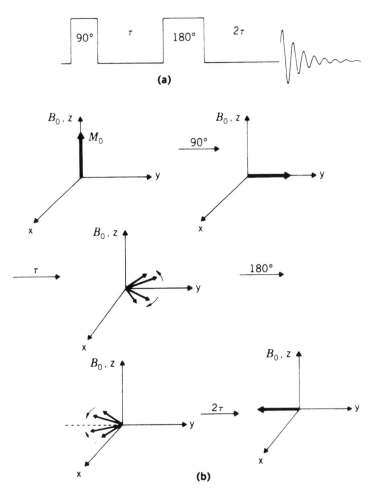

**Figure 3** Concept of a spin echo: (a) the pulse sequence and (b) the corresponding magnetizaiton vectors. Initially the spins are at equilibrium, aligned along the z axis. A 90° rf pulse brings the magnetization vector into the xy plane, where the spins begin to dephase. After a time τ, a 180° pulse is applied to reverse the sense of the dephasing, causing refocusing. After a time τ again the spins have refocused. [Reprinted from Bovey, F. A., and Jelinski, L. W. (1987). Nuclear magnetic resonance, in *Encyclopedia of Polymer Science and Engineering*, Vol. 10, 2nd ed., John Wiley & Sons, New York, p. 254.]

allows for acquisition during the center of a gradient period. The gradient echo is preferable for more rapid detection of the signal when a delay period is not wanted before echo formation. Gradient echoes refocus gradient inhomogeneities only; dephasing induced by local magnetic susceptibility or field inhomogeneity is not refocused and leads to intensity losses in the gradient echo image.

Also, gradient pulses are susceptible to the adverse effects of eddy currents during gradient switching, if the gradient coils are not well shielded.

## 3. Selective Excitation, Selective Saturation, and Slice Selection

In the presence of a one-dimensional gradient, spins located within a plane perpendicular to the gradient will resonate at a unique Larmor frequency. The 90° excitation pulse can be tailored to selectively irradiate at that frequency, with a very narrow bandwidth, while the gradient is on. It then excites only the spins in that plane. This is the concept of *slice selection* [15]. The thickness of the slice is determined by the gradient strength as well as the bandwidth of the pulse. Most selective excitation schemes involve a *shaped* rf pulse. For example, a pulse whose time dependence is a sinc $(\sin(x)/x)$ function has a frequency spectrum that is nearly rectangular. An alternative concept is that of *selective saturation*. A series of rf pulses can be made to destroy sample magnetization in all but the desired frequency range. It is common to counter unwanted dephasing that may have occurred during the pulse of finite duration by following the selective excitation pulse with a gradient reversal.

## 4. Frequency Encoding and (Two-Dimensional) Phase Encoding

A gradient that is on during the acquisition of the FID or echo is called a *readout* gradient; the signal is *frequency encoded* with spatial information in the direction of that gradient, as discussed above. Fourier transformation of the transient signal yields the distribution of spins along the gradient axis. An orthogonal gradient can be turned on prior to but not during the acquisition—for example, during the dephasing period between the 90° and 180° pulses of an echo. The amount of time that this gradient is on or, alternatively, the strength of this gradient, determines the *phase* of the signal at the beginning of the acquisition. A series of spectra with varying gradient strengths are generally collected. This method of *phase encoding* yields spatial (or functional) information in a second dimension. A two-dimensional Fourier transform creates a two-dimensional frequency spectrum from the series of FIDs. Most images represent magnitude calculations, and phase information is not preserved. However, phase information is useful for detecting flow.

## 5. Line Narrowing

Line narrowing is not an issue for most biological specimens, where isotropic reorientation of water produces sharp spectral lines. However, techniques have been developed in the field of high resolution solid state NMR spectroscopy [4,16] to average or effectively remove the orientation-dependent interactions that cause spectral broadening in solid samples. Two of the possible options are magic angle spinning (MAS) or multiple pulse line-narrowing techniques. These are necessary to enable imaging of true solids.

## D. The Imaging Experiment

Now it can be seen how a three-dimensional image can be created. A $z$ slice is selected, for example, and then a series of phase-encoded $y$ gradients is applied prior to acquiring a two-dimensional image in the presence of an $x$ readout gradient. Subsequent slices can be obtained to build up a three-dimensional image.

### 1. General Form

An imaging experiment can be divided into two parts. The first period, called the *preparation* period, determines the *contrast* of the image (i.e., the gray scale or the ordinate). The contrast can be based on, in the simplest case, spin density, or otherwise $T_1$, $T_2$, or functional parameters such as flow rate and diffusivity. The second period is the actual formation of the spatial image. This may consist of a slice selection (placed either before or after the preparation period), and/or an evolution period, and of course it includes a detection period:

**preparation** $\longrightarrow$ **image formation**
**slice selection – evolution – detection**

An important part of the imaging process is the actual display of the image. Figure 4 shows three images, each representing a slice, to demonstrate the ways in which the image contrast can be displayed as a function of two spatial dimensions. In the first image, areas of lower spin density in a plane are shown with increased darkness. The second image uses the method of Lauterbur [17] and others to present early images—a contour plot shows the regions of constant spin density. In the last, a three-dimensional depiction shows the contrast, in this case increasing fluid velocity, along the third axis or ordinate (pointing upward).

### 2. Types of Preparation

*Spin Density and Relaxation Rates*

Control over image contrast comes from the preparation period preceding formation of the image. In the simplest case, implicit in what we have been discussing until this point, the preparation is a 90° pulse applied to a fully relaxed sample, and the contrast is determined primarily by spin density. It is then desirable to minimize the time between preparation and detection so that relaxation effects can be minimized.

Contrast on the basis of *relaxation rates* can augment spin density contrast in many applications of NMR imaging, including the medical identification of disease states. For example, some tumorous tissue exhibits longer $T_1$ and $T_2$ relaxation times than normal tissue, possibly because of the decrease in the degree

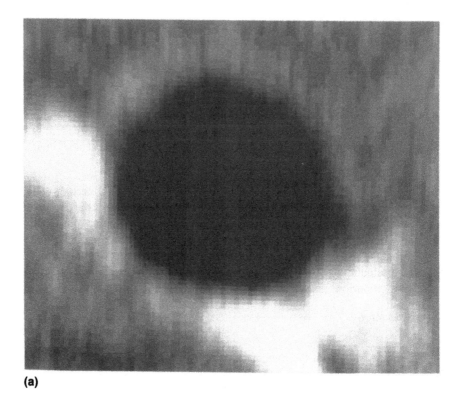

**(a)**

**Figure 4** Three ways to present image data: (a) a magnitude image of a transverse section through the carotid artery of a rat, (b) a contour plot of a rat artery, with thresholding to pick out the edges, and (c) a wire mesh drawing of blood flow through the artery, taken from phase-sensitive data.

of ordering of intracellular water in the tumorous tissue [18]. Relaxation contrast can be based on either $T_1$ or $T_2$. In its simplest form, a $T_1$-weighted image can be obtained by varying the repetition time between experiments, $T_R$. $T_2$ contrast can be achieved by preceding the image formation period with a spin echo and varying the length of the echo delay.

If the image formation is preceded by a spin inversion (180° pulse) followed by a recovery delay, the image can be made to show long or short $T_1$ regions selectively, based on whether the delay is long or short. This concept has been extended [19] by incorporating a *storage* period during which the magnetization is stored either along the $z$ axis or in the $xy$ plane for variable time. $T_1$ relaxation governs signal loss for storage along $z$, while *rotating frame relaxation* $(T_{1\rho})$ governs signal loss for storage in the $xy$ plane. An image of a sample of adamantane inside a neoprene rubber cylinder taken with $z$ storage and a short repetition delay shows predominantly the short $T_1$ component (rubber), while

**(b)**

**(c)**

the image taken with a long *xy* storage time and a long repetition delay shows only the component with long $T_{1\rho}$ (adamantane), as shown in Fig. 5.

*Function and Flow*

One of the major advantages of the technique of NMR imaging is that it can provide *functional* information *noninvasively* and without ionizing radiation.

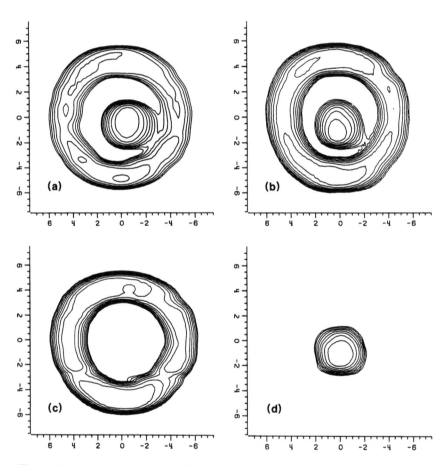

**Figure 5**  Multiple-pulse proton NMR images of a sample of a 5 mm tube of adamantane surrounded by a 12 mm OD neoprene hose section: (a, c) magnetization stored along *z* (Zeeman storage) and (b, d) magnetization stored in the *xy* plane (spin-lock storage). In (a) and (b) both species are seen because the repetition times and storage times were balanced. In (c), taken with a short repetition time, only the rubber image is seen, while the adamantane has been saturated. In (d), where the spin-lock time was long, the neoprene signal vanishes because it has a short $T_{1\rho}$. (From Ref. 19.)

Function can be studied by creating contrast based on velocity or diffusivity, and this information can be obtained in situ in combination with steady state NMR parameters, discussed above. Particularly useful is the ability to *noninvasively* image fluid flow. Other noninvasive techniques, such as ultrasound Doppler measurements, suffer from such drawbacks as the attenuation of signal from obstructing tissue, and interference from intervening gases and bones. Another advantage that NMR imaging has over techniques such as laser Doppler velocimetry is that it allows the simultaneous observation of flow information in more than one dimension. In addition, moving spins can be isolated from static spins, as discussed briefly below.

The subject of flow imaging by NMR has been reviewed [20,21]. Different classifications of flow measurement methods include time-of-flight, velocity encoding to obtain one-dimensional velocity profiles, and two-dimensional phase angle imaging.

The time-of-flight method involves selecting a slice and observing an echo at a later time. For a selective refocusing pulse, fluid flowing out of the slice leads to attenuation of the echo signal. If the 180° pulse is nonselective and a readout gradient is applied during the echo formation, the flow of fluid out of the originally excited slice is manifested as a loss of intensity in the Fourier-transformed echo signal.

A simple method to observe flow via velocity encoding is by the imposition of a *bipolar* gradient [22], consisting of two successive gradient pulses of equal strength and duration but opposite sign, with no delay between them, as shown in Fig. 6. Dephasing followed by complete refocusing occurs if no flow is present, whereas if spins have flowed out of their initial pixels during the time of application of the two *flow encoding* gradients, the observed echo signal will be attenuated. A more rigorous analysis follows.

For a gradient pulse of duration $T$ imposed along the flow direction ($z$) during the dephasing period in a spin echo experiment, the phase of the echo signal, measured in time $t$ from the center of the echo, or equivalently, from the beginning of the gradient, can be expressed [following Eq. (6)] as follows:

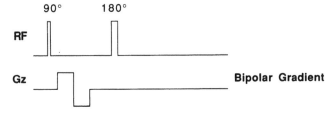

**Figure 6**  Simple spin echo pulse sequence with imposition of a bipolar gradient for measuring flow.

$$\phi(z,t) = \gamma \int_0^t \mathbf{G}(s) \cdot \mathbf{r}(s) \, ds = \gamma \, \mathbf{G}_z \, zT + \phi(v) \tag{7}$$

where $s$ is a dummy time variable measured from the start of the gradient pulse and $\phi(v)$ is a velocity-dependent phase correction:

$$\phi(v) = \gamma \int_0^T \mathbf{G}_z(s) \int_0^s v(p) dp \, ds$$

$$= \gamma \, v \int_0^T s \, \mathbf{G}_z(s) ds \quad \text{if } v \text{ is constant} \tag{8}$$

where $p$ is another dummy time variable. $\phi(v)$ is independent of position and is zero for stationary spins. For a simple (unipolar) gradient pulse of duration $T$, the total phase would be:

$$\phi(z,T) = \gamma \, \mathbf{G}_z \, z \, T + \frac{\gamma \, \mathbf{G}_z v \, T^2}{2} \tag{9}$$

One way to think of it is that the *apparent* position would be $z_{app} = z + vT/2$. (This is a valid expression for motion artifacts during spatial imaging sequences.) For a bipolar $z$ gradient, as shown in Fig. 6, with each gradient pulse being of duration $T$, the zeroth moment (area) is zero:

$$\int_0^{2T} \mathbf{G}_z(s) ds = 0 \tag{10}$$

The total phase is proportional to the first moment of the gradient and the velocity:

$$\phi(v,T) = -\gamma \, \mathbf{G}_z v T^2 \tag{11}$$

The velocity information contained in the phase is in this case independent of spin position. An orthogonal gradient can be applied to frequency encode position as well. For a frequency-encoding gradient along $x$, the image intensity will then be

$$S(t) = \int \rho(x) \exp[-i\phi(v,T)] \exp[-i\gamma x \mathbf{G}_x t] dx \tag{12}$$

where $\rho(x)$ is the spin density distribution. The velocity profiles are obtained from $\phi(v,T)$. Again, the observed echo signal is completely refocused with no attenuation from the phase term only if no flow has occurred.

Subtraction of a signal from a subsequent experiment performed with the polarity of the bipolar gradient reversed yields signal only from the moving spins. This can be generalized: serial application of two or more bipolar gradients will cause static spins to refocus after every bipolar gradient; spins moving at constant velocity will be caused to refocus after every second (even) bipolar gradient. Thus signal subtraction allows removal of signal emanating from non-moving spins.

*Flow compensation* refers to juxtaposition of a second bipolar gradient of opposite polarity in the same sequence to partially refocus the flowing spins that have dephased during the first bipolar gradient. Flow compensation can be done after the slice selection gradient pulses, so that the area and first moment of the gradients in the velocity encoding direction (usually parallel to the slice selection gradients) are zero before the start of the velocity encoding period.

A velocity-encoding scheme alternate to the bipolar gradient sequence involves application of two gradients of the same polarity separated by a period during which a 180° rf pulse is applied. In this case, flow compensation is achieved by repeating the sequence: gradient pulse – 180° rf pulse – gradient pulse.

In either scheme, the velocity-encoding gradient strengths can be incremented in a series to *phase-encode* the velocity (i.e., stronger gradients cause larger echo attenuation). The signal can be expressed as follows:

$$S = \int \rho(v_z,x) \exp[-i\phi(v,T)] \exp[-i\gamma x G_x t] \, dx \, dv_z \qquad (13)$$

Fourier transformation with respect to the gradient strength yields the velocity *distribution* function $\rho(v_z,x)$, which is the number of spins between $x$ and $x + dx$ having velocity between $v$ and $v + dv$. Again, the signal phase gives the velocity profile, $v(x)$. (Alternatively, the gradient durations can be incremented with the equivalent result.) If a second orthogonal position-encoding gradient of varying strength is added to the sequence, the phase information will give the two-dimensional velocity profile, $v(x,y)$.

Figure 7 shows how these parts fit together in a pulse sequence designed to yield a velocity profile. First, slice selection is followed by flow compensation. The flow encoding is combined with a spin echo experiment which uses an orthogonal frequency-encoding (readout) gradient during the dephasing period, while the bipolar gradient is on, and during the refocusing period of the spin echo. The flow information in one direction is superimposed on the spatial information in the orthogonal direction, and a velocity distribution profile is obtained.

Flow can also be measured at every voxel (volume element) by utilizing the inherent phase information in an image. Equation (8) showed that the phase accumulated by a flowing spin in a magnetic field gradient is directly related to its velocity. In a *phase angle* image, the phase angle of *each voxel* is calculated as the inverse tangent of the ratio of the imaginary to the real frequency–space data point:

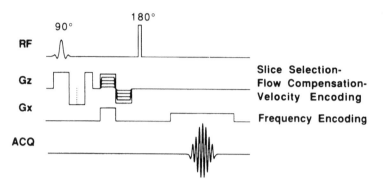

**Figure 7** Complete pulse sequence for measuring flow showing slice selection, flow compensation, and velocity and frequency encoding.

$$\phi_i = \tan^{-1} \left[ \frac{I_i(f)}{R_i(f)} \right] \tag{14}$$

Another method for obtaining velocity profiles via phase-encoded NMR imaging uses spatial modulation of magnetization (SPAMM) [23,24]. In the preparation period, two nonselective 90° rf pulses are separated by a time during which a gradient is imposed. The spins dephase in the *xy* plane under the influence of the gradient, but then are flipped back to the *z* axis by the second 90° pulse. The oscillatory nature of the transverse magnetization, which encodes position, is preserved when it is converted to *z* magnetization for a time on the order of $T_1$. (Residual transverse magnetization can be spoiled with a second gradient pulse, or is spoiled automatically during the slice selection stage of image formation.)

The image formation experiment that follows shows a pattern of "stripes" of light and dark in a plane parallel to the direction of the applied gradient. The gradient strength and amplitude control the width and spacing of the stripes. The time interval between rf pulses is minimized to prevent unwanted dephasing from local interactions. If the 'labeled' spins in a given "bright" pixel (corresponding to positive *z* magnetization) have *flowed* out of their pixel in the interval between the creation of modulated *z* magnetization and the formation of the image, the bright spot will be displaced to another pixel in the image. The stripe pattern therefore directly *shows* the spin displacements. Figure 8 shows a pulse sequence used to measure velocity profiles using SPAMM.

This method can be extended to a second dimension to create a two-dimensional array of stripes, or a *grid* pattern, of tagged magnetization. It has also been shown [25] that a series of nonselective saturation pulses can be used

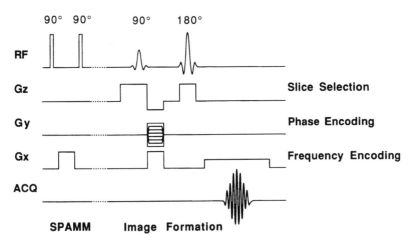

**Figure 8** Complete pulse sequence for a SPAMM experiment with a spin echo used for image formation.

to produce stripes that are more rectangular (sharp-edged) in their intensity profile than the sine-shaped stripes produced with a selective saturation pulse. The amplitude of sequential nonselective rf pulses in the pulse train is varied according to the binomial coefficients (e.g., 1:3:3:1 intensity ratios), with the gradient pulses placed in between the rf pulses. A higher density of stripes and thus a finer resolution of motion is attainable, thereby enhancing the feasibility of creating two-dimensional grid patterns. Another way of generating the grid pattern has been presented [26]—it makes use of a special train of pulses, called a DANTE sequence, in the presence of a continuous frequency-encoding gradient. Its advantages are superior line resolution and reduced eddy current formation.

It is worth noting at this point that consideration of relaxation effects is of practical importance in performing flow imaging experiments. A long $T_1$ places an upper limit on the velocity that can be measured, since the spins must be aligned in the magnetic field before they reach the coil. On the other hand, if $T_1$ is short, return of the magnetization to equilibrium competes with flow to attenuate the observed signal. A second consideration is that the experimental time must be short enough to ensure that the spins have not flowed out of the coil before the end of the acquisition. When phase-encoding methods are used to determine velocities, rapid flow leads to signal-to-noise limitations. Another concern is the suppression of signal from nonmoving spins. Several schemes in addition to the subtraction of signals acquired under opposite bipolar gradients, as mentioned above, accomplish this. They include selective saturation, selective excitation, and cancellation excitation [20].

*Diffusion*

Diffusion as it pertains to an imaging experiment can be thought of as intravoxel incoherent motion [21]. Spin diffusion will be neglected, and we refer to molecular displacements only. Diffusion processes can be studied by the pulsed-gradient spin echo (PGSE) technique, invented by Stejskal and Tanner [27] long before the advent of NMR imaging. An imaging pulse sequence that incorporates this technique to achieve diffusion contrast [28,29] is shown in Fig. 9. In this case, slice selection precedes the preparation period. The preparation consists of two gradient pulses, each of strength $g$ and duration $\delta$, separated by a time $\Delta$. The spatially dependent dephasing caused by the first gradient is inverted by a 180° rf pulse placed between the two gradients. The second gradient refocuses the magnetization and an echo is observed, provided the spins have not been displaced in space during the time period $\Delta$. On the other hand, if diffusion has occurred during the time $\Delta$, the peak echo amplitude can be expressed as follows:

$$E(\mathbf{g}, \Delta) = \int \rho(r) \int Ps(\mathbf{r}|\mathbf{r}',\Delta) \exp[i\gamma\delta\mathbf{g}\cdot(\mathbf{r}' - \mathbf{r})]d\mathbf{r}' \, d\mathbf{r} \tag{15}$$

where $P_s(\mathbf{r}|\mathbf{r}',\Delta)$ is the probability that a spin initially at position $\mathbf{r}$ will have moved to position $\mathbf{r}'$ in time $\Delta$. The function $\rho(r)$ is the initial spin density distribution. Alternatively, this expression can be written as follows:

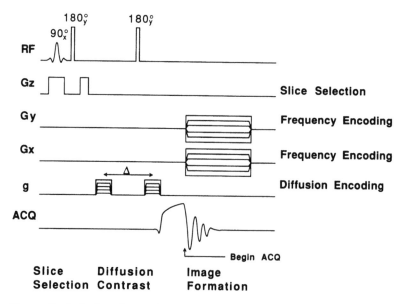

**Figure 9** Pulsed-gradient spin echo (PGSE) sequence with slice selection, frequency encoding, and phase encoding for providing a spatial map of diffusivity.

$$E(\mathbf{g}, \Delta) = \int P_s(\mathbf{R}, \Delta) \exp[i\gamma\delta\mathbf{g} \cdot \mathbf{R}]d\mathbf{R} \tag{16}$$

where $P_s(\mathbf{R}, \Delta)$ is an *average displacement propagator*, which represents the ensemble-averaged probability that a spin will move a distance $\mathbf{R} = (\mathbf{r} - \mathbf{r}')$ in the diffusion time $\Delta$. As the entire sample contributes to each pixel in $\mathbf{R}$-space, the spectrum of spin displacements can be obtained with high resolution. The gradient strength is varied, and the echo signal is Fourier transformed to $\mathbf{q}$-space, defined by $\mathbf{q} = \gamma\delta\mathbf{g}$, to give the displacement profile. For *free* diffusion, the function $P_s(\mathbf{r} \,|\mathbf{r}', \Delta)$ is a Gaussian. The diffusion coefficient can be determined from the displacement profile. For *restricted* diffusion, the geometry (i.e., cell size) is reflected in the dependence of $P_s(\mathbf{r} \,|\mathbf{r}', \Delta)$ on $\Delta$ [30–32]. The displacement spectrum is the autocorrelation function of the particular geometry along the gradient axis. The diffusion time can be varied to show the time dependence of the displacement profile. With the PGSE method, *indirect* information on average structures of porous materials is attainable with higher resolution than is possible with conventional NMR imaging.

A variation of the PGSE technique is the pulsed-gradient *stimulated* echo technique, where the $[\delta - \Delta/2 - 180° - \Delta/2 - \delta]$ rf sequence is replaced with $[90° - \delta - -90° - \Delta - 90° - \delta]$ so that magnetization is stored along $z$ during the diffusion time $\Delta$. The experiment time is then limited by $T_1$ instead of $T_2$ relaxation. Merboldt et al. [33] have demonstrated the use of the stimulated echo preparation sequence along with an imaging sequence to obtain spatially resolved molecular self-diffusion coefficients via the image contrast.

The PGSE sequence can be combined with an image formation sequence and slice selection, as depicted in Fig. 9; systematic variation of the strengths of the gradient pulses creates phase and amplitude modulations that give spatial displacement profiles for *each pixel* of the image. Spatially resolved velocity (or *coherent* flow) and diffusion maps can thus be simultaneously obtained in this technique, referred to as "dynamic NMR microscopy" [29,34]. Callaghan and Xia have used it to study capillary flow, achieving a 20 μm transverse resolution. To correctly calculate diffusion coefficients, one must take care to consider the effect of "cross terms" between parallel diffusion and imaging gradients when the two are of comparable magnitude [35].

## Chemical Shift Imaging

In a conventional imaging experiment, spins at different positions along the direction of the applied gradient resonate at different frequencies. One cannot distinguish between signals arising from spins with different *chemical shifts* and those from spins at different positions. In many cases it is desirable to obtain the spatially localized, volume-selective chemical shift spectra. The important issues are optimizing the signal-to-noise ratio and resolving the complexity arising from overlapping resonances. Methods used for separating the chemical shift effects include (1) varying the gradient strength so that the spatially dependent frequency offset scale is varied [36], (2) multiple phase-encoding methods in which no frequency-encoding gradient is used during the signal acquisition [37], (3) stimulated echo acquisition mode (STEAM) [38], and (4) the volume-selective spatial and chemical-shift-encoded excitation (SPACE) sequence [39–41].

The first option is time-consuming because it requires many experiments to obtain a frequency spectrum with a sufficient number of points to resolve chemical shift differences. With the phase-encoding methods, the principal difficulty lies in shimming the static field well enough to achieve high resolution in a number of contiguous voxels simultaneously. The stimulated echo localization resembles a spin echo method, with the 180° pulse replaced by two 90° pulses with a variable delay between them. The pulses are slice selective, and the gradients that determine the slices are orthogonal to each other as well as to the original slice selection gradient, to specify a volume of interest. This method has inherent signal losses due to relaxation effects and because the pulse sequence causes refocusing of only half of the originally available magnetization. SPACE requires very precise slice selection—the pulses generate dephasing of spins not in the specific slice-selected region. Chemical shift offsets and inhomogeneities are refocused, but for successful implementation very homogeneous and strong ($\gamma \mathbf{B}_1 \sim 10$ kHz) rf pulses are required—this places severe demands on rf probe design [41].

In some cases, one may wish to select a certain spectral line from a complex spectrum for the image: for example, water versus lipid resonances. Two ways to accomplish this are with a frequency-selective refocusing pulse in an echo and by selective saturation of the undesired lines before the image formation part of the experiment [36]. Chemical shift selective (CHESS) imaging [42] is a preparation period method that involves application of a selective 90° pulse followed by dephasing under a slice selection gradient to destroy a specific unwanted magnetization component prior to collection of an image. For example, the water and fat (methylene) resonances whose chemical shifts are separated by 3 ppm can be easily resolved [42]. It is an easily applied method that can be placed ahead of a sequence such as STEAM. In another selection method, Sotak and coworkers [43,44] have modified the STEAM sequence with a method of "zero-quantum coherence spectral editing" to remove lipid peaks that overlap lactate peaks, for example.

*Multiple Quantum Filtering*

Multiple quantum filtering [8] can be used to provide an additional source of contrast in an image. For example, with the use of a double-quantum filter [45] in $^{23}$Na imaging, one can distinguish the isotropically mobile spins from those in anisotropic environments (i.e., bound or physically confined). This has been demonstrated by Cockman et al. [46] for NaCl solutions of physiological concentrations where sodium spins are either free or entrapped in agarose gels. The double-quantum-filtered images show selectively the sodium in its bound state, whereas the single-quantum image shows both the isotropic and bound sodium. A drawback of this technique is that long signal acquisition times are required because the double- and triple-quantum signals are inherently insensitive.

*Chemical Preparation*

Contrast agents are chemical compounds that can be added to a specimen to increase relaxation rates and therefore increase the relative intensity in $T_1$ and $T_2$-weighted images. They are used to help identify and delineate neurological diseases such as brain lesions, blood–brain barrier leaks, and tissue abnormalities. An example is Gd HP-DO3A [ProHance, Squibb Diagnostics], a nonionic gadolinium chelate.

## 3. Methods of Image Formation

Now that we have seen how different methods can be used, separately or in combination, to prepare the magnetization or the chemical state of a sample for displaying the desired contrast, we turn to a discussion of methods of forming the actual spatial images. Image formation experiments may be classified according to the number of experiments used to form an image of an object divided into $n_x \times n_y \times n_z$ volume elements, as represented in Fig. 10. Methods in which each volume element is sampled by one and only one experiment are termed *sequential point* techniques. The conceptually simplest way to do this is to use a small surface coil and perform the NMR experiment on a selected spatial region defined by the field of the coil. If one simultaneously observes volume elements along an entire line, say in the $x$ direction, then only $n_y \times n_z$ experiments are required. Such methods are termed *sequential line* methods. Similarly, in *sequential plane* methods, an entire plane may be imaged in one experiment. The resolution of $n_x \times n_y$ elements is experimentally demanding. A variation involves simultaneously sampling from all elements in a plane, while resolving only elements along a line. In *three-dimensional* methods all volume elements

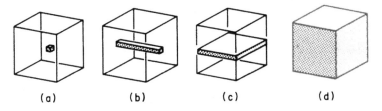

**Figure 10** Depiction of four image formation techniques: (a) sequential point, (b) sequential line, (c) sequential plane, and (d) three-dimensional.

are sampled simultaneously. These techniques have the highest sensitivity, since the signal emanates from all spins in the sample. The tradeoff is in resolution— in going from point to three-dimensional techniques, the attainable resolution in general decreases, while the sensitivity increases and total experimental time decreases. We discuss several of the techniques most commonly used.

*Projection–Reconstruction*

Projection–reconstruction is the method Lauterbur [17] used to obtain the first NMR image. He acquired a series of one-dimensional projections of the object (two capillaries of water) in the presence of a gradient that was rotated about the object a complete 180°, in discrete steps. The two-dimensional image, a projection of the spin density of the object onto a plane, was reconstructed using back-projection. An alternative to electronically rotating the gradient is to mechanically rotate the sample. The projection–reconstruction technique, depicted in Fig. 11, is straightforward, and the acquisition times are short, meaning minimal relaxation effects. Projection–reconstruction is therefore an ideal technique for obtaining images of solids. Several methods are available for the image reconstruction [8], including filtered back-projection, iterative reconstruction, and Fourier reconstruction. Some of these methods result in image distortions or artifacts. Slice selection can be used to obtain two-dimensional maps of a plane rather than projections onto the plane.

*Fourier Imaging*

Fourier imaging, also referred to as zeugmatography, was first reported in 1975 [47]. The technique is depicted in Fig. 12. An excitation pulse is followed by

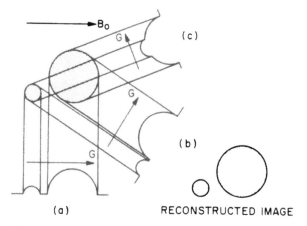

**Figure 11** Illustration of the projection–reconstruction technique: (a–c) projections obtained with gradients in three different orientations with respect to the external magnetic field. The reconstructed image is also shown.

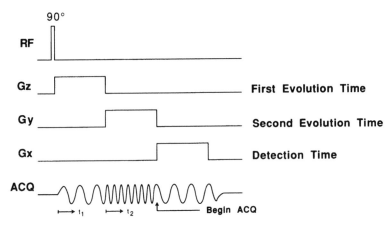

**Figure 12** Three-dimensional Fourier imaging sequence. The spins dephase under the influence of the gradients $G_x$, $G_y$, and $G_z$ during the times $t_1$, $t_2$, and the acquisition time, respectively. The parameters $t_1$ and $t_2$ are incremented in a series of experiments to produce the three-dimensional data set.

two or three evolution periods in the presence of orthogonal gradients. The first one or two serve the purpose of phase encoding. A series of FIDs is acquired during the last frequency-encoding gradient period, for varying duration(s) of the first gradient period(s). A two- or three-dimensional Fourier transform directly yields the spatial map of the spin system.

*Spin Warp Imaging*

In spin warp imaging, depicted in Fig. 13, a slice is chosen with a selective 90° rf pulse in the presence of a gradient, say in the $z$ direction ($G_z$). Subsequently the spins dephase in phase-encoding gradient, ($G_y$), which is incremented in strength for each experiment. A selective 180° pulse, again during $G_z$, refocuses the dephasing that has been caused by field inhomogeneities and chemical shift dispersions. A third gradient ($G_x$) is positioned symmetrically about the point of maximum echo formation for encoding spatial information during the acquisition. The readout gradient $G_x$ is also on during the phase-encoding period to ensure that the spatial information will be properly refocused following the echo peak. The time between the excitation and refocusing pulses (the echo time $T_E$) is typically several times shorter than $T_2$. Spin warp imaging was first described in 1980 [48]. A variation is the gradient echo (Fig. 14), where a negative gradient accomplishes the refocusing. The advantage here is a shorter echo time.

*Echo Planar Imaging*

Echo planar imaging (Fig. 15) is a cleverly devised [10,49–51] technique that allows simultaneous imaging of an entire plane in one experiment with high resolution. It is therefore very fast and very sensitive. Following the slice selection,

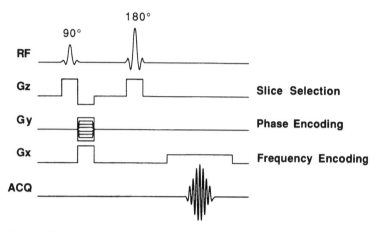

**Figure 13**  Spin warp pulse sequence with slice selection along $z$, phase encoding along $y$, and frequency encoding along $x$.

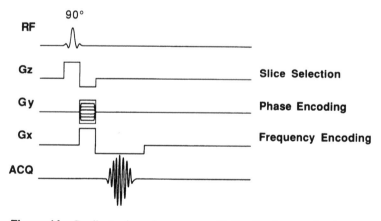

**Figure 14**  Gradient echo pulse sequence. Notice the shorter echo time compared to the spin warp sequence in Fig. 13.

an $x$ gradient is turned on. Meanwhile, a switched $y$ gradient of alternating polarity is turned on and the signal is acquired. For each cycle, dephasing and rephasing will occur under the $y$ gradient. Dephasing under the $x$ gradient meanwhile occurs throughout many cycles. The signal implicitly contains two temporal dimensions: $nT$, the time since the beginning of the acquisition at each echo peak, which frequency-encodes the $x$ dimension, and $t_2$, the time since the beginning of the $n^{th}$ echo, which also frequency-encodes the $y$ dimension. The signal is the same as in a Fourier imaging experiment:

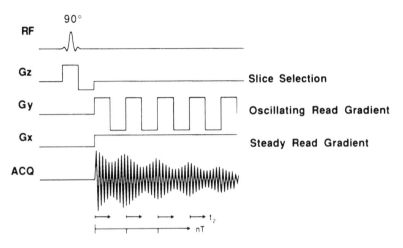

**Figure 15** Echo planar imaging pulse sequence.

$$S(nT,t_2) = \int \rho(x,y)\exp[-i\gamma G_x nT]\exp[-i\gamma G_y t_2]dx\ dy \qquad (17)$$

A two-dimensional Fourier transform directly yields the two-dimensional spatial distribution function, $\rho(x,y)$.

## III. RANGE AND LIMITATIONS

### A. Nuclei and Sensitivity

One could theoretically obtain images of essentially all the NMR-active nuclei. However, in practice, sensitivity, relaxation time, and line width considerations generally limit the application of NMR imaging to a small subset of possibilities. Protons, by virtue of their sensitivity and abundance, have been extensively imaged. In biological specimens, water, the protons on fat, the methyl group of lactic acid, and the acetyl groups of various neuromodulators are all present in sufficient quantities to provide acceptable images in a matter of minutes. Deuterated water, which has been added exogenously to biological specimens, has also been imaged. Much of the imaging work on true solids (i.e., materials below their glass transition temperature) has been performed on the protons of various polymers.

Other nuclei, which are simultaneously present in sufficient quantities in biological specimens and can be readily subjected to magnetic resonance imaging (MRI) (see Table 1) include lithium, sodium, potassium, and phosphorus. For example, sodium MRI is particularly useful for imaging the brain, where it can be used to pinpoint edema and regions of stroke damage. Some phosphorus-containing species are present in sufficient concentrations in biological samples to provide images. These include brushite and hydroxyapatite in bone, and inorganic phosphate, creatine phosphate, and some phosphate esters in muscle.

The natural abundance of $^{13}$C (1.1%) makes it too insensitive for imaging without isotopic enrichment. It is likely that the indirect detection of carbon through the protons may ultimately provide the most sensitive way to detect specific carbon species.

## B. Resolution, Sensitivity, and Acquisition Time

There are a number of factors that limit the resolution ultimately obtainable with NMR microscopy. These include a desired signal-to-noise ratio, the effects of self-diffusion of the species being imaged, and magnetic susceptibility effects arising from inhomogeneities in the sample. These factors depend, in turn, on a myriad of experimental variables, including the coil dimensions, its quality factor $Q$, the magnetic field strength, and relaxation parameters. A sobering realization is that the time required to obtain an image goes as the inverse sixth power of the resolution [52]. The theoretical limit on resolution presently attainable is about $10 \times 10 \times 10 \ \mu m^3$; this theoretical limit, which has not been experimentally verified, makes a number of assumptions concerning sample perfection. A practical resolution limit at present for real-world, living, samples is about $20 \times 20 \times 100 \ \mu m^3$. This resolution limit is consistent with a straightforward calculation based on a present-day sensitivity of about $10^{17}$ spins, and calculating that the smallest box they would fit into is one with dimensions about $30 \times 30 \times 40 \ \mu m^3$. Since between three and five pixels are needed to define a feature in an image, these ballpark numbers suggest that the lower limit on the size of an object that can be imaged is of the order of $50 \ \mu m$.

Some examples of attainable resolution are provided in Section IV. It should be noted that these are usually *computed* resolutions, determined simply by the strengths of the gradients, the number of points in both dimensions, and the spectral width. These computed resolutions have been verified in very few cases. The *actual* resolution is determined by the considerations discussed below.

The signal-to-noise ratio in a free induction decay is given by [1,53]:

$$S/N = \alpha^{-1/2} \left[ \frac{V_s}{2V_c} \right] \left[ \frac{x_0}{\gamma} \omega_0 \right] \left[ \frac{\mu_0 \, Q \, \omega_0 \, V_c}{4kT \, \Delta f} \right]^{1/2} \frac{1}{F_n} \tag{18}$$

where $\alpha$ depends on coil geometry, $V_s$ and $V_c$ are the sample and coil volumes, respectively, $Q$ is the quality factor, $\Delta f$ the frequency bandwidth of the receiver, and $F_n$ the spectrometer noise figure. It can be shown that the imaging time for a volume $(\Delta x^3)$ is given by [54]:

$$t = \left[\frac{S}{N}\right]^2 a^2 \left[\frac{T_1}{T_2}\right]\left[\frac{2.8 \times 10^{-15}}{f^{7/2}}\right]\left[\frac{1}{\Delta x}\right]^6 \tag{19}$$

where $a$ is the coil radius and $f$ the spectrometer frequency in megahertz. Allowing a 1000-second imaging time, an $S/N$ of 20, a $T_2$ of 0.5 second, a field strength of 400 MHz, a $T_1$ of 0.5 second, and a coil of radius 2 mm, one calculates a resolution of $1.4 \times 1.4 \times 1000 \ \mu m^3$. The limiting resolution is, of course, much less than this.

## C. Diffusion, Susceptibility, and Motion Effects on Resolution

Resolution calculated according to Eq. (19) is the optimum resolution, or $\Delta x_{opt}$. The transverse resolution can be degraded by diffusion and magnetic susceptibility effects. Broadening by diffusion arises because motion of the molecule in the presence of the imaging gradient causes loss of phase coherence of the transverse magnetization [55]. The new (degraded) resolution is calculated by:

$$\Delta x = 1.34 \left[\Delta x_{dif} \Delta x_{opt}^2\right]^{1/3} \tag{20}$$

where $\Delta x_{dif}$ is the diffusion-limited resolution. Figure 16 shows the effects of diffusion limits on resolution [11].

Magnetic susceptibility effects comprise an additional factor affecting resolution. These arise from inhomogeneities in the sample and can originate from sources such as tissue–air interfaces and organic–water interfaces [56]. Such interfaces are often found in biological samples, especially in the nasal cavities and in plants. The effect of susceptibility-limited resolution is dependent on the strength of the external magnetic field (the higher the field strength, the more serious the problem), the gradient strengths (the weaker the gradient, the more serious the problem), the sample shape, and the orientation of this shape with respect to the magnetic field.

Although these effects can cause resolution loss, they can also be used to advantage. An example of this is discussed in the blood oxygenation level dependent (BOLD) contrast technique [57]. This method takes advantage of the fact that hemoglobin that is carrying oxygen is diamagnetic, whereas deoxyhemoglobin is paramagnetic. The paramagnetic deoxyhemoglobin produces intensity losses in the vicinity of capillaries, thereby producing enhanced contrast in regions that are deoxygenated.

As opposed to the considerations discussed above, which can be readily calculated, there are other more difficult-to-quantify effects that tend to adversely affect resolution. These effects generally pertain to sample motion. Breathing or motion artifacts can rapidly degrade resolution. Furthermore, even for plant

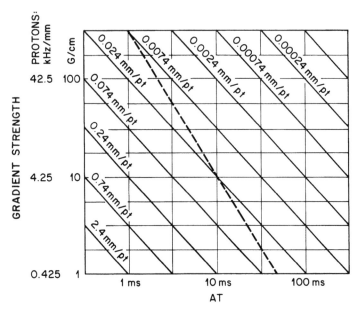

**Figure 16** Log–log plot of gradient strength versus acquisition time (AT). The solid diagonal lines define the spatial resolution and the dashed line corresponds to the resolution where approximately 15% intensity loss occurs because of water diffusion. (From Ref. 11.)

samples, it is conceivable that sample growth or elongation during long imaging experiments could adversely affect resolution.

The foregoing discussion has focused on the ultimate resolution obtainable for the proton resonance. Similar arguments pertain to other nuclei. For example, at present, the resolution limit for $^{31}P$ chemical shift imaging is about 3 cm$^3$.

The limitations of MRI for resolution and sensitivity have been well studied and well established. In contrast, the use of MRI to assess biological function is still its infancy, and the scope and limitations have not been carefully delineated. Various parts of this chapter describe state-of-the-art experiments on flow, diffusivity, and other parameters. For example, we show that one can generally measure flow ranging from about 25 cm/s [58] to 6 μm/s [59].

## IV. APPLICATIONS

## A. Introduction

In almost all biological systems water is a primary constituent. For example, water comprises approximately 80% of the human body. If this water is free, or exchanges rapidly with proteins or other cellular components, then isotropic averaging of local interactions results in a narrow NMR line width. The narrow line, in combination with the abundance and high sensitivity of protons, make the body ideally suited for study by NMR imaging. This is borne out by the tremendous impact the imaging technique has had on the medical community. The observed sensitivity [18] of the proton relaxation rates on cellular environments in biological systems, mentioned in Section II.D.2 above, adds yet another dimension to the utility of proton imaging. Many novel applications of imaging focus on a new dimension, the elucidation of biological structure and function.

Water or solvent taken up by a nonbiological material also lends itself easily to study by NMR imaging. By examining the properties of the saturating fluid, one can infer information about the solid host. For example, the distribution of water in polymer composites having polar functional groups, such as epoxies, is an important characteristic of the materials that can be studied in this way [60,61].

Solid systems in general are more difficult to study by NMR imaging, for the reasons mentioned above. However, the feasibility of imaging in solid systems has been demonstrated by pioneering researchers in studies of pore size distributions [30], polymer microphase separation [62], and ceramic structures [63].

## B. Ultra-High-Resolution Images

The greatest image resolution that has been claimed as of the time of this writing is 4.5 $\mu$m in the transverse direction in a slice that is 100 $\mu$m thick [64]. Homebuilt gradient coils were used in a spectrometer operating at a proton frequency of 500 MHz, with an 89 mm bore magnet. A spin warp imaging sequence was used, where the phase-encoding gradients spanned only half of phase–space (0 to $\pi$) to improve the apparent signal-to-noise ratio. The investigators applied the high resolution imaging sequence to study plant cell structure using a section of a geranium stem. An image in which individual cells can be differentiated is shown in Fig. 17. It was found that magnetic susceptibility differences between air in very vacuolar cells and water create the image contrast; near the air spaces, extra susceptibility-induced gradients that may be as large as 20 T/m cause dephasing and lead to signal distortion and loss of intensity.

In another study, a resolution of $10 \times 13 \ \mu m^2$ in the transverse plane with a 250 $\mu$m slice was obtained to resolve features in a *single cell* [65]. The cytoplasm was found to be distinguishable from the free fluid surrounding the cell as well as from the cell nucleus, as a result of either spin density or $T_1$ relaxation, in the egg cell of an African clawed toad (*Xenopus laevis*). In addition, the heterogeneity of the cell cytoplasm could be qualitatively observed.

## C. Mechanical Motion

One of the important recent developments in the study of biological function is the use of imaging to study *mechanical* motion, such as that occurring in the

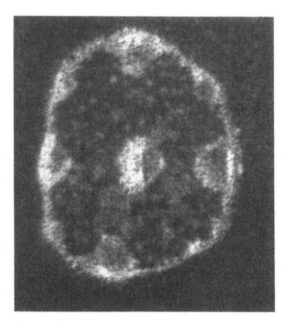

**Figure 17**   Image of a geranium stem. The in-plane resolution was 4.5 μm, and the slice thickness was 100 μm. The magnetic field strength was 11.7 T. (From Ref. 64.)

human heart, arteries, and spinal cord. In addition, kinematic changes in anatomical features, such as joint motion, could potentially be detected. Behling et al. [66] have used microimages of the rat carotid artery to measure features of its motion that are relatable to local effects of vasoconstriction and vasodilation throughout the entire heart cycle. From images obtained with gating to the electrocardiogram (EKG), they measured the cross-sectional area of the arteries, ranging from 600 to 800 μm, during the various stages of the cardiac cycle. A series of these images is shown in Fig. 18. These values were used along with the pressure exerted by the blood on the arterial wall at systole and diastole, measured with a pressure transducer, to determine the average Young's modulus, a measure of elasticity, of the artery. Although measurements of this kind can easily be made *in vitro*, the measurements made from NMR imaging data are unique in that they are noninvasive, hence can be conducted in vivo.

To observe heart motion with NMR imaging, the image acquisition can be EKG-gated, and successive images recorded from different points in different cardiac cycles. For example, this has been done using the SPAMM technique [23] to study heart wall motion, as well as the flow of cerebrospinal fluid. Axel and Dougherty [25] have also used the two-dimensional SPAMM technique to *quantitatively* study the deformation of the heart throughout the cardiac cycle,

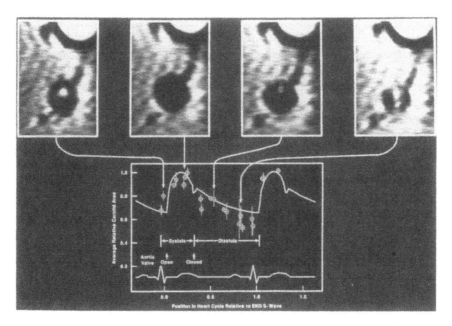

**Figure 18** EKG-gated images of the carotid artery of a rat along with arterial cross-sectional areas, shown as a function of the phase in the heart cycle. (From Ref. 66.)

by analysis of the strain tensor that can be obtained from the stripe patterns in orthogonal planes. Translation, rotation, and cardiac twisting during the contractile phase of the heart cycle have also been observed [67].

Recently, very rapid imaging techniques [49,51,69] have been used to image the heart motion in a *single* heart cycle. Frahm et al. [69] report heart images obtained in 0.3 second using the fast low angle shot (FLASH) pulse sequence [68]. The idea is to use pulses that are less than 90° (in this case 10°) and repeat the image formation sequence, incrementing the phase-encoding gradients, with very short recovery delays (4.8 ms) and echo times (2.8 ms). With this short experiment time, artifacts due to motion of the subject are almost completely absent. In addition, Frahm et al. used very strong gradients (10 mT/m) and a large receiver bandwidth to lessen chemical shift artifacts, an echo sequence to refocus susceptibility effects, and flow suppression via saturation of adjacent slices in the preparation period to gain contrast between blood and muscle. These researchers showed, in selected images from *different* heartbeats obtained without electrocardiogram gating, the opening and closing of the mitral valve. It is possible to obtain images from different points in a *single* cardiac cycle.

Pykett and Rzedzian [70] reported scan times of 40 ms per image using an echo planar sequence with a spin echo, to obtain ''instant'' images of the body.

They created an eight-frame "movie" (Fig. 19) of peristaltic motion of the intestine from a series of EKG-gated images acquired in one breath-hold using this very rapid technique. Their experimental parameters resulted in a voxel size of 0.08 cm$^3$, using a 4.7 mm thick slice. In addition, by making the 180° pulse of the spin echo frequency selective, they were able to separate the water resonances from lipid resonances, and they presented the combined data as single images in color, with two hues for the two spin density maps.

Levy et al. [71] have used phase-encoded flow imaging sequences, with electrocardiogram gating, to observe pulsatile motion of the spinal cord in human subjects who were either healthy or afflicted with various abnormalities of the spinal cord. The stationary cord is distinguishable from the flowing cerebrospinal fluid in a phase image acquired at the time of the gating, with no delay, whereas after variable delays on the order of 100 ms (corresponding to early systole), the spinal cord displayed motion. The capability to monitor this motion may prove to be clinically useful in evaluating certain diseases of the spinal cord.

## D. Flow and Diffusion

The literature on applications of imaging to study flow and diffusion is vast, demonstrating its utility for providing rapid, noninvasive information that is otherwise unavailable. Here we present a synopsis of a few exemplary applications in diverse systems. We begin with a comparison. Rittgers et al. [72] have used a time-of-flight imaging method of measuring velocities and have compared the results to those obtained by laser Doppler velocimetry (LDV). The

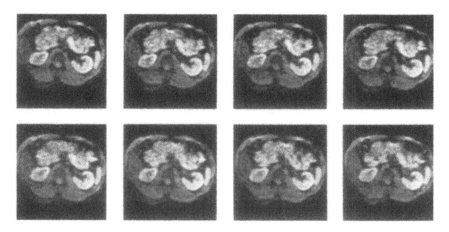

**Figure 19**   Eight-frame "movie" of magnetic resonance images showing normal peristaltic motion. The complete sequence was acquired in a single breath-hold. (From Ref. 70.)

velocities from NMR imaging were frequency encoded, and the fluid displacement profiles were obtained as projections across the tube diameter. The centerline velocities were measured from the leading edge of the profiles. Comparison of these values with the velocities measured from LDV gave a good correlation ($R = 0.991$). The investigators obtained velocity profiles downstream of tube stenoses for flow rates ranging from 5 to 170 cm/s, with Reynolds numbers ranging from 500 to 4200. Differences between the two techniques were observed under conditions approaching turbulence; whereas LDV measures local average velocities, the NMR imaging technique employed measures that temporally average local maximum velocities. The ability to observe changes in the flow profiles for different degrees of stenosis may prove to be very important in applications to living systems.

## 1. Blood Flow

Aspects of blood flow are of interest to medical researchers as well as researchers in biophysics who study the factors that affect biological function. NMR imaging has been used to study vascular flow, capillary flow, and perfusion. Blood flow measurements may be obtained in the presence of stenoses, near occlusions, and in the vicinity of the carotid bifurcation to provide important physiological information [58].

Weeden et al. [73] have imaged blood flow with EKG gating along with the technique of subtraction of successive signals from opposite bipolar gradients to observe the signal emanating from only those vessels that carry pulsatile flow. Thus the three-dimensional vascular system was projected onto a plane with virtually no background signal. The images were obtained of a human chest, thigh, and knee. An image showing the vessels in a human thigh is presented in Fig. 20.

NMR imaging experiments have provided the first direct visualization of flow in submillimeter-sized arteries. Behling et al. [58] have used the spin echo technique to obtain velocity fields of flowing blood in the carotid arteries of live rats. They used the spin warp imaging technique to obtain phase angle images, where the complex echo signal reflects how much phase the spins have accumulated by moving coherently in a magnetic field gradient that is parallel to the flow. Under the experimental conditions employed, the upper limit on the flow rate that could be detected was 23 cm/s. The acquisitions were gated to part of the electrocardiogram signal so that flow during the different periods of the heart cycle (systolic and diastolic) could be distinguished. Representative velocity fields are shown in Fig. 21.

Arai et al. [74] attempted to study cerebral blood flow by following in time the *oxygen-17* signal in images obtained after injection of isotopically enriched $H_2[^{17}O]$. They were able to obtain $^{17}O$ images of a rabbit brain in vivo, but with insufficient resolution in space as well as in time for quantification. Nevertheless, use of such isotopic experiments to measure flow appears to be promising.

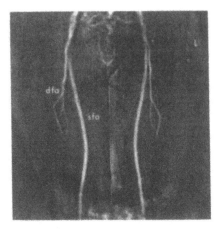

**Figure 20**  Blood flow image of the thigh of a normal human subject, showing parts of the vascular system. The superficial femoral (sfa) and deep femoral (dfa) arteries are labeled. (From Ref. 73.)

**Figure 21**  Blood flow velocity fields for the carotid artery of a rat, shown for fractional positions in the heart cycle (measured from the opening of the aortic valve) of 0.49, 0.77, and 0.91. The vertical axis corresponds to velocity; the maximum velocity for the image on the left is about 8 cm/s. (From Ref. 58.)

## 2.  Water in Plants

NMR imaging studies of plants have focused on understanding how physiological conditions affect the rate of water transport and the tissue water content. $T_2$ relaxation times of water in plants are typically less than in animal tissue (10 ms vs. 50 ms) because more of the water is near solidlike surfaces, which provide mechanisms for enhanced relaxation. Consequently, the pulse sequences must be more compact in time and the imposed gradients stronger, to encompass the broader line widths. Strong rectangular gradient pulses with short rise times can be achieved with low inductance gradient coils.

The first noninvasive measurement of flow in plants was by van As and Schaafsma [75], who used a time-of-flight technique [76,77] involving successive application of a series of equally spaced rf pulses while a gradient is applied in the flow direction. In their technique, the sample is in a Helmholtz coil, and the tip angle of the rf pulses varies from 180° at the center of the coil to 0° outside the coil. If no motion is present, the spins symmetrically positioned in either direction from the center of the coil will experience the same tip angle and opposite gradients; the magnetization components in the $xy$ plane will precess in opposite directions and no signal will be observed. If flow is present, some spins will be moving into the coil, while the spins moving out of the coil will not be detected, and a net magnetization will be generated. Application of this technique demonstrated the ability to discriminate flowing from stationary water. Water flowing through the stem of an intact cucumber plant was observed. The measured velocity was in good agreement with the measured weight loss due to evaporation. The velocity was observed to increase after water was added to the soil. Although a distribution of vessel sizes was known to exist, it was postulated that the signal from the largest vessels dominated the image.

Bottomley et al. [78] have studied water distribution and transport in plant root systems in situ. They used spin warp imaging and $T_2$ contrast to distinguish mobile from bound water, such as that bound to cellulose. Figure 22 shows an image of the *Vicia faba* root system, showing the lower stem, cotyledon, and root. The investigators compared different soil types and found that ferromagnetic particles in certain soils created image distortion and signal loss. Also, they used $Cu^{2+}$ nuclei in 0.04% $CuSO_4 \cdot 5H_2O$ as a paramagnetic tracer; this refinement decreases $T_1$ relaxation times and therefore enhances contrast in images taken with short repetition delays. Water leaving the cotyledon area was observed upon exposure of the plants to light sufficient to cause wilting. Following a recovery period in a cool, dark place, NMR imaging revealed that the water from an injected bolus was restored to the root tissues. Decay of the cotyledon area and in situ germination of seeds were also observed.

Johnson et al. [79] have studied transpiration in plant stems via NMR imaging. They obtained proton density and $T_1$ values, indicative of water content and the degree to which water is bound, which allowed them to discern different parts of stem tissue in plants. These parameters were studied as the plants transpired at low or high rates. $T_1$ decreased, indicating a smaller percentage of free water, as the total water content decreased in plants undergoing rapid transpiration. In tissue regions with small cellular structure and more water per volume, the effects were more pronounced than in tissues with more intercellular spaces.

## 3. Diffusion

The length scales that can be probed by NMR imaging measurements of diffusion are shorter than the obtainable resolution of the NMR image itself. Microstructure can therefore be studied by measuring, rather than spin positions, spin displacements over a fixed time scale. The factor limiting the resolution of features is not the gradient strength but the diffusivity, as shown in Fig. 16. For

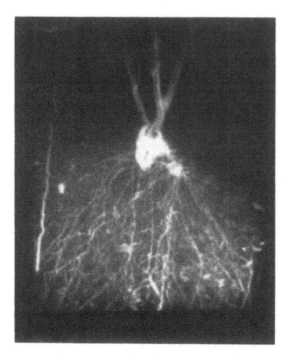

**Figure 22** Image of a *Vicia* faba root system with 0.56 mm resolution. The plant is viewed from the side, with the lower stems (top), cotyledons, and roots easily distinguished. The soil–air interface is just above the cotyledons. (From Ref. 78.)

example, features ranging from 0.1 to 25 μm, which create diffusion boundaries, can be characterized [30]. Fields of potential applications include fluid in living cells, liquids in polymers, and colloids and emulsions.

In situ water flow and diffusion in developing grains of wheat have been studied by Jenner et al. [80]. They used the PGSE technique with a stimulated echo for achieving contrast, and projection reconstruction for formation of the microscopic image. The sequence they used is similar to that shown in Fig. 9, with the spin echo replaced by a stimulated echo for magnetization storage along $z$. Figure 23 shows the resulting two-dimensional spin displacement (velocity and diffusion) and spin density distributions, with an in-plane pixel resolution of 100 μm × 100 μm, as well as the velocity *distribution* profile for one pixel. Volumetric flow rates of $6.2 \times 10^{-3}$ mm$^3$/s were measured, as well as diffusion coefficients of $8.3 \times 10^{-10}$ m$^2$/s. The observed mass flow is much larger than it would be if it were due to phloem transport alone; the authors postulate that recirculation occurs as nutrients are loaded and unloaded in the vicinity of the vascular system. The same group [81] has imaged water self-diffusion normal to

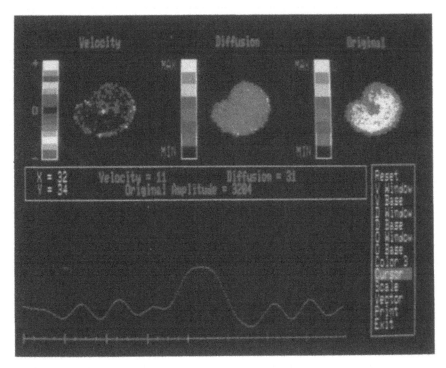

**Figure 23** Spatial maps of velocity, diffusion, and spin density ("original") for water in a transverse section of attached wheat grain. Also shown is a displacement profile showing the *distribution* of velocities in a given pixel ($X = 32$, $Y = 34$). (From Ref. 80.)

the imaging plane in various distinct structural regions of the wheat grain with 150 µm resolution. The measured diffusion coefficients range from $\sim 5 \times 10^{-10}$ to $\sim 10 \times 10^{-10}$ m$^2$/s.

## 4.   Diffusion in Porous Media

Studies of diffusion throughout a pore system can reveal structural features in a porous solid. Spatial characteristics such as morphology, granularity, and porosity in a bulk sample can be easily obtained in times shorter than those required to obtain an NMR image with comparable resolution [82]. The analogy to diffraction experiments was first proposed by Mansfield and Grannell [83]. Consider a PGSE experiment to measure diffusivity, with the echo intensity defined as in Eq. (8). In the limit of a diffusion time $\Delta$, which is long enough to permit a spin to sample all locations in the pore, $P_s (\mathbf{r} \,|\mathbf{r}',\Delta)$ becomes $\rho(\mathbf{r}')$. The experiment is then analogous to optical diffraction through a single slit [30]. If diffusion *between* pores is also possible, the resulting signal is analogous to that arising from multislit diffraction. The probability that a spin will migrate to a different pore that is a distance $Z$ away in a time $\Delta$, where $\Delta$ is long enough for intrapore diffusion to completely occur, can be defined as $C(Z,\Delta)$. Then $P_s(\mathbf{r} \,|\mathbf{r}',\Delta)$ becomes $\rho(\mathbf{r}')C(Z,\Delta)L(Z)$, where $L(Z)$ is a lattice correlation function that describes the relative positions of the pore centers [32]. A caveat is that one cannot distinguish a system of uniformly shaped pores or cells of different sizes from a system of pores or cells of varying shape but uniform size [30].

Callaghan et al. [32] have demonstrated these concepts with a test system of monodisperse polystyrene spheres in a random loose packing arrangement, where the sphere diameters were about 16 $\mu$m. The system consists of short-range order and long-range orientational disorder. Callaghan et al. performed PGSE experiments and observed a *coherence peak* in the echo amplitude, shown in Fig. 24 as a function of $\mathbf{q} = \gamma g \delta / 2\pi$, at a value of $1/\mathbf{q} \sim 16$ $\mu$m, the approximate pore separation distance. They then conducted the experiment with orthogonal gradient directions and used the projection reconstruction algorithm on the Fourier-transformed echo intensity versus $\mathbf{q}$ data to create a three-dimensional depiction of the lattice correlation function $L(Z)$.

Cory and Garroway [30] have studied pore structures using diffusion imaging. They obtained displacement profiles for several different diffusion times. The displacement profiles will continue to broaden as the diffusion time increases for free diffusion, but will reach a final profile for bound diffusion. Critical dimensions and times can be examined. Figure 25 shows displacement profiles for self-diffusion occurring in water and dimethyl sulfoxide (DMSO), which serve as model unbound systems. The self-diffusion coefficients were calculated from the data and were in good agreement with literature values. For water in yeast cells, a limiting width of the displacement profile was achieved by 20 ms. The diffusivity of water in the cells and the characteristic cell size (about 5 $\mu$m) were calculated from the profiles.

## 5. Perfusion

In terms of an imaging experiment, "perfusion" refers to flow in which spins change direction several times within a given voxel during the measurement time. Tissue perfusion and blood flow have been studied via a method involving a bolus administration of $D_2O$ as a freely diffusible tracer. After an intravenous, intra-arterial, or intratissue injection to a specific organ or tissue site, the deuterium concentration was monitored using an NMR signal picked up from a surface coil. In the simplest case, the decay can be modeled as an exponential to determine the volumetric rate of flow and perfusion [84]. In further experiments, it was postulated that recirculation of the $D_2O$ tracer, as in a multicompartment model, was responsible for observed results that were not well fit to the single exponential model [85].

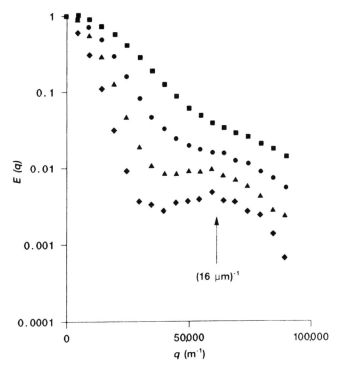

**Figure 24** Echo intensities as a function of gradient wave vector $\mathbf{q}$, from a PGSE experiment performed on a system of monodisperse polystyrene spheres, for a series of diffusion times $\Delta$: squares, 20 ms; circles, 40 ms; triangles, 70 ms; diamonds, 110 ms. The feature marked at $(16 \ \mu m)^{-1}$ corresponds to a pore separation distance of 16 $\mu$m, consistent with a random loose packing arrangement. An effective diffusivity of $2 \times 10^{-9} m^2/s$ was calculated from the low $\mathbf{q}$ data. (From Ref. 32.)

## E.   Chemistry and Mobile Materials

NMR imaging can be used to elucidate spatially dependent physical processes that involve mobile materials (i.e., materials above their glass transition temperature). Applications are as diverse as food processing, polymer processing, and hydrocarbon recovery. One can noninvasively monitor the effects of time, temperature, or other variables on processes.

## 1.   Food Processing

NMR imaging can be applied in food processing to study transport conditions during drying, freezing, crystallization, and other processes [86]. McCarthy and Kauthen [87] have utilized a tube rheometer as a flow system inside the imaging

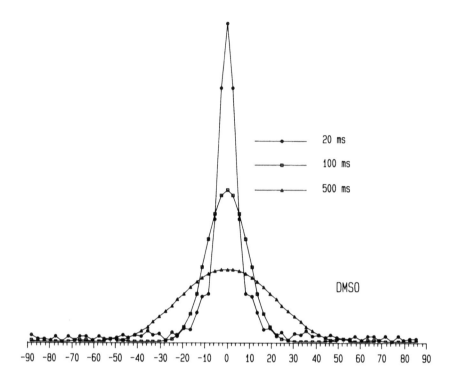

DMSO

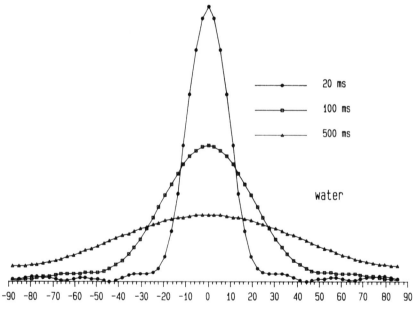

water

displacement microns

apparatus to study flow of tomato juice. They employed a time-of-flight method, where velocity profiles are built up as contours from two-dimensional images of a series of detection slices, displaced along the flow direction from the original excitation slice. The tomato juice was a suspension containing 6.1% solids, traveling at an average velocity of 11.2 cm/s. Non-Newtonian behavior was observed, and plug flow observed in the inner regions of the tube was in good agreement with theoretical predictions. The results of studies such as this have implications for improving technologies in aseptic food processing.

Attard et al. [88] reported the results of a comparison of fast versus slow thermal cooling of chocolate. Fast cooling leads to a mixture of low melting polymorphs from the cocoa butter and changes in the composition of milk-fat triglycerides. The intensities of the signals emanating from the two vials of chocolate that had undergone the different cooling protocols reflect the corresponding differences in $T_2$.

## 2. Polymer Systems

In addition to studies of diffusion in porous media, discussed above, studies of the properties of porous media by observation of *static* images of adsorbed fluids have been performed. Rothwell et al. [61] have studied the distribution of water adsorbed in glass-reinforced epoxy–resin composites. These belong to a class of polymer materials that have polar functional groups and interact strongly with water. Understanding the chemical changes that the water causes is important in the design of moisture-resistant materials. The protons from adsorbed water are preferentially observed over the protons from the polymer itself, which typically have much shorter $T_2$s. Water was imaged in two polymer rods that had been cured with different curing agents. Though the concentration of water was comparable in the two samples, its distribution was very different. In the sample cured with an aromatic amine, which acquired cracks during the fabrication process, the water was observed to be distributed fairly uniformly about the center of the rod. In the sample cured with an anhydride, the water was nonuniformly distributed about the rod's perimeter, as shown in Fig. 26. The authors surmise that the cracks provide the mechanism for macroscopic transport of the fluid to the center of the rod.

## 3. Hydrocarbon Systems

Quantification of the properties of liquid-saturated rock systems is important for studies of hydrocarbon recovery. Gummerson et al. [89] have obtained profiles

**Figure 25** Displacement profiles for water and DMSO at diffusion times of 20, 100, and 500 ms. The profiles are essentially Gaussian-shaped. Self-diffusion coefficients were calculated to be $2.03 \pm 0.15 \times 10^{-5}$ cm$^2$/s for water and $4.93 \pm 0.36 \times 10^{-6}$ cm$^2$/s for DMSO, at ambient conditions. (From Ref. 30.)

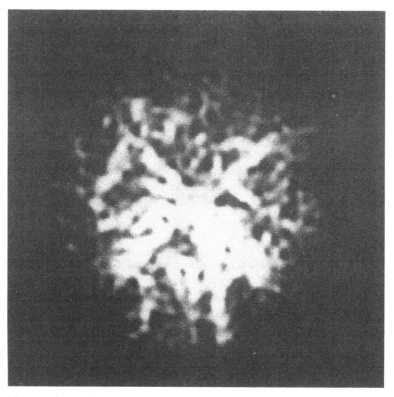

**(a)**

**Figure 26** NMR images of cross sections of glass-fiber-reinforced epoxy–resin composites: (a) sample cured with an aromatic amine, and (b) sample cured with an anhydride, both following exposure to water at 93°C. It can be seen that in (b) the water is distributed nonuniformly about the perimeter. (From Ref. 61.)

of water content as a function of distance during capillary adsorption into different porous materials, such as limestones and plasters, to determine hydraulic diffusivities. Attard et al. [88], who observed different $T_1$ values for water in a shale band compared to the 3% brine-saturated core of sandstone, obtain bimodal $T_1$ distributions from the image data. They are attempting to correlate $T_1$ with porosity, determined from a three-dimensional spin density image, which may allow calculation of permeability.

Baldwin and Yamanashi [90] have studied water and oil in reservoir rocks with NMR imaging. An important issue in improving the efficiency of oil recovery is how various fluids are distributed inside reservoir rocks. Dye injection

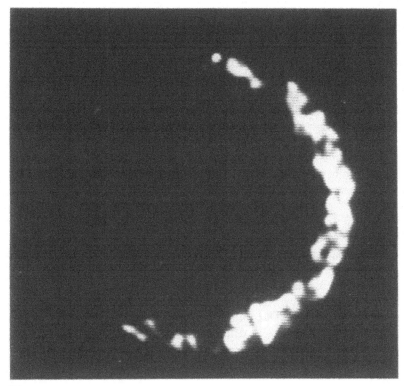

**(b)**

techniques can be used to study this effect, but they preclude multiple experiments on a single core. In this study, the flow of water and oil in a core was followed by observations of images as a function of percent water uptake. In the experiment, water was taken up by the core via capillary action at one end of the sample, a Berea sandstone core, and oil was expelled at the other. To distinguish the two phases, the water was doped with $Mn^{2+}$ ions (which are insoluble in the oil) to reduce its $T_2$ relative to that of the oil. The investigators found that in a previously water-wet core, the water front moved through in a pluglike fashion, bypassing some oil areas. In a previously oil- and water-wet core, the water traveled peripherally, leaving oil behind in the center of the core.

## 4. Colloidal Suspensions

Filtration of colloidal suspensions has been studied by imaging [91]. The buildup of solids on a filter surface during pressure-driven filtration was studied in situ using $T_1$ contrast, where a calibrated functionality of $T_1$ on surface-area-to-volume ratio of the particulate suspensions allowed determination of the liquid-to-solid volume ratio as a function of distance from the filter. This information can be useful in prediction of loss of filtrate.

Majors et al. [92] have demonstrated the applicability of magnetic resonance imaging to the study of multiphase flow. They measured velocities between 1.6 and 31 cm/s in solid suspensions, composed of polyethylene spheres in viscous oil, with solid concentrations up to 10% by volume. Their method involved frequency encoding along the direction of flow and phase encoding along the other two orthogonal directions to obtain concentration and velocity profiles. Their results show nonhomogeneous particle distributions and fluid velocity distributions for the suspensions, compared to homogeneous distributions for pure fluids.

Similar results were obtained by Altobelli et al. [93] for particles consisting of divinyl–benzene styrene copolymer in a viscous oil, with solid volume fractions up to 0.39 and average fluid velocities up to 25 cm/s. Some of their data were presented in terms of the maximum velocity divided by the average velocity (a measure of the "bluntness" of the velocity profile) versus the average solid volume fraction. They concluded that at high velocities the particles form a concentrated core in the tube center while the velocity profile flattens.

Another group of researchers [94] have also studied pure fluids and solid suspensions in Poiseuille flow, using a similar imaging technique. For single-phase Newtonian *and* non-Newtonian fluids undergoing Poiseuille flow, their imaging results were in good agreement with calculated velocity profiles. For concentrated suspensions, they obtained results that could be interpreted consistently with shear-induced particle migration. They hypothesized possible effects of flow-induced changes in $T_2$.

Flow-induced particle migration in concentrated suspensions of neutrally buoyant polymethacrylate spheres in Newtonian fluids undergoing Couette flow (between rotating concentric cylinders) has been studied [95,96]. The angular velocity was phase-encoded along the diameter of the cylinders. For pure Newtonian fluids, the measured velocities showed good agreement with the exact analytical solutions for wide gap Couette flow. For the suspensions, it was found that migration of particles occurs *irreversibly* under creeping-flow conditions from the high shear area near the inner wall to the low shear area of the outer wall. Figure 27 shows representative images of the fluid concentration fields. From these the authors obtained fluid fraction profiles, as shown in Fig. 28. They varied the parameters of the flow system and found that the migration rate depends on the shear rate, the particle diameter, and the polydispersivity of the particulate phase, not on the viscosity of the fluid nor on the rate of strain. Hydrodynamic diffusion is postulated to be the driving force for the particle migration.

## 5. Bioreactors

Flow imaging has also been applied to the study of hollow-fiber bioreactors. Heath et al. [97] have studied convection (leakage flow) in the extracapillary

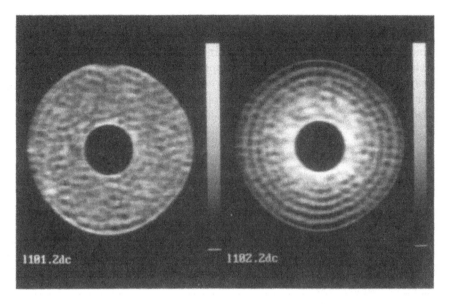

**Figure 27** NMR images of a cross section of a suspension of 50% polydisperse poly-methacrylate spheres between concentric cylinders. The left-hand image represents the initially well-dispersed suspension, while the right-hand image represents the steady state achieved upon rotation of the inner cylinder. The darker area near the inner cylinder represents a higher fluid fraction, or fewer particles, indicative of particle migration to the outer wall. (From Ref. 95.)

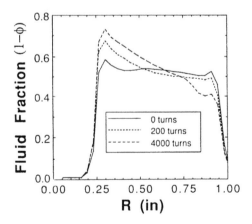

**Figure 28** Radial concentration profile, determined from the average value of signal intensity as a function of radial position for a suspension similar to that described in Fig. 27. The parameter $(1 - \phi)$ is the fluids volume fraction. (From Ref. 95.)

spaces of membrane bioreactors not containing cells. Their method was to apply a two-dimensional spin warp sequence with a bipolar gradient to phase-encode spins along the flow direction; the bipolar gradient was incremented to produce velocity images. Their results were compared with a mass transfer model and may provide very important information for future design of reactors regarding, for example, nutrient transport, packing of cells, membrane characteristics, and flow rates.

## F. Chemical Shift Imaging

In 1982 Hall and Sukumar [98] demonstrated the ability to select species possessing distinct chemical shifts in images, where the chemical shifts do not overlap, using capillaries of water, acetone, benzene, and methylene chloride. Since then, volume-localized spectroscopy and chemical shift imaging have been applied to a number of medical and nonmedical problems.

Frahm et al. [99a] have applied localized in vivo high resolution proton spectroscopy, using the STEAM technique described in Section II.D.2, with CHESS in the preparation period, to the human brain. They have resolved the spectral lines of cerebral metabolites in a voxel of dimensions $4 \times 4 \times 4$ cm$^3$. They optimized the gradient pulses in the sequence with respect to generating the spin echo while dephasing the magnetization that leads to unwanted echoes and minimizing the effects of motion and diffusion. Improvements were made [99b] in the technique to obtain a resolution of 1 cm$^3$, using water suppression and shorter echo times. Figure 29 shows a further improved [99c] spectrum of an 18 mL localized volume of the gray matter of a human brain in which *Scyllo*-inositol was identified. The same group has also measured [100] the concentrations, $T_1$s, and $T_2$s of the major cerebral metabolites in the insular and occipital areas, thalami, and cerebella of the brains of normal human volunteers. The $T_2$s ranged from 900 to 1850 ms, and the $T_2$s from 112 to 1200 ms. Variations in the relaxation times may be used to elucidate structure and dynamics in the different regions of the brain.

Using STEAM with water suppression, others [101] have performed localized spectroscopy of the brain on 18 multiple sclerosis (MS) patients and 17 control subjects. Their most prominent finding was a significant decrease in the signal intensity of *N*-acetyl aspartate in the MS patients. They also found that this decrease is not affected by the extent of progression of the disease, nor by the density of MS plaque. Another group [102] has used a similar technique, along with positron emission tomography, to obtain metabolic information in human gliomas. The magnetic resonance results indicated a relative decrease in the concentration of *N*-acetyl aspartate and an increase in the choline concentration in tumorous tissue.

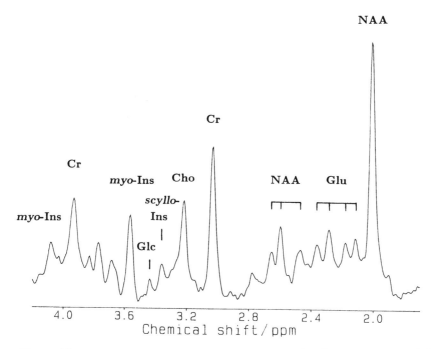

**Figure 29**   Localized proton spectra with water suppression from gray matter of the brain of a normal human volunteer showing the metabolites labeled as follows: *myo*-inositol (*myo*-Ins), *scyllo*-inositol (*scyllo*-Ins), creatin (Cr), glycine (Glc), choline-containing compounds (Cho), *N*-acetyl aspartate (NAA), and glutamate (Glu). The spectrum represents 64 scans on an 18 cm³ localized volume with an echo time of 20 ms and a repetition time of 6 seconds. (From Ref. 99c.)

Doddrell et al. [41] have obtained a spectrum from 0.2 cm³ of a rat's brain in vivo using the SPACE technique described in Section II.D.2, along with water suppression. They used a home-built rf probe with a 5.2 cm diameter resonator, giving a $\pi/2$ pulse time of 27 μs. The gradient strengths were 17 mT/m, in a 4.7 T magnet with a 13 cm bore. They localized the spectra in a cubic voxel where 0.19 cm³ encompassed ~85% of the signal. The selective spectrum displays up to 13 resolved peaks.

One of the goals in many volume-selective and chemical shift imaging studies of the brain is to select resonances from certain metabolites (e.g., lactate). Crozier et al. [39] use the SPACE sequence along with two-dimensional, multiple-quantum editing to target glutamate and glutamine, two excitatory neurotransmitters. While providing enhancements in the resolution, editing increases the time requirement—their spectra require 100 minutes of signal averaging, compared to 20 minutes in the nonedited study cited above.

Image-selected in vivo spectroscopy (ISIS) is a method to localize a volume element in combination with an acquired image. For plane selection, the spins in a plane are selectively inverted (with a 180° pulse) prior to image formation. The image signal is then subtracted from a signal obtained in the same way but without inversion. Ordidge et al. [103] have demonstrated the extension of this concept to three dimensions to achieve volume selection. The total experiment consists of eight subexperiments involving combinations of selective inversions in three orthogonal planes, with the appropriate addition of the resulting signals. The location of the selective volume can be controlled with control over the location of the gradient zero-crossing, or more practically, by adjusting the rf frequencies, with the offsets determined by reference to an image acquired using the same image formation protocol. This technique is very insensitive to non-uniformity of the rf field. $T_2$ effects are minimal, although some $T_1$ effects may be observed during the delay imposed between the selective inversions and the image formations to allow for stabilization following gradient switching. The technique was demonstrated using a phantom sample consisting of three tubes: water, methanol, and toluene. Cubes with a volume of 4.1 cm$^3$ were selected with a 40-second acquisition time, to isolate the spectra from the different compounds. The same volume cubes were selected in an image of a cross-sectional slice through a human's lower leg. Figure 30 shows that in the selected muscle region, only one peak attributed to water was observed, whereas in the selected bone marrow region four peaks from water and fatty acids were distinguished. The selected bone marrow spectrum acquired in vivo, compared very well to a spectrum of fat tissue acquired in vitro.

Gassner and Lohman [104] used ISIS to follow the development of the locust embryo. They acquired images followed by localized spectra every 12 hours and observed changes in both the water and lipid content of the embryo. Their images were acquired in 27 minutes, with $100 \times 100 \times 500$ µm$^3$ resolution. For the localized spectra, the spectral volume was 0.000125 cm$^3$. Figure 31 shows selected regions that were focused on to obtain the ISIS spectra.

## G. Chemical Exchange Imaging

Wolff and Balaban [105] have demonstrated the ability to obtain images of *chemical exchange* in reactions in solution involving labile protons. Their model system was a dilute ammonium chloride solution. They examined exchange between protons of water and those of ammonia, whose chemical shift difference is about 2.5 ppm. They used the *saturation transfer* concept, whereby selective irradiation at the ammonium resonance frequency decreased the intensity of the water resonance peak. Pseudo-first-order rate constants for the exchange

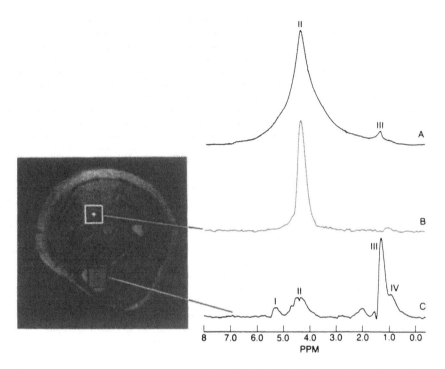

**Figure 30** Proton image of a cross-sectional slice through a human lower leg, midway between the knee and ankle: (A) overall spectrum, (B) ISIS spectrum of soleus muscle, and (C) ISIS spectrum of bone marrow in the tibia, where the individual resonances are labeled I, fatty acid (—HC=CH—); II, water; III, fatty acid, (—CH₂—)n; and IV, fatty acid (—CH₃). The intensity scale on plots (B) and (C) is two orders of magnitude larger than on plot (A). (From Ref. 103.)

process $k$ were determined from the ratio of the water peak intensity with ammonia saturation to that without, $M_{sat}/M_0$:

$$k = \frac{1 - M_{sat}/M_0}{T_{1sat}} \tag{21}$$

where $T_{1sat}$ is the $T_1$ of the water protons when the ammonium protons are saturated. To obtain a spatial map of $k$ values, an image is acquired with selective irradiation before the slice selection step and again after the phase-encoding step. The image thus obtained is compared to a control image obtained without selective irradiation at the appropriate frequency, to measure $M_{sat}/M_0$ at each pixel, and then combined with a separate $T_1$- weighted image to construct the $k$

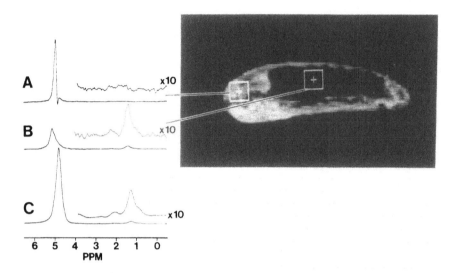

**Figure 31** Midsagittal slice water image of a locust egg, oriented with the anterior end to the left, at 66% development. Also shown are the volume-localized (ISIS) spectra from two regions, as indicated by the boxes, corresponding to different types of tissue. (A) Spectrum shows only a water peak. (B) Spectrum shows a broader water peak as well as a lipid peak. (C) The overall spectrum. (From Ref. 104.)

map from Eq. [21]. The effectiveness was demonstrated for a phantom sample consisting of an inner compartment filled with water and an outer compartment filled with dilute ammonium chloride. Two possible applications of the exchange imaging technique are to determine the rate constants and to enhance the effective signal-to-noise ratio to dilute species in solution.

Hsieh and Balaban [106] have demonstrated the feasibility of spatially mapping reaction rates with $^{31}$P imaging using the principle of saturation transfer. They studied the forward reaction catalyzed by creatine kinase:

$$PCr + ADP + H^+ <=> ATP + creatine \tag{22}$$

with the rate constant expressible as follows:

$$k = \frac{1 - M_s/M_0}{T_{1app}} \tag{23}$$

They obtained $^{31}$P images of PCr, with and without the $\gamma$-ATP resonance saturated, to monitor the exchange. With a 3 mm $\times$ 3 mm $\times$ 3 cm voxel size, they mapped $M_s/M_0$; comparison with a proton density image confirmed that the reaction activity was focused in the skeletal muscle.

# H.  Solids

Imaging of solids has not been exploited to the same extent as imaging of liquids. However, some important information is available uniquely from magnetic resonance imaging of some solid samples, as discussed in a review on solids imaging [107]. The materials that can be studied cannot be highly conductive, or else rf fields will not penetrate the samples. Polymers represent a broad class of materials to which imaging may be applied and may provide such information as the degree of mixing of components or the distribution of adsorbed fluids and fillers [108].

## 1.  Methods

As mentioned earlier, the most severe limitation posed by a solid sample is usually its very large line width. Typical proton line widths determined by homonuclear dipolar interactions can be tens of kilohertz in amorphous polymers or 50–100 kHz in rigid lattice solids; for quadrupolar nuclei, such as deuterium, line widths may be even larger. In contrast, liquid line widths are usually only several hertz. To obtain a reasonable resolution with such broad lines using conventional liquids, imaging techniques would require unrealistically large gradients. Therefore, the methods used to image solid samples rely on *line-narrowing* techniques. These include multiple pulse techniques, magic angle spinning (MAS), or a combination of the two: combined rotational and multiple pulse sequence (CRAMPS) [109].

MAS [110] averages chemical shift anisotropy as well as dipolar interactions, but it can be difficult to implement because the sample must be contained in a spinning rotor, and some samples may deform during spinning. For imaging, one must simultaneously rotate the gradients.

Of the multiple pulse line-narrowing sequences, the WHH4 and MREV8 sequences [16] average the *homonuclear* dipolar energy to 0. A "side effect" is the reduction in the magnitude of linear interactions; the gradient strength is effectively reduced by $\sqrt{3}$ for the WHH4 sequence and by $3/\sqrt{2}$ for the MREV8 sequence. These pulse sequences are repeated cyclically during the times that the gradients are on. Multiple pulse sequences act uniformly over narrow frequency ranges but typically do not work well off-resonance, so the applied magnetic field gradients cannot be too large. This limits the attainable spatial resolution. A maximum resolution of 50 μm has been obtained in solid polymers using proton MAS, and about 100 μm resolution has been obtained using line-narrowing sequences.

A third approach to imaging solids is the use of multiple-quantum resonance [111]. In a gradient, an $n$-quantum coherence evolves $n$ times more rapidly than a single-quantum coherence. This phenomenon can be used to effectively increase the gradient strength and enhance the resolution.

Wind and Yannoni [112] have used a line-narrowing technique with a combination of rf irradiation and field or frequency modulation to select "sensitive" frequency regions, making use of the fact that the efficiency of the line-narrowing technique is a strong function of the frequency offset. They demonstrate the technique with the ability to selectively observe either one of two tubes of adamantane separated by 3.2 mm, each of diameter 1 mm.

Miller and Garroway [113,114] have used a different modified multiple pulse line-narrowing sequence with a surface coil. Variation of the rf field strength selects parallel planes at varying distances from the coil. One sequence used is similar to the MREV-8 sequence, with the 90° pulses replaced by 180° pulses. A further improvement is obtained by replacing the 180° pulses with composite inversion pulses that average the dipolar interaction during irradiation. This relieves the problem of dipolar dephasing during the rf irradiation period, when the total irradiation time is on the order of twice $T_2$ or longer. The test sample used by Miller and Garroway consisted of three polyacrylic rectangles, each 1.2 mm thick, separated by 1 mm thick Teflon spacers. Figure 32 shows the image obtained, with a total rf irradiation time of 560 μs. As expected, the resolution varies nonlinearly with rf strength; the best resolution, about 200 μm, is attained nearest to the coil.

Miller et al. [115] have used a modification of the MREV-8 line-narrowing sequence in which the gradient is applied in the form of pulses during windows after every four pulses in the rf sequence, rather than continuously. Figure 33 shows the increased resolution attained this way on a sample consisting of two Mylar film pieces, each 130 μm thick, separated by 560 μm of Teflon.

MAS has been used [109], with the gradient rotating synchronously with the sample, to average chemical shift anisotropies to their isotropic values. During the acquisition, an MREV-8 line-narrowing sequence was used to average the homonuclear dipolar interaction. Two-dimensional images of polyethylene were obtained.

## 2. Other Nuclei

The *heteronuclear* dipolar coupling of nuclear spins other than protons (e.g., $^{31}P$) to any neighboring proton spins can be removed by employing proton *decoupling*. If the spins are magnetically dilute, homonuclear couplings are not contributors to the line width. Nuclei other than protons are therefore attractive for solids imaging, even though sensitivities may be significantly lower.

Szeverenyi and Maciel [116] have imaged $^{13}C$ occurring in natural abundance. They used a well-known technique called *cross-polarization* to enhance the carbon magnetization using the proton magnetization. For acquiring images, they let the cross-polarized spins evolve with the gradient on, then turned the gradient off but continued decoupling the proton spins during detection. The evolution of the spins before detection contained chemical shift and gradient information, whereas the detection time contained only chemical shift informa-

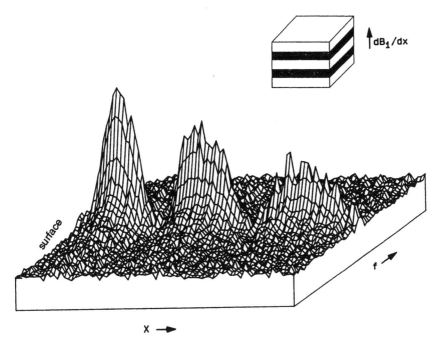

**Figure 32** One-dimensional ($x$) image of the polyacrylic–Teflon sample (inset), obtained using a line-narrowing sequence with a surface coil. The sample dimensions are $8 \times 8$ mm$^2$, with three 1.2 mm thick polyacrylic plates separated by 1.0 mm thick Teflon spacers. The $f$ dimension is the Fourier transform of the signal. Resolution nearest the coil is better than 200 μm. (From Ref. 114.)

tion; hence the two could be deconvoluted. The lines were further narrowed by detecting with a series of $\pi$ pulses to generate echoes and collecting data points at the center of every echo to form a slowly decaying "pseudo-FID." These techniques were demonstrated on a sample consisting of a Delrin cylinder with slits filled with adamantine, camphor, and hexamethylbenzene.

Suits and White [117] have studied the ion motion of $^{23}$Na in the solid ionic conductor β-alumina. They used the projection–reconstruction technique to image a single crystal, where sodium ions were exchanged for potassium ions by dipping one end of the crystal in molten KNO$_3$. They observed a more uniform distribution of the sodium ions after the sample had been annealed.

A remarkable demonstration of the ability to obtain structural or dynamic information from images of solids comes from the use of *deuteron* MAS imaging [118]. These authors made use of the fact that the quadrupolar interaction is refocused, whereas the influence of the gradient is invariant under rotation. They obtained two-dimensional images where one dimension encodes position, determined by phase modulation with respect to the gradient strength, and the other

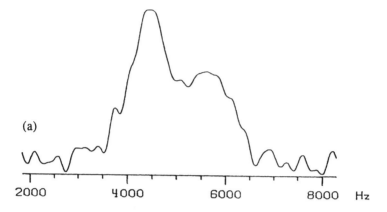

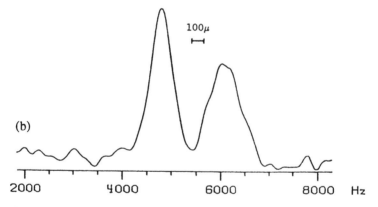

**Figure 33** One-dimensional images of two rectangular pieces of Mylar film, each 130 μm thick, separated by a 560 μm thick Teflon spacer: (a) obtained with the MREV-8 sequence and a static magnetic field gradient and (b) obtained with the MREV-8 sequence and a pulsed-field gradient. (From Ref. 115.)

dimension is the spectral frequency, determined by evolution under MAS conditions during the detection period. They obtained a resolution of better than 100 μm with a spectral width of 20 kHz. Figure 34 shows a two-dimensional data set for a sample of polycarbonate and polystyrene, both deuterated at the phenyl ring, with the polycarbonate possessing a much longer $T_2$. The spatial resolution is very clear, and the expected deuterium powder pattern, determined by the interaction of the quadrupolar nucleus with external electric field gradients, is very well defined.

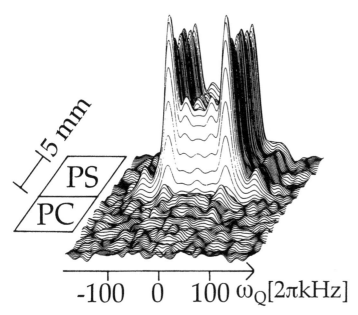

**Figure 34** Deuterium MAS one-dimensional image of a phantom sample consisting of polystyrene (PS) and polycarbonate (PC). The polycarbonate is not seen because of its longer $T_2$. The horizontal axis is the frequency axis, showing the characteristic deuterium powder line shape. The spinning speed was 4.7 kHz, and the total acquisition time was less than 4 hours. (From Ref. 118.)

## 3. Applications

The examples presented above, and indeed most of the applications of solids imaging to date, have used "phantom" samples to demonstrate the effectiveness of various techniques. Because these have been largely successful, applications to "real" systems are likely to increase in the future. One application to a polymer system is presented below.

Cory et al. [62] have used proton MAS to study polybutadiene–polystyrene blends. They used back-projection–reconstruction with phase encoding to form the images, with better than 50 μm resolution. The phase-encoding time was in integral multiples of the rotational period of the rotor; the time was incremented along with the gradient. The gradient was spun synchronously with the sample, and the phase shift between the spinner and the gradient provided the variation in effective gradient orientation. Images were obtained of 50/50 blends of the two polymers. One sample was a mechanical mixture and the other was cast from toluene. Sample deformation from spinning could clearly be seen. $T_2$ differences were used to select the signal from polybutadiene. Differences between the two samples are seen in Fig. 35, and a nonuniform distribution of the two components appears in both cases.

## V.  PROGNOSIS

In 1980 a review article was written entitled "NMR in Chemistry: An Evergreen" [119]. The title was intended to convey the constantly fresh, rapidly growing applications of nuclear magnetic resonance. As we reach the twentieth an-

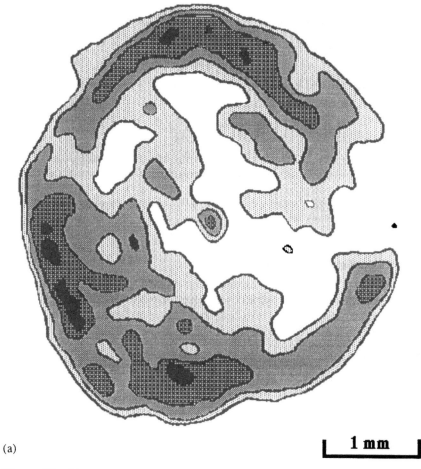

(a)                                                                    **1 mm**

**Figure 35**  [1]H MAS images of polybutadiene in (a) a mechanical blend with polystyrene and (b) a blend with polystyrene cast from toluene. Both samples were 3.5 mm in diameter and 0.75 mm thick. (From Ref. 62.)

niversary of the discovery of NMR imaging [17], it appears that new developments in imaging are also coming forward at a rapid pace. For example, one is
now able to record a series of images *during* a human heartbeat; one can use
NMR imaging to routinely diagnose disease and pathological conditions in patients; and NMR imaging is being investigated as an alternative to angiography.
To date, many of the advances in NMR imaging have been motivated by the
exceptional promise it holds for medical applications. It is likely that new health-
driven applications will continue to be developed at a rapid pace. However, we
are beginning to see several major themes that suggest future potential for NMR
imaging in other areas, as well.

One important theme for the future is the use of NMR imaging to understand
biological *function*. In this chapter, we have described the application of MRI to
the measurement of the biological functions of flow and diffusivity. It is likely

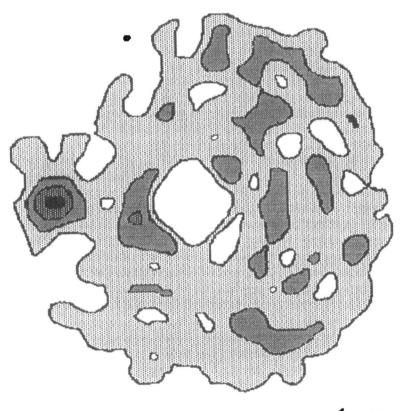

(b)

1 mm

that these applications will continue to flourish and expand. For example, imaging based primarily on differences between oxy- and deoxyhemoglobin has been used to pinpoint regions of active metabolism in the rat brain [57]. Furthermore, changes in blood flow have been used to distinguish between resting and active cognitive states of the human brain [120]. It is likely in the future that MRI will augment the information available from positron emission tomography (PET), eventually enabling us to understand the neurobiophysics of cognitive and higher order brain functions.

Another future theme for NMR imaging involves its ability to follow, noninvasively, the time course of development [104]. For example, imaging has been used in the field of developmental biology to follow individual cell lineages in a developing embryo [121].

A further theme is in the application of NMR imaging in the understanding of plant science. We have already seen examples of water transport and circulation (Section IV.D) in plant materials. Future developments are likely to involve chemical shift imaging and/or localized spectroscopy to understand the transport of nutrients or important metabolites. Furthermore, NMR imaging and localized spectroscopy would be very useful for selection of seeds with enhanced oil content.

Chemical shift imaging and localized spectroscopy (i.e., the application of NMR imaging to understand spatial variations in chemistry and biochemistry in tissues) is another area of intense interest. One would ultimately like to have NMR-measurable markers, particularly in the areas of disease and tumor diagnoses. Enthusiasm for this area of research is modulated somewhat by our knowledge that NMR is an inherently insensitive technique and that chemical markers for disease states may be present in very low concentrations. Nevertheless, it is likely that saturation transfer, indirect detection schemes, and other methods for signal amplification will contribute to the future of this area of research.

Another topic that holds much potential for the future is the application of NMR imaging to materials science. However, it is unlikely that the applications will supplant invasive light microscopic analyses except in situations calling for noninvasive methods. The major applications are therefore more likely to be in the areas of investigating tissue–material interfaces, understanding bioerosion and the controlled delivery of drugs, and following wear and fatigue of artificial joints or heart valves.

A presently underutilized application of NMR imaging involves veterinary medicine. The potential to diagnose joint problems, particularly in expensive animals such as race horses, has not been thoroughly investigated. Veterinary medicine applications will probably require the development of substantially different magnet geometries.

Another as-yet underutilized application of NMR imaging is in the area of food processing. The advent of rapid imaging techniques now makes feasible the possibility of real-time, on-line monitoring of food production.

Since NMR spectroscopy and NMR imaging continue to be driven by important applications that affect the human condition—health care, agriculture, materials science, and food processing—it is likely that this field will remain an "evergreen" for a long time to come.

## REFERENCES

1. Abragam, A. (1961). *Principles of Nuclear Magnetism*, Clarendon Press, Oxford.
2. Bovey, F. A. (1988). *Nuclear Magnetic Resonance Spectroscopy*, 2nd ed., Academic Press, New York.
3. Ernst, R. R., Bodenhausen, G., and Wokaun, A. (1987). *Principles of Nuclear Magnetic Resonance in One and Two Dimensions*, Clarendon Press, Oxford.
4. Gerstein, B. C., and Dybowski, C. R. (1985). *Transient Techniques in NMR of Solids: An Introduction to Theory and Practice*, Academic Press, Orlando, FL.
5. Slichter, C. P. (1990). *Principles of Magnetic Resonance*, 3rd ed., Springer-Verlag, New York.
6. Bottomley, P. A. (1982). NMR imaging techniques and applications: A review, *Rev. Sci. Instrum. 53*(9):1319.
7. Callaghan, P. T., and Xia, Y. (1991). Velocity and diffusion imaging in dynamic NMR microscopy, *J. Magn. Reson. 91*:326.
8. Ernst, R. R. (1987). Methodology of magnetic resonance imaging, *Q. Rev. Biophys. 19*(3/4):183.
9. Jelinski, L. W., Behling, R. W., Tubbs, H. K., and Cockman, M. D. (1989). NMR imaging: From whole bodies to single cells, *Am. Biotech. Lab. 34* (April).
10. Mansfield. P., and Pykett, I. L. (1978). Biological and medical imaging by NMR, *J. Magn. Reson. 29*:355.
11. Listerud, J. M., Sinton, S. W., and Drobny, G. P. (1989). NMR: Imaging of materials, *Anal. Chem. 61*(1):23A.
12. Wehrli, F. W., Shaw, D., and Kneeland, J. B., Eds., (1988). *Biomedical Magnetic Resonance Imaging*, VCH Publishers, New York.
13. Hahn, E. L. (1950). Spin echoes, *Phys. Rev. 80*:580.
14. Carr, H. Y., and Purcell, E. M. (1954). Effects of diffusion on free precession in nuclear magnetic resonance experiments, *Phys. Rev. 94*:630.
15. Garroway, A. N., Grannell, P. K., and Mansfield, P. (1974). Image formation in NMR by a selective irradiative process, *J. Phys. C. 7*:L457.
16. Haeberlen, U. (1976). High resolution NMR in solids: Selective averaging, in *Advances in Magnetic Resonance*, Supplement 1 (Waugh, J. S., Ed.), Academic Press, New York.
17. Lauterbur, P. C. (1973). Image formation by induced local interactions: Examples employing nucler magnetic resonance, *Nature, 242*:190.
18. Damadian, R. (1971). Tumor detection by nuclear magnetic resonance, *Science, 171*:1151.
19. Chingas, G. C., Miller, J. B., and Garroway, A. N. (1986). NMR images of solids, *J. Magn. Reson. 66*:530.
20. Caprihan, A., and Fukushima, E. (1990). Flow measurements by NMR, *Phys. Rep. 198*:195.

21. Moonen, C. T. W., van Zijl, P. C. M., Frank, J. A., Le Bihan, D., and Becker, E. D. (1990). Functional magnetic resonance imaging in medicine and physiology, *Science, 250*:53.

22. Redpath, T. W., Norris, D. G., Jones, R. A., and Hutchinson, M. S. (1984). A new method of NMR flow imaging, *Phys. Med. Biol. 29*:891.

23. Axel, L., and Dougherty, L. (1989). MR imaging of motion with spatial modulation of magnetization, *Radiology, 171*:841.

24. Hoatson, G. L., Cockman, M. D., and Jelinski, L. W. (1990). Measurement of slow anisotropic diffusion with microscopic MRI, *Soc. Magn. Reson. Med. Works Prog. 15*:1287.

25. Axel, L., and Dougherty, L. (1989). Heart wall motion: Improved method of spatial modulation of magnetization for MR imaging, *Radiology, 172*:349.

26. Mosher, T. J., and Smith, M. B. (1990). A DANTE tagging sequence for the evaluation of translational sample motion, *Soc. Magn. Reson. Med. Works Prog. 15*:334.

27. Stejskal, E. O., and Tanner, J. E. (1964). Spin diffusion measurements: Spin echoes in the presence of a time-dependent field gradient, *J. Chem. Phys. 42*:288.

28. Callaghan, P. T. (1984). Pulse field gradient nuclear magnetic resonance as a probe of liquid state molecular organization, *Aust. J. Phys. 37*:359.

29. Callaghan, P. T. (1991). *Principles of Nuclear Magnetic Resonance Microscopy*, Clarendon Press, Oxford.

30. Cory, D. G., and Garroway, A. N. (1990). Measurement of translational displacement probabilities by NMR: An indicator of compartmentation, *Magn. Reson. Med. 14*:435.

31. Callaghan, P. T., MacGowan, D., Packer, K. J., and Zelaya, F. O. (1990). High-resolution q-space imaging in porous structures, *J. Magn. Reson. 90*:177.

32. Callaghan, P. T., Coy, A., MacGowan, D., Packer, K. J., and Zelaya, F. O. (1991). Diffraction-like effects in NMR diffusion studies of fluids in porous solids, *Nature, 351*:467.

33. Merboldt, K., Hänicke, W., and Frahm, J. (1985). Self-diffusion NMR imaging using stimulated echoes, *J. Magn. Reson. 64*:479.

34. Callaghan, P. T., Eccles, C. D., and Xia, Y. (1988). NMR microscopy of dynamic displacements: k-space and q-space imaging, *J. Phys. E, 21*:820.

35. Neeman, M., Freyer, J. P., and Sillerud, L. O. (1990). Pulsed-gradient spin-echo diffusion studies in NMR imaging. Effects of the imaging gradients on the determination of diffusion coefficients, *J. Magn. Reson. 90*:303.

36. Axel, L., and Dougherty, L. (1986). Chemical-shift-selective magnetic resonance imaging of multiple-line spectra by selective saturation, *J. Magn. Reson. 66*:194.

37. Brown, T. R., Kincaid, B. M., and Ugurbil, K. (1982). NMR chemical shift imaging in three dimensions, *Proc. Natl. Acad. Sci. (USA) 79*:3523.

38. Frahm, J., Merboldt, K., and Hänicke, W. (1987). Localized proton spectroscopy using stimulated echoes, *J. Magn. Reson. 72*:502.

39. Crozier, S., Brereton, I. M., Rose, S. E., Field, J., Shannon, G. F., and Doddrell, D. M. (1990). Application of volume-selected, two-dimensional multiple-quantum editing in vivo to observe cerebral metabolites, *Magn. Reson. Med. 16*:496.

40. Doddrell, D. M., Brooks, W. M., Bulsing, J. M., Field, J., Irving, M. G., and Baddelly, H. (1986). Spatial and chemical-shift-encoded excitation. SPACE, a new technique for volume-selected NMR spectroscopy, *J. Magn. Reson. 68*:367.

41. Doddrell, D. M., Field, J., Crozier, S., Brereton, I. M., Galloway, G. J., and Rose, S. E. (1990). High-field localized in vivo proton spectroscopy on micro volumes, *Magn. Reson. Med. 13*:518.

42. Haase, J., Frahm, J., Hänicke, W., and Matthaei, D. (1985). $^1$H NMR chemical shift selective (CHESS) imaging, *Phys. Med. Biol. 30*:341.

43. Sotak, C. H., and Freeman, D. M. (1988). A method for volume-localized lactate editing using zero-quantum coherence created in a stimulated-echo pulse sequence, *J. Magn. Reson. 77*:382.

44. Sotak, C. H., and Alger, J. R. (1991). A pitfall associated with lactate detection using stimulated-echo proton spectroscopy, *Magn. Reson. Med. 17*:533.

45. Jelicks, L. A., and Gupta, R. K. (1989). Observation of intracellular sodium ions by double-quantum-filtered $^{23}$Na NMR with paramagnetic quenching of extracellular coherence by gadolinium tripolyphosphate, *J. Magn. Reson. 83*:146.

46. Cockman, M. D., Jelinski, L. W., Katz, J., Sorce, D. J., Boxt, L. M., and Cannon, P. J. (1990). Double-quantum-filtered sodium imaging, *J. Magn. Reson. 90*:9.

47. Kumar, A., Welti, D., and Ernst, R. R. (1975). NMR Fourier zeugmatography, *J. Magn. Reson. 18*:69.

48. Edelstein, W. A., Hutchison, J. M. S., Johnson, G., and Redpath, T. (1980). Spin warp NMR imaging and applications to human whole-body imaging, *Phys. Med. Biol. 25*:751.

49. Howseman, A. M., Stehling, M. K., Chapman, B., Coxon, R., Turner, R., Ordidge, R. J., Cawley, M. G., Glover, P., Mansfield, P., and Coupland, R. E. (1988). Improvements in snap-shot nuclear magnetic resonance imaging, *Br. J. Radiol. 61*:822.

50. Mansfield, P. (1977). Multi-planar image formation using NMR spin echoes, *J. Phys. C. 10*:L55.

51. Stehling, M. K., Turner, R., and Mansfield, P. (1991). Echo-planar imaging: Magnetic resonance imaging in a fraction of a second, *Science, 254*:43.

52. Brunner, P., and Ernst, R. R. (1979). Sensitivity and performance time in NMR imaging, *J. Magn. Reson. 33*:83.

53. Eccles, C. D., and Callaghan, P. T. (1986). High resolution imaging. The NMR microscope, *J. Magn. Reson. 68*:393.

54. Callaghan, P. T., and Eccles, C. D. (1987). Sensitivity and resolution in NMR imaging, *J. Magn. Reson. 71*:426.

55. Callaghan, P. T., and Eccles, C. D. (1988). Diffusion-limited in nuclear magnetic resonance microscopy, *J. Magn. Reson. 78*:1.

56. Callaghan, P. T. (1990). Susceptibility-limited resolution in nuclear magnetic resonance microscopy, *J. Magn. Reson. 87*:304.

57. Ogawa, S., Lee, T.-M., Kay, A. R., and Tank, D. W. (1990). Brain magnetic resonance imaging with contrast dependent on blood oxygenation, *Proc. Natl. Acad. Sci. (USA) 87*:9868.

58. Behling, R. W., Tubbs, H. K., Cockman, M. D., and Jelinski, L. W. (1990). Stroboscopic nuclear magnetic resonance microscopy of arterial blood flow, *Biophys. J. 58*:267.

59. Xia, Y., Jeffrey, K. R., and Callaghan, P. T. (1991). Imaging the velocity profiles for water flow through an abrupt contraction and an abrupt enlargement,'' in *Abstracts of the International Conference on NMR Microscopy*, Heidelberg, Germany, September.

60. Hoh, K. P., Perry, B., Rotter, G., Ishida, H., and Koenig, J. L. (1989). *J. Adhes.* 27:245.

61. Rothwell, W. P., Holecek, D. R., and Kershaw, J. A. (1984). NMR imaging: Study of fluid absorption by polymer composites, *J. Polym. Sci.: Polym. Lett. Ed. 22*:241.

62. Cory, D. G., de Boer, J. C., and Veeman, W. S. (1989). Magic angle spinning $^1$H NMR imaging of polybutadiene/polystyrene blends, *Macromolecules, 22*:1618.

63. Ellingson, W. A., Wong, P. S., Dieckman, S. L., Ackerman, J. L., and Garrido, L. (1989). Magnetic resonance imaging: a new characterization technique for advanced ceramics, *Am. Ceram. Soc. Bull. 68*:1180.

64. Bowtell, R. W., Brown, G. D., Glover, P. M., McJury, M., and Mansfield, P. (1990). Resolution of cellular structures by NMR at 11.7 T, *Phil. Trans. R. Soc. London A, 333*:457.

65. Aguayo, J. B., Blackband, S. J., Schoeniger, J., Mattingly, M. A., and Hintermann, M. (1986). Nuclear magnetic resonance of a single cell, *Nature, 322*:190.

66. Behling, R. W., Tubbs, H. K., Cockman, M. D., and Jelinski, L. J. (1989). Stroboscopic NMR microscopy of the carotid artery, *Nature, 341*:321.

67. Zerhouni, E. A., Parish, D. M., Rogers, W. J., Yang, A., and Shapiro, E. P. (1988). Human heart: Tagging with MR imaging—A method for noninvasive assessment of myocardial motion, *Radiology, 169*:59.

68. Haase, A., Frahm, J., Matthaei, D., Hänicke, W., and Merboldt, K. D. (1986). FLASH imaging. Rapid NMR imaging using low flip-angle pulses, *J. Magn. Reson. 67*:258.

69. Frahm, J., Merboldt, K. D., Bruhn, H., Gyngell, M. L., Hänicke, W., and Chien, D. (1990). 0.3-Second FLASH MRI of the human heart, *Magn. Reson. Med. 13*:150.

70. Pykett, I. L., and Rzedzian, R. R. (1987). Instant images of the body by magnetic resonance, *Magn. Reson. Med. 5*:563.

71. Levy, L. M., Di Chiro, G., McCullough, D. C., Dwyer, A. J., Johnson, D. L., and Yang, S. S. L. (1988). Fixed spinal cord: Diagnosis with MR imaging, *Radiology, 169*:773.

72. Rittgers, S. E., Fei, D., Draft, K. A., Fatouros, P. P., and Kishore, P. R. S. (1988). Velocity profiles in stenosed tube models using magnetic resonance imaging, *Trans. ASME: J. Biomech. Eng. 110*:180.

73. Weeden, V. J., Meuli, R. A., Edelman, R. R., Geller, S. C., Frank, L. R., Brady, T. J., and Rosen, B. R. (1985). Projected imaging of pulsatile flow with magnetic resonance, *Science, 230*:946.

74. Arai, T., Mori, K., Nakao, S., Watanabe, K., Kito, K., Aoki, M., Mori, H., Morikawa, S., and Inubushi, T. (1991). In vivo oxygen-17 nuclear magnetic resonance for the estimation of cerebral blood flow and oxygen consumption, *Biochem. Biophys. Res. Commun. 179*:954.

75. van As, H., and Schaafsma, T. J. (1984). Noninvasive measurement of plant water flow by nuclear magnetic resonance, *Biophys. J. 45*:469.

76. de Jager, P. A., Hemminga, M. A., and Sonneveld, A. (1978). Novel method for determination of flow velocities with pulsed nuclear magnetic resonance, *Rev. Sci. Instrum.* 49:1217.

77. Hemminga, M. A., and de Jager, P. A. (1980). The study of flow by pulse nuclear magnetic resonance. II: Measurement of flow using a repetitive pulse method, *J. Magn. Reson.* 37:1.

78. Bottomley P. A., Rogers, H. H., and Foster, T. H. (1986). NMR imaging shows water distribution and transport in plant root systems in situ, *Proc. Natl. Acad. Sci. (USA)* 83:87.

79. Johnson, G. A., Brown, J., and Kramer, P. J. (1987). Magnetic resonance microscopy of changes in water content in stems of transpiring plants, *Proc. Natl. Acad. Sci. (USA)* 84:2752.

80. Jenner, C. F., Xia, Y., Eccles, C. D., and Callaghan, P. T. (1988). Circulation of water within wheat grain revealed by nuclear magnetic resonance micro-imaging, *Nature*, 336:399.

81. Eccles, C. D., Callaghan, P. T., and Jenner, C. F. (1988). Measurement of the self-diffusion coefficient of water as a function of position in wheat grain using nuclear magnetic resonance imaging, *Biophys. J.* 53:77.

82. Barrall, G. A., Frydman, L., and Chingas, G. C. (1992). NMR diffraction and spatial statistics of stationary systems, *Science*, 255:714.

83. Mansfield, P., and Grannell, P. K. (1973). NMR "diffraction" in solids? *J. Phys. C*, 6:L422.

84. Ackerman, J. J. H., Ewy, C. S., Becker, N. N., and Shalwitz, R. A. (1987). Deuterium nuclear magnetic resonance measurements of blood flow and tissue perfusion employing $^2H_2O$ as a freely diffusible tracer, *Proc. Natl. Acad. Sci. (USA)* 84:4099.

85. Kim, S., and Ackerman, J. J. H. (1988). Multicompartment analysis of blood flow and tissue perfusion employing $D_2O$ as a freely diffusible tracer: A novel deuterium NMR technique demonstrated via application with murine RIF-1 tumors, *Magn. Reson. Med.* 8:410.

86. McCarthy, M. J., and Kauten, R. J. (1990). Magnetic resonance imaging applications in food research, *Trends Food Sci. Technol.* 1(6):134.

87. McCarthy, K. L., Kauten, R. J., McCarthy, M. J., and Steffe, J. F. (1993). Flow profiles in a tube rheometer using magnetic resonance imaging, *J. Food Eng.* in press.

88. Attard, J., Hall, L., Herrod, N., and Duce, S. (1991). Materials mapped with NMR, *Phys. World, 4* (July):41.

89. Gummerson, R. J., Hall, C., Hoff, W. D., Hawkes, R., Holland, G. N., and Moore, W. S. (1979). Unsaturated water flow within porous materials observed by NMR imaging, *Nature, 281*:56.

90. Baldwin, B. A., and Yamanashi, W. S. (1988). NMR imaging of fluid dynamics in reservoir core, *Magn. Reson. Imaging*, 6:493.

91. Horsfield, M. A., Fordham, E. J., Hall, C., and Hall, L. D. (1989). $^1H$ NMR studies of filtration in colloidal suspensions, *J. Magn. Reson.* 81:593.

92. Majors, P. D., Givler, R. C., and Fukushima, E. (1989). Velocity and concentration measurements in multiphase flows by NMR, *J. Magn. Reson.* 85:235.

93. Altobelli, S. A., Givler, R. C., and Fukushima, E. (1991). Velocity and concentration measurements of suspensions by nuclear magnetic resonance imaging, *J. Rheol. 35*:721.

94. Sinton, S. W., and Chow, A. W. (1991). NMR flow imaging of fluids and solid suspensions in Poiseuille flow, *J. Rheol. 35*:735.

95. Abbott, J. R., Tetlow, N., Graham, A. L., Altobelli, S. A., Fukushima, E., Mondy, L. A., and Stephens, T. S. (1991). Experimental observations of particle migration in concentrated suspensions: Couette flow, *J. Rheol. 35*:773.

96. Graham, A. L., Altobelli, S. A., Fukushima, E., Mondy, L. A., and Stephens, T. S. (1991). Note: NMR imaging of shear-induced diffusion and structure in concentrated suspensions undergoing Couette flow, *J. Rheol. 35*:191.

97. Heath, C. A., Belfort, G., Hammer, B. E., Mirer, S. D., and Pimbley, J. M. (1990). Magnetic resonance imaging and modelling of flow in hollow-fiber bioreactors, *AICHE J, 36*:547.

98. Hall, L. D., and Sukumar, S. (1982). Chemical microscopy using a high-resolution NMR spectrometer. A combination of tomography/spectroscopy using either $^1$H or $^{13}$C, *J. Magn. Reson. 50*:161.

99. (a) Frahm, J., Bruhn, H., Gyngell, M. L., Merboldt, K. D., Hänicke, W., and Sauter, R. (1989). Localized high-resolution proton NMR spectroscopy using stimulated echoes: Initial applications to human brain in vivo, *Magn. Reson. Med. 9*:79. (b) Frahm, J., Michaelis, T., Merboldt, K. D., Bruhn, H., Gyngell, M. L., and Hänicke, W. (1990). Improvements in localized proton NMR spectroscopy of human brain. Water suppression, short echo times, and 1 mL resolution, *J. Magn. Reson. 90*:464. (c) Michaelis, T., Helms, G., Merboldt, K. D., Hänicke, W., Bruhn, H., and Frahm, J. (1993). Identification of *Scyllo*-inositol in proton NMR spectra of human brain *in vivo, NMR Biomed.* in press.

100. Frahm, J., Bruhn, H., Gyngell, M. L., Merboldt, K. D., Hänicke, W., and Sauter, R. (1989). Localized proton NMR spectroscopy in different regions of the human brain in vivo. Relaxation times and concentrations of cerebral metabolites. *Magn. Reson. Med. 11*:47.

101. Van Hecke, P., Marchal, G., Johannik, K., Demaerel, P., Wilms, G., Carton, H., and Baert, A. L. (1991). Human brain proton localized NMR spectroscopy in multiple sclerosis, *Magn. Reson. Med. 18*:199.

102. Alger, J. R., Frank, J. A., Bizzi, A., Fulham, M. J., DeSouza, B. X., Duhaney, M. O., Inscoe, S. W., Black, J. L., van Zijl, P. C. M., Moonen, C. T. W., and Di Chiro, G. (1990). Metabolism of human gliomas: assessment with H-1 MR spectroscopy and F-18 fluorodeoxyglucose PET, *Radiology, 177*:633.

103. Ordidge, R. J., Connelly, A., and Lohman, J. A. B. (1986). Image-selected in vivo spectroscopy (ISIS). A new technique for spatially selective NMR spectroscopy, *J. Magn. Reson. 66*:283.

104. Gassner, G., and Lohman, J. A. B. (1987). Combined proton NMR imaging and spectral analysis of locust embryonic development, *Proc. Natl. Acad. Sci. (USA) 84*:5297.

105. Wolff, S. D., and Balaban, R. S. (1990). NMR imaging of labile proton exchange, *J. Magn. Reson. 86*:164.

106. Hsieh, P. S., and Balaban, R. S. (1987). $^{31}P$ imaging of in vivo creatine kinase reaction rates, *J. Magn. Reson. 74*:574.

107. Jezzard, P., Attard, J. J., Carpenter, T. A., and Hall, L. D. (1991). Nuclear magnetic resonance imaging in the solid state, *Prog. Nucl. Magn. Reson. Spectrosc. 23*:1.

108. Jelinski, L. W., and Cockman, M. D. (1990). Materials imaging, *Polym. Prepr., Am. Chem. Soc. Div. Polym. Chem. 31*:100.

109. Cory, D. G., Reichwein, A. M., Van Os, J. W. M., and Veeman, W. S. (1988). NMR images of rigid solids, *Chem. Phys. Lett. 143*:467.

110. Veeman, W. S., and Cory, D. G. (1989). $^1H$ nuclear magnetic resonance imaging of solids with magic-angle spinning, *Adv. Magn. Reson. 13*:43.

111. Garroway, A. N., Baum, J., Munowitz, M. G., and Pines, A. (1984). NMR imaging in solids by multiple-quantum resonance, *J. Magn. Reson. 60*:337.

112. Wind, R. A., and Yannoni, C. S. (1979). Selective spin imaging in solids, *J. Magn. Reson. 36*:269.

113. Miller, J. B., and Garroway, A. N. (1988). NMR imaging of solids with a surface coil, *J. Magn. Reson. 77*:187.

114. Miller, J. B., and Garroway, A. N. (1989). Dipolar-decoupled inversion pulses for NMR imaging of solids, *J. Magn. Reson. 85*:432.

115. Miller, J. B., Cory, D. G., and Garroway, A. N. (1989). Pulsed field gradient NMR imaging of solids, *Chem. Phys. Lett. 164*:1.

116. Szeverenyi, N. M., and Maciel, G. E. (1984). NMR spin imaging of magnetically dilute nuclei in the solid state, *J. Magn. Reson. 60*:460.

117. Suits, B. H., and White, D. (1984). NMR imaging in solids, *Solid State Commun. 50*:291.

118. Günther, E., Blümich, B., and Spiess, H. W. (1991). Spectroscopic imaging of solids of deuteron magic-angle-spinning NMR, *Chem. Phys. Lett. 184*:251.

119. Jonas, J., and Gutowsky, H. S. (1980). NMR in chemistry—An evergreen, *Annu. Rev. Phys. Chem. 31*:1.

120. Belliveau, J. W., Kennedy, D. N., McKinstry, R. C., Buchbinder, B. R., Weiskoff, R. M., Cohen, M. S., Vevea, J. M., Brady, T. J., and Rosen, B. R. (1991). Functional mapping of the human visual cortex by magnetic resonance imaging, *Science, 254*:621.

121. Fraser, S., and Jacobs, R. (1992). What, when and where. Applications of MRI microscopy to problems in developmental biology, in *Abstracts of the 33rd Experimental Nuclear Magnetic Resonance Conference*, Pacific Grove, CA, April.

These references pertain to the Figure captions only:

*Bovey, F. A. and Jelinski, L. W. (1987). Nuclear magnetic resonance, *Encycl. Polym. Sci. Eng.*, John Wiley & Sons, New York, p. 254.

*Jelinski, L. W. (1984). Modern NMR spectroscopy, *Chem. Eng. News., 62*:26.

# Index